Physics of Strength and Plasticity

The M.I.T. Press *Cambridge, Massachusetts, and London, England*

Physics of Strength and Plasticity

EDITED BY ALI S. ARGON

PHYSICS

Copyright © 1969 by The Massachusetts Institute of Technology

Designed by Dwight E. Agner.

Set in Monotype Times Roman. Printed and bound in the
United States of America by The Maple Press Company.

SBN 262 01030 5

Library of Congress catalog card number: 70–86606

Dedicated to Egon Orowan in acknowledgement of the debt we owe him for his contributions to this field, both in details and in perspective, both in phenomenology and in physics
(U. F. Kocks)

EGON OROWAN

Preface

The original manuscripts making up this book were presented as a handsomely bound volume to Professor Egon Orowan during a special celebration at the M.I.T. Endicott House, on the occasion of his retirement.

The idea of publishing a special volume summing up the present state of affairs in the field of the physics of strength and plasticity, to which Orowan so significantly contributed, came quite naturally to several people at once. In this respect I have merely implemented a widespread common wish.

Orowan came to M.I.T. in 1950 at the height of a successful career that started at Berlin-Charlottenburg, and continued at Birmingham and at the Cavendish Laboratory in Cambridge. The developments which led Orowan to introduce dislocations into crystal plasticity are discussed in detail in the following laudatory address of E. Kröner, delivered on the occasion of the presentation of the Gauss Medal to Orowan by the Braunschweigische Wissenschaftliche Gesellschaft in April 1968.

At M.I.T. in the Department of Mechanical Engineering, where he held the George Westinghouse Professorship, Orowan taught a graduate course entitled "Physics of Strength and Plasticity" (after which this book is named) in which he presented to the M.I.T. student a clear and simple

vii

picture of the basic concepts in crystal plasticity and the mechanics of fracture of materials. Uncomplicated by the absence of a heavy mathematical crust, the course strove to give the student "vitamins instead of calories." These lectures were always strongly flavored with Orowan's current preoccupations —whether these were in geotectonics or in the role of iron in the military posture of the Hittite State. In his relations with his doctoral students Orowan rarely taught directly; he merely presented a series of challenges. With his careful skepticism he quickly imparted to his students the necessity to penetrate to the essence of the problem without getting lost in the peripheral details. No elegant theory was worth the effort if it did not describe accurately the physical phenomenon. In the criticism of the works of his colleagues who sought his views he was always penetrating, and gave unvarnished but welcome advice.

In the preparation of this book I received prompt and full cooperation from all the contributors, for which I am deeply thankful. The production of the book, however, would never have become a reality without the generous financial help of the U.S. Army Mechanics and Materials Research Agency, the Air Force Office of Scientific Research, the Boeing Scientific Research Laboratory, the General Electric Research Laboratory, and the Monsanto Research Laboratory. For this help I am especially grateful to Drs. J. Burke, J. Pomerantz, G. L. Hollingsworth, H. A. Dewhurst, and R. Buchdahl, respectively. I am also grateful to Professors Ashby, Berg, and McClintock for their help in reading the manuscripts.

Ali S. Argon

Cambridge, Massachusetts
December, 1968

Contributors

A. S. Argon
Massachusetts Institute of Technology, Cambridge, Massachusetts, U.S.A.

M. F. Ashby
Harvard University, Cambridge, Massachusetts, U.S.A.

C. A. Berg
Massachusetts Institute of Technology, Cambridge, Massachusetts, U.S.A.

M. J. Bomford
University of Cambridge, Cambridge, England.
(Present address: The International Nickel Company, Suffern, New York, U.S.A.)

R. Bullough
Atomic Energy Research Establishment, Harwell, England.

J. W. Christian
Oxford University, Oxford, England.

M. Cohen
Massachusetts Institute of Technology, Cambridge, Massachusetts, U.S.A.

A. H. Cottrell
Christ's College, Cambridge, England.

J. E. Dorn
University of California, Berkeley, California, U.S.A.

W. M. Elsasser
University of Maryland, College Park, Maryland, U.S.A.

J. D. Eshelby
University of Sheffield, Sheffield, England.

J. Friedel
University of Paris, Orsay, France.

J. J. Gilman
Allied Chemical Corporation, Morristown, New Jersey, U.S.A.

P. Guyot
University of California, Berkeley, California, U.S.A.

A. K. Head
Commonwealth Scientific and Industrial Research Organization, University of Melbourne, Australia.

P. B. Hirsch
Oxford University, Oxford, England.

J. P. Hirth
Ohio State University, Columbus, Ohio, U.S.A.

F. J. Humphreys
Oxford University, Oxford, England.

T. Ishii
University of Tokyo, Tokyo, Japan.

A. Kelly
National Physical Laboratory, Teddington, England.

U. F. Kocks
Argonne National Laboratory, Argonne, Illinois, U.S.A.

D. Kuhlmann-Wilsdorf
University of Virginia, Charlotesville, Virginia, U.S.A.

N. Levy
Brown University, Providence, Rhode Island, U.S.A.

J. C. M. Li
E. C. Bain Laboratory of the United States Steel Corporation, Monroe-ville, Pennsylvania, U.S.A.
(Present address: Allied Chemical Corporation, Morristown, New Jersey, U.S.A.)

J. Lothe
Oslo University, Oslo, Norway.

F. A. McClintock
Massachusetts Institute of Technology, Cambridge, Massachusetts, U.S.A.

F. R. N. Nabarro
University of the Witwatersrand, Johannesburg, South Africa.

M. S. Paterson
The Australian National University, Canberra, Australia.

J. R. Rice
Brown University, Providence, Rhode Island, U.S.A.

A. Seeger
Max-Planck Institut für Metallforschung, and University of Stuttgart, Stuttgart, Germany.

J. A. Simmons
National Bureau of Standards, Washington, D.C., U.S.A.

T. Stefansky
University of California, Berkeley, California, U.S.A.

T. Suzuki
University of Tokyo, Tokyo, Japan.

M. R. Vukcevich
Case Western Reserve University, Cleveland, Ohio, U.S.A.

J. Weertman
Northwestern University, Evanston, Illinois, U.S.A.

Z. Wesolowski
Polish Academy of Sciences, Warszawa, Poland.

T. Yokobori
Research Institute for Strength and Fracture, and Tohoku University, Sendai, Japan.

Contents

4. Geology

The laudatio of E. Kröner to the Braunschweigische Wissenschaftliche Gesellschaft on the occasion of the presentation of the Gauss Medal to Egon Orowan on April 30, 1968.

Gauss — Medaille für Egon Orowan

Meine sehr verehrten Damen, meine Herren!

Zu den einschneidendsten Ereignissen in der Entwicklungsgeschichte der Menschheit gehört die etwa 5000 Jahre v.Chr. allmählich einsetzende Nutzbarmachunge der Metalle für alle Arten von Gebrauchsgegenständen. Die Auswirkungen dieser Entwicklung auf die Lebensweise des Menschen können gar nicht überschätzt werden. Ich erwähne als einziges Beispiel, daß ohne die Möglichkeit der Verwendung metallischer Leiter die Nutzung der Elektrizität für unseren täglichen Bedarf auf ein Minimum beschränkt wäre.

Was macht nun eigentlich die Sonderstellung der Metalle unter den Werkstoffen aus? Die Antwort auf diese Frage sehen wir—ganz abgesehen von der elektrischen Leitfähigkeit, die uns heute nicht interessieren soll— in der bei den Metallen besonders glücklichen Vereinigung einer hohen Festigkeit mit einer auch im festen Zustand bestehenden Verformbarkeit, der sogenannten Plastizität.

In größtem Ausmaß wird die Plastizität der Metalle in den Hüttenwerken verwertet, deren Aufgabe es ist, das im allgemeinen in sehr heterogener Form anfallende Rohmaterial in einen für den Verbraucher—

etwa die Autoindustrie oder die Elektroindustrie—geeigneten Zustand zu bringen. In der Regel geschieht das so, daß das geschmolzene Rohmaterial zu großen Blöcken gegossen wird, die dann im festen Zustand plastisch verformt werden.

Viele von Ihnen, meine Damen und Herren, sind vielleicht schon staunend vor einer der großen Walzstraßen der modernen Hüttenwerke gestanden, auf denen riesige Metallblöcke in immer wiederkehrenden Arbeitsgängen zu gebrauchsfertigen Blechen heruntergewalzt werden. Oder Sie haben gesehen, wir dicke Eisenstangen durch immer kleiner werdende Düsen gepreßt oder gezogen werden, bis schließlich noch lange feine Drähte übrig bleiben, wie sie etwa in der Elektrotechnik so viel gebraucht werden.

Natürlich könnte man auch Bleche und Drähte direkt durch Gießen herstellen. Es zeigt sich dann jedoch, daß diese für die meisten Zwecke wegen schlechter Festigkeitseigenschaften unbrauchbar sind. Ich erinnere nur an die bekannte Erscheinung, daß Gußeisen bei vergleichsweise schwachen Stößen zerspringt.

Tatsächlich ist die Verformung im festen Zustand nicht nur ein bequemes Mittel, um das Material in die gewünschte Form zu bringen. Bei der plastischen Formgebung verbessern sich vielmehr die mechanischen Eigenschaften, insbesondere die Festigkeit, des Werkstoffs in einer Weise, die dessen Einsatz in der Regel erst ermöglicht.

Mit diesen Ausführungen ist die oben gestellte Frage nach der Sonder-stellung der Metalle für den Normalverbraucher sicherlich ausreichend beantwortet. Nicht dagegen wird sich der Naturforscher mit einer solchen Auskunft begnügen. Er kann Befunde, wie sie die beobachteten Eigen-schaften der Metalle darstellen, nicht einfach zur Kenntnis nehmen, sondern er muß—sozusagen von Berufs wegen—weiter fragen: "Warum haben die Metalle diese besonderen Eigenschaften?"

Um einen Ansatzpunkt zur Behandlung dieses Problems zu finden, kann man allgemeiner fragen: "Was ist es überhaupt, das die Eigen-schaften der verschiedenen Stoffe bestimmt?"

Es kann heute keinen Zweifel geben, daß hierfür der atomare, bzw. molekulare, Aufbau—präziser ausgedrückt: die Anordnung der Atom-kerne und Elektronen, die den festen Körper aufbauen—maßgebend ist. Die Aufgabe des Naturwissenschaftlers bei dem Versuch einer Erklärung der Stoffeigenschaften ist es demnach, die atomare Struktur der Körper zu untersuchen und ihren Zusammenhang mit den beobachteten Eigen-schaften herauszufinden.

Die hierzu notwendigen Untersuchungen wurden zum erstenmal möglich, als im Jahre 1912 M. von Laue, rasch danach dann auch W. H. und W. L. Bragg, P. Debye, P. Scherrer und andere, Methoden entwickelten, den atomaren Aufbau der festen Körper mit Hilfe von Röntgenstrahlen aufzuklären. Nur angedeutet sei, daß Röntgenstrahlen elektromagnetische

Wellen mit einer dem Atomabstand vergleichbaren Wellenlänge sind; im Gegensatz zum sichtbaren Licht, dessen Wellenlänge ungleich viel größer ist, das deshalb Strukturen auf der atomaren Skala nicht mehr erkennen läßt.

Die röntgenographischen Untersuchungen an zahlreichen Festkörpern führten tatsächlich einen großen Schritt vorwärts. Es wurde nicht nur nachgewiesen, daß alle Metalle kristallin aufgebaut sind, sondern auch, daß andere kristalline (und nur solche) Festkörper die Eigenschaft der Plastizität in der bei den Metallen beobachteten Form haben können. Damit war der Zusammenhang zwischen kristallinem Zustand und Plastizität eindrucksvoll belegt.

Unter einem Kristall versteht man in der Wissenschaft einen Körper, dessen Atomkerne regelmäßig in einem dreidimensionalen Raumgitter angeordnet sind (Figure 1). Große Materiestücke enthalten in der Regel eine Vielzahl kleiner, gegeneinander verdrehter Kriställchen, sogenannter Kristallite, die mit dem Auge oft gar nicht erkennbar sind, im atomaren Maßstab gemessen aber doch riesengroß erscheinen, zum Beispiel einen Durchmesser von einer Million Atomabständen haben.

In einem solchen vielkristallinen Aggregat — auch Vielkristall genannt — ist immer noch der geordnete Atomaufbau das beherrschende Strukturmerkmal. Deshalb sollten sich die Eigenschaften des Vielkristalls nicht allzusehr von denen des einzelnen Kristalls — auch Einkristall genannt — unterscheiden. Es bestand berechtigte Hoffnung, die Eigenschaften der Vielkristalle zu verstehen, wenn man erst einmal die einfacheren Verhältnisse bei den Einkristallen hinreichend durchschaut haben würde. In den Jahren nach von Laues großer Entdeckung entwickelten J. Czochralski,

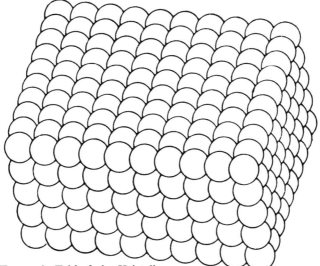

FIGURE 1. Fehlerfreier Kristall

P. Bridgman und andere raffinierte Methoden, um die in der Natur nur selten anzutreffenden Einkristalle künstlich zu züchten. Durch diese Entwicklung wurden die experimentellen Grundlagen für die Erforschung der Einkristalleigenschaften geschaffen.

Entscheidende Experimente und theoretische Überlegungen hierzu wurden in den zwanziger Jahren insbesondere in England und in Deutschland angestellt. Ein Zentrum von internationaler Geltung auf dem Gebiet der Metalle war das Kaiser-Wilhelm-Institut für Metallforschung in Berlin-Dahlem, in dem eine Gruppe tatkräftiger junger Wissenschaftler durch präzise mechanische und röntgenographische Experimente die schon vermutete Tatsache sicherstellte, daß die plastische Verformung durch Gleitung entlang von Atomgitterebenen erfolgt; etwa so, als ob dünne, parallel geschnittene Scheibchen übereinander hinwegglitten. Wie E. Schmid zeigte, setzt diese Gleitung bei einer bestimmten, in der betreffenden Ebene wirkenden Schubspannung ein. Diese kritische Schubspannung wurde für viele Stoffe exakt gemessen und mit theoretischen Betrachtungen von J. Frenkel, M. Polanyi und E. Schmid verglichen.

Das Ergebnis sar höchst bemerkenswert: Man glaubte, das Wesentliche verstanden zu haben und mußte feststellen, daß die gemessene kritische Schubspannung um mehrere Zehnerpotenzen unter der theoretischen Spannung lag, daß also die realen Materialien erheblich nachgiebiger waren, als man es nach den theoretischen, auf der Kristallvorstellung fußenden, Überlegungen erwarten sollte. Die Diskrepanz zwischen Theorie und Experiment war sogar noch größer: Die Überlegungen der genannten Forscher ließen den Schluß zu, daß die einsetzende Gleitung fortdauern sollte, bis der ganze Kristall in zwei Hälften abgeschert war. Ein Kristall sollte sich also gar nicht plastisch verformen, sondern bei Überschreiten einer bestimmten Belastung sofort zu Bruch gehen.

Nach dieser Erkenntnis mußte man einen grundsätzlichen Fehler in der bisherigen Konzeption Plastizität — Kristall annehmen. Alles konzentrierte sich nun immer mehr auf die besonders von Schmid und Polanyi herausgearbeitete Fragestellung: "Was ist es, das entgegen der Erwartung einen einen Kristall plastisch macht?"

Es war etwa um die gleiche Zeit, um die zweite Hälfte der zwanziger Jahre, als sich die Aufmerksamkeit der Festkörperphysiker auf ein Phänomen richtete, von dem bisher gar nicht die Rede gewesen war, nämlich auf die möglichen Störungen des regelmäßigen Gitterbaus, auch Gitterfehler genannt. So führten J. Frenkel, W. Schottky und C Wagner die Vorstellung ein, daß vielleicht hier und da ein regulärer Gitterplatz unbesetzt sein könnte (sogenannte Leerstelle), oder daß sich ein Atom auch einmal in einer der Lücken zwischen den regulären Gitterplätzen aufhalten könnte (sogenanntes Zwischengitteratom).

A. Smekal äußerte im Jahre 1926 die Ansicht, daß die Rolle der Gitterfehler für manche Kristalleigenschaften geradezu entscheidend sein könnte und versuchte eine Klassifizierung in störungsempfindliche und störungsun-empfindliche Eigenschaften. Zu den störungsempfindlichen Eigenshaften sollte auch die Plastizität gehören; doch fehlten noch begründete Vorstel-lungen über die Natur der Fehler. L. Prandtl und U. Dehlinger wiesen in den Jahren 1928/29 auf die Möglichkeit der Existenz bis dahin nicht diskutierter Gitterfehler hin. Nach Dehlingers atomtheoretischen Rech-nungen sollten die Verhakungen, wie er seine Gitterstörungen nannte, einen erheblichen Einfluß auf den Ablauf der als Rekristallisation bekannten Erscheinung besitzen.

Meine Damen und Herren! Ich habe versucht, Ihnen eine Einsicht in die Situation zu vermitteln, wie sie ausgangs der zwanziger Jahre auf dem Gebiet der Kristallplastizität herrschte. Man kannte eine große Zahl von experimentellen und theoretischen Zusammenhängen. Es war aber nicht gelungen, diese Erfahrungen zu einem wierspruchslosen Gesamtbild der Plastizität zu vereinigen.

Genau in dieser Zeit meldete sich bei dem Ordinarius für theoretische Physik der Technischen Hochschule Charlottenburg, Professor R. Becker, ein junger Mann zur Diplomarbeit. Es war der 1902 in Budapest geborene Egon Orowan, der bislang, zuerst in Wien, dann in Berlin Maschinenbau und Elektrotechnik studiert hatte. Becker hatte eine für einen theoretischen Physiker ungewohnte Forschungsrichtung. In der großen Zeit der Entwicklung der Quantenmechanik arbeitete er — ebenfalls sehr erfol-greich — über Probleme der Plastizität, Festigkeit und Rekristallisation. Gerade hatte er eine Theorie entwickelt, die den Einfluß der Wärme-bewegung der Atome auf das plastische Fließen erklären sollte. Insbeson-dere konnten auch in der Beckerschen Theorie Hinweise für die Existenz gewisser Materialfehler gefunden werden.

Die Vorstellung von Gitterfehlern irgendwelcher Art als Ursache des plastischen Verhaltens der Kristalle war zu jener Zeit keineswegs allgemein anerkannt. Orowan glaubte, in den im Rahmen seiner Diplomarbeit durchgeführten Experimenten neue Anhaltsphnkte für die Mitwirkung von Gitterstörungen gefunden zu haben. Mit Hilfe des damals wie heute vielverwendeten Polanyischen Dehnungsapparates verformte er Zink-kristalle und fand unter gewissen besonderen Bedingungen, daß die Probe bei konstanter Belastung vielleicht minutenlang unbeweglich blieb, um sich dann sprunghaft im Bruchteil einer Sekunde plastisch zu verformen, wonach dann wieder für eine Weile Stillstand eintrat. Orowan führte diese sprunghafte Dehnung auf eine Schwierigkeit bei der Entstehung einer ersten "lokalen Gleitung" zurück: Hatte die Gleitung einmal ein-gesetzt, so sollte sie sich längs der ganzen Atomgitterebene wellenartig ausbreiten.

Orowan war überzeugt, daß die sprunghafte Dehnung dem elementaren Gleitprozess entsprach, den er durch seine besondere Versuchsführung isoliert hatte. Die normalerweise stetig ablaufende plastische Verformung deutete er als die gleichzeitige Aktion von sehr vielen elementaren Gleitungen.

Mit diesen Überlegungen kam Orowan zu jener Zeit — seine Diplom-arbeit wurde 1929 abgeschlossen — des Rätsels Lösung am nächsten. Die noch zu klärende Frage war es, worin nun eigentlich die lokale Gleitung bestand. Auch die Antwort hierauf findet sich im wesentlichen schon in der Diplomarbeit. Veröffentlicht wurde sie erst im Jahre 1934 in der Zeitschrift für Physik.

Aus der Vorstellung, daß die Gleitung lokal anfängt und sich dann über die ganze Gleitebene ausbreitet, folgt die Existenz einer streifen- oder linienhaften Übergangszone, die die beiden Bereiche der Gleitebene trennt, in denen die Gleitung schon erfolgt, bzw. noch nicht erfolgt ist. Diese Zone muß aus geometrischen Gründen eine Störung im normalen Gitteraufbau darstellen. Man nennt sie eine Versetzungslinie oder Versetzung schlechthin. Figure 2 zeigt, wie man sich die Atomkonfigura-tion um eine Versetzung herum vorzustellen hat. Charakteristisch hierfür ist, daß eine der Gitterebenen im Innern des Kristalls endet. Das gerade macht die Gitterstörung aus.

Fortschreiten der Gleitung bedeutet auch Vorrücken der Versetzung. Die Figur vermittelt einen gewissen Eindruck davon, daß die Atome in der Gitterstörung leichter beweglich sein sollten, als die Atome im ungestörten Gitter. Dies bedeutet, daß sich die Versetzung bei Spannungen bewegen kann, die weit unter der von Frenkel angegebenen theoretischen

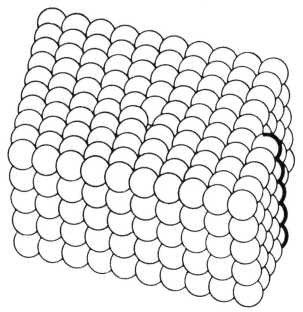

FIGURE 2. Kristall mit Stufenversetzung

kritischen Schubspannung liegen, bei der zwei benachbarte Gitterebenen starr übereinander abgleiten.

Orowan hat diese Überlegungen in äußerst scharfsinniger Weise im Detail ausgeführt. Die früher gestellte Frage "Was ist es, das entgegen der Erwartung einen Kristall plastisch macht?" war damit beantwortet: Es ist die Versetzung. Ihr Vorrücken stellt den elementaren Gleitschritt dar.

Die große Leistung Orowans bei der Aufklärung der physikalischen Hintergründe der Kristallplastizität wird weder durch den Befund geschmälert, daß sich seine Versetzungen also nahe verwandt mit den von Prandtl und Dehlinger betrachteten Gitterfehlern erweisen, noch dadurch, daß gleichzeitig und unabhängig auch M. Polanyi und G. I. Taylor zu ähnlichen Schlüssen kamen.

Die Entdeckung der Rolle, die die Versetzung in der Kristallplastizität, somit auch in der ganzen Metallindustrie, spielt, bildete den Abschluß einer langen Reihe schwierigster theoretischer und experimenteller Untersuchungen, an denen sich viele hervorragends Wissenschaftler beteiligten. Das Erscheinen der Versetzung bedeutete zugleich den Anfang einer neuen Aera in der Festkörperphysik, insbesondere in der Festkörpermechanik. Stellte sich doch heraus, daß die Versetzungen für zahllose Phänomene, die an festen Körpern beobachtet werden, verantwortlich ist. Eine Fülle von Problemen, für deren Lösung bisher jeder Ansatzpunkt fehlte, hatte sozusagen nur auf diesen Augenblick. gewartet.

Es würde zu weit führen, auf die nach 1934 folgenden Entwicklungen im einzelnen einzugehen. Eine nur Beispiele gebende Aufzählung mag genügen.

Vielfach unter Orowans Führung wurden tiefe Einsichten in die technisch so bedeutsamen Probleme der bei der plastischen Verformung eintretenden Materialverfestigung, der durch gewisse Wärmebehandlungen zusätzlich hervorgerufenen Härtung, der bei Dauerbelastung oft beobachteten fatalen Erscheinung des Kriechens, der nicht minder unerwünschten Erscheinung der Materialermüdung — bekanntlich einer Ursache so mancher Flugzeugabstürze — und der für unser Leben so bedeutsamen Erscheinung des Bruches gewonnen.

Die Festkörpermechanik ist keineswegs ein abgeschlossenes Gebiet. Man kann den Naturwissenschaftler in mancher Hinsicht mit einem Bergsteiger vergleichen, der sich vorgenommen hat, einen Gipfel zu ersteigen. Kommt er nach großen Mühen dort an, so stellt er vielleicht fest, daß sein Gipfel in Wirklichkeit nur ein Plateau ist, hinter dem sich weitere Berge türmen, die es zu ersteigen gilt.

Alle bisherigen Ausführungen haben eine Frage nicht beantwortet, nämlich die Frage: "Gibt es denn die Versetzung wirklich?" Diese Frage darf gestellt werden, waren doch die Überlegungen, die zur Versetzungshypothese führten, eher ein Indizienbeweis, als eine echte Überführung.

Dieses Stadium der Unsicherheit wurde in den fünfziger Jahren überwunden, als geschickte Experimentatoren Methoden entwickelten, Versetzungen mehr oder weniger direkt sichtbar zu machen. Sehr wirksame Methoden basieren auf der Elektronen- und Röntgenbeugung, die auf die Gitterverzerrungen in der Umgebung der Versetzung ansprechen.

Eine vielbenützte Methode, an deren Entwicklung auch Orowan Anteil hat, besteht darin, die Oberfläche eines Kristalls mit geeigneten Essenzen anzuätzen. Hierbei werden die relativ locker gebundenen Atome, die sich nahe des Zentrums der auf der Oberfläche mündenden Versetzung befinden, herausgelöst. Die so entstehenden erweiterten Mündungen der Versetzungen können im Mikroskop photographiert werden [Figure 3].

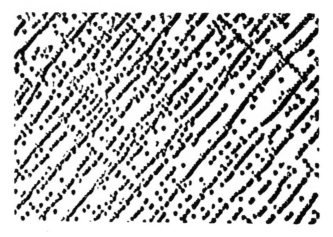

FIGURE 3. Zur Sichtbarmachung von Versetzgungen. 700 × (Nach Livingston)

Die Methoden der direkten Beobachtung der Versetzungen haben nicht nur die Richtigkeit der Versetzungshypothese eindrucksvoll belegt, sondern sich auch als äußerst wirkungsvoll bei der Aufklärung weiterer Versetzungsprobleme erwiesen.

Orowan hat die große Entwicklung der Versetzungstheorie nicht mehr in Deutschland erlebt. Nach dem Doktorat ging er nach Budapest, wo er die große Versetzungsarbeit (1934) schrieb und die Errichtung der Kryptonfabrik der Vereinigten Glühlampen- und Elektrizitätsgesellschaft überwachte. 1937 folgte er einem Ruf an die Universität Birmingham, 1939 einem Ruf an die Universität Cambridge, England. Im Jahre 1950 wurde Orowan Professor für Mechanical Engineering am Massachusetts Institute of Technology in Cambridge, U.S.A., einer der Hochburgen der internationalen wissenschaftlichen Forschung.

Hier wagte Orowan den kühnen Schritt vom Institutslaboratorium in das großartige Laboratorium der Natur: Vielbeachtete Arbeiten erschienen zur Theorie der Kontinente, zur Bildung der Gebirge und zur Entstehung der Erdbeben. Seine profunde Kenntnis des Wesens der Verformungsvorgänge kam ihm auch hier sehr zustatten. In der Tat darf man annehmen, daß die meisten Erdbeben durch plötzliche Verformungen in Erdinnern entstehen, wenn es nämlich unter dem gewaltigen Druck der Gravitationskräfte an exponierten Stellen innerhalb der Erde lokal zu einer spontanen Gleitung kommt.

Die Fakultät für Bergbau und Hüttenwesen der Technischen Universität Berlin hat im Jahre 1965 das weitreichende schöpferische Wirken des früheren Studenten Egon Orowan durch die Verleihung der akademischen Würde eines Doktor-Ingenieur ehrenhalber anerkannt. Damals wie heute wurde ein Mann geehrt, der in glücklicher Weise die Fähigkeiten zu großen Leistungen auf dem Gebiet der Naturwissenschaften in sich vereinigt, nämlich eine scharfe Beobachtungsgabe, experimentelles Geschick, theoretische Tiefgründigkeit und die zähe Beharrlichkeit, die ihn auch vor unüberwindlich scheinenden Schwierigkeiten nicht kapitulieren ließ.

Mit der Aufstellung der Versetzungshypothese hat Egon Orowan nicht nur eine uralte Frage der Menschheit in gewissem Sinn abschließend beantwortet. Mit seiner Antwort hat er zugleich gezeigt, daß es, wie auch in anderen Bereichen menschlichen Lebens, so auch im Naturgeschehen nicht die größte Vollkommenheit—hier widergespiegelt in der größtmöglichen Symmetrie—ist, die unsere Welt regiert. Vielmehr wurde klar, daß erst die Fehler in der Vollkommenheit des strukturellen Aufbaus das innere Leben der Materie ermöglicht, ein Befund, der weit über die ursprüngliche Fragestellung hinausgehend die tiefsten Wurzeln unseres Daseins berührt.

E. Kröner

Braunschweig, Germany
April 30, 1968

I. Physics of Plasticity

I. Individual Dislocations and
Basic Deformation Mechanisms

ABSTRACT. Dislocation motions are related to the concept of fluid viscosity. In this way it is possible to take advantage of existing viscosity theory to interpret dislocation behavior and to unify the description of dislocation mobility. The resulting viewpoint places the behavior of plastic solids between elastic solids at one extreme and fluids at the other.

1. A Unified View of Flow Mechanisms in Materials

J. J. GILMAN

1.1 Introduction

Deformational flow in solids is often viewed as a special process because it occurs by means of the movements of dislocations. This is not a necessary viewpoint, however. If deformation is considered as a transport process, the flow mechanisms in solids can be unified with those in gases and liquids. There is considerable advantage in this concept because it allows a more free exchange of ideas among persons who study these various material types.

The idea that flow in crystalline solids is microscopically inhomogeneous and can be described in terms of the movements of specific configurations (crystal dislocations) has been crucial in developing understanding of the process. It is equally important to realize that flow in noncrystalline solids does not, in general, result from a sequence of random local shears between molecules (or atoms). The molecular shear events tend to be correlated, and if the correlation is high the process can be described in terms of generalized dislocation lines. That is, if a shear event occurs at a certain place in a solid, the probability that another event will occur adjacent to the first is considerably greater than the probability that the second event will occur at some random position. In other words, once a region of shear

3

is nucleated it tends to grow, and as it grows it is surrounded by a dislocation line-loop.

On the other hand, the overall behavior of a plastic crystal is intermediate between that of a solid and that of a liquid. This is because the perfect regions of a crystal have substantial shear rigidity, but the centers of the dislocated regions do not. These central regions (cores) behave essentially like liquids, sometimes with high viscosity and sometimes with low. The viscosity is not necessarily constant along the length of a dislocation if the material is heterogeneous.

The above comments indicate that high-viscosity "liquids" have some of the flow characteristics of solids, and plastic "solids" behave in part like liquids. Therefore, certain connecting bridges exist between these differing materials, and the following paragraphs are intended to describe some of their common features

1.2. Dislocations in High-Viscosity Liquids (Noncrystalline Solids)

The description of flow in crystals is greatly simplified by the fact that Burgers vectors are constant and well conserved in them for given glide systems. However, the concept of a dislocation line remains useful even if the Burgers vector does not have a fixed value, as may be the case for noncrystalline solids such as glasses.

In a noncrystalline solid a dislocation line will have a somewhat variable

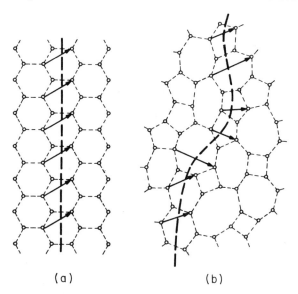

(a)   (b)

FIGURE 1.1. Dislocation lines in crystalline and noncrystalline silica compared (only silicon atoms are drawn). (a) plane of crystal structure; (b) projection of sheet of glassy network.

Burgers vector along its length, as suggested in Figure 1.1 for the prototype material: silica glass. Projections onto the plane of the drawing of the positions of the silicon atoms of a single sheet in the structure are shown. The oxygen atoms are not shown, but each silicon atom is bonded to one oxygen atom that lies just above it, plus to three others that lie just below it parallel to the plane of the drawing. If parts of the upper oxygen layers are translated while their positions are not, dislocation lines can be formed at positions indicated by the dashed lines in Figure 1.1. The arrows represent the translations that move oxygen atoms in the next higher layer from initial sites to equivalent final sites during an elementary motion of the dislocation line.

It may be seen that by allowing the magnitude and direction of the Burgers vector to fluctuate about mean values, the concept of a dislocation line can be retained for glassy structures (Gilman 1968a). This analogy is not forced but is desirable because it allows the flow properties of these structures to be discussed in a more organized way than is otherwise possible. That is, it provides a simple means for describing the correlations that must exist between adjacent elementary shear processes. In existing treatments of the flow of glasses (Eyring et al. 1964), it is assumed that the elementary flow events are independent of one another. This is not justified except at very high temperatures.

If an elementary flow event occurs somewhere locally, both the chemical and the stress state become changed in the immediate vicinity. Therefore, at ordinary temperatures, the probability that another event will occur in that vicinity is enhanced. That is, a dislocation line is created and continues to exist and move until it becomes annihilated by another line, or by a free surface.

The Burgers vector of a dislocation in a glass will not have a constant value but will fluctuate about a mean value that is determined by the average network dimensions. However, for small fluctuations of **b** the increase in the self-energy is small. Suppose a unit length of dislocation has a Burgers displacement of $(b + \delta)$ along one-half of its length, and $(b - \delta)$ along the other half, where δ is a small increment. Then, since the self-energy is proportional to the square of b, the ratio of its energy to that of the same length without the fluctuations is $1 + \delta^2/b^2$. Thus fluctuations as large as 30 percent cause only a 10 percent energy increase.

In the noncrystalline case of Figure 1b, the mean Burgers displacement has a definite value that is determined by the network dimensions, but there are fluctuations in both its magnitude and its direction along the line. In order to minimize the energy of such a dislocation, it is necessary for the mean **b** to be conserved over long distances; so although the local **b**'s may fluctuate, there are long-range correlations (occasional large energy densities may cause this condition to be relaxed). Furthermore, there will be little tendency for the line to lie on a single plane, and its local structure will change as its moves. Nevertheless, it is expected that such dislocations will exist in noncrystalline solids, especially under flow conditions. When they are viewed with a somewhat fuzzy microscope (resolution of approximately 10A), their behavior should resemble that of dislocations in crystals.

Most of the usual techniques for observing dislocations in solids are ineffective for noncrystals. One that might be used is the high-resolution observation of surface steps which would reveal the egress of dislocations from a material (Gilman 1968a). An effective method for observing monomolecular surface steps is the gold decoration technique discovered by Bassett and developed by Bethge (1962).

Evidence that the ideas of dislocation dynamics that have been developed to describe crystalline solids can be applied to noncrystals (or partial crystals) has been obtained by Dey (1967) for the case of flow in nylon. He measured the velocities of reorientation (Luders) fronts as a function of stress and temperature, and showed that the behavior is consistent with the behavior of crystals.

1.3 Viscous Resistance to Dislocation Motion

It is well known that dislocation motion is a very dissipative process with most of the plastic work being converted into heat and some of it into structural defects within the material. Because so much dissipation occurs, there is no unique way of describing the details of the process, but there is a distinct advantage in using the language and ideas associated with the behavior of fluids. One reason is that this tends to unify discussions of crystalline and noncrystalline solids. Another reason is that the theory of transport in fluids has deep traditions and a highly developed status.

Since dislocations have both micro-aspects and macro-aspects, it is necessary to describe the viscous resistance to their motion in at least two stages: first in terms of semi-macromechanics, where viscous effects are described by means of a viscosity coefficient that is treated as a continuous parameter, and second, in terms of the molecular mechanisms that determine the local value of the viscosity coefficient. A complication is introduced by the special structure at the core of a dislocation. This structure can be expected to have a different viscosity coefficient than the remainder of the material. Fortunately, the core region actually plays a dominant role, as will be demonstrated shortly, so attention can be focused on it and its local viscosity.

The net effect of various linear loss-mechanisms integrated over the entire flow field of a dislocation is described by means of a "damping constant" which conventionally has the symbol B. Oftentimes nonlinear losses are larger than the linear ones, and the damping depends on the stress and/or the velocity in a nonlinear way.

Another source of drag on moving dislocations is the anelastic relaxation that can occur if impurities or other defects are present and can move to cause stress relaxation. This type of drag depends strongly on the dislocation velocity (Schoek and Seeger, 1959) and has a relatively small magnitude. It will not be discussed further here because it is absent in pure materials, and point defects usually cause other effects that are larger in magnitude.

As a dislocation moves along its glide plane the elastic strains at points remote from its center undergo changes. Thus the moving dislocation is surrounded by a strain-rate field. In addition, at the very center the atoms on the top side of the glide plane slide over those on the bottom side. Thus a velocity gradient exists across the glide plane. For a narrow core, its magnitude can be very large compared with the other velocity gradients (strain-rates) in the system. Whenever a velocity gradient exists in a material (gas, liquid, or solid) it tends to become decreased as momentum is transported from the higher velocity regions to those with lower velocities. The viscosity coefficient measures the efficiency of this transport.

Mason (1960) first emphasized the usefulness of this viewpoint in considering dislocation losses, and showed how to calculate the power loss in the strain-rate field. However, he arbitrarily excluded the core region when he integrated over the field. Gilman (1968b) showed how the core region can be included in the calculation and that most of the loss occurs there for a given viscosity coefficient.

Consider a screw dislocation line that lies parallel to the z axis and moves with velocity v_x along the xz glide plane. Outside the core region, the strain-rate field is given by

$$\dot{\varepsilon}_{13} = \left(\frac{b}{2\pi}\right) \frac{v_x \cos \theta}{r^2} \tag{1.1}$$

and if the viscosity coefficient is called η, the power loss dP in a differential volume dV is

$$dP = \eta(\dot{\varepsilon}_{13})^2 \, dV. \tag{1.2}$$

If the separation distance at the glide plane is a, then the elastic approximation can be used in the regions $(x_2 < -a/2)$ and $(x_2 > +a/2)$, which lie outside a slab of thickness a centered on the glide plane. The power loss in these regions is

$$P_0 = 2\eta \left(\frac{bv_x}{2\pi}\right)^2 \int_{a/2}^{\infty} \int_{-\infty}^{+\infty} x_1^2 (x_1^2 + x_3^2)^{-3} \, dx_1 \, dx_3 = \frac{\eta}{2\pi} \left(\frac{bv_x}{2a}\right)^2, \tag{1.3}$$

and this is essentially the same as Mason's previous result.

In the region within the slab $(-a/2 < x_2 < +a/2)$, the velocity gradient can be obtained from the rate of relative displacement (sliding) across the glide plane. To a good approximation (and exactly in some cases), the relative displacement is given by

$$u_3(x_1) = -\frac{b}{\pi}\tan^{-1}\left(\frac{2x_1}{w}\right), \tag{1.4}$$

where w is the " width " of the dislocation core. Then the velocity gradient in the glide plane region is

$$\dot{\varepsilon}_g = \frac{1}{a}\left(\frac{du_3}{dt}\right) = \frac{v_x}{a}\left(\frac{\partial u_3}{\partial x_1}\right). \tag{1.5}$$

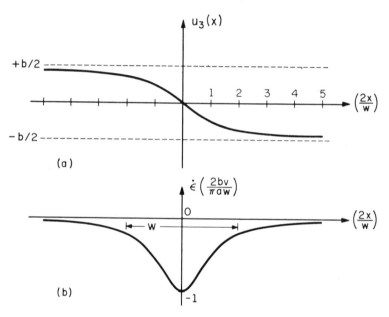

FIGURE 1.2. Relation of velocity gradient ($\dot{\varepsilon}$) to displacement along the glide plane of a dislocation. (a) displacement function of stationary dislocation; (b) velocity gradient for moving dislocation (neglects perturbations caused by local atomic interactions)

The relations just above are illustrated by Figure 1.2. Substitution of this into Equation 1.2 followed by integration yields the power loss

$$P_g = \frac{\eta v_x^2 b^2}{\pi a w} \tag{1.6}$$

Equations 1.3 and 1.6 combined give the total power loss

$$P \simeq \frac{\alpha \eta v_x^2 b^2}{\pi a w}\left[1 + \frac{1}{8}\left(\frac{w}{a}\right)\right], \tag{1.7}$$

where α is a numerical factor that equals unity for a screw dislocation; and $3/4(1 - v)$ for an edge dislocation. The width is usually comparable with a in size so the second term in Equation 1.7 is small compared with the first. Thus most of the power loss occurs on the glide plane near the core.

Because of the localization of the power loss, viscous processes that occur at the glide plane are most important in determining the dislocation damping constant, B, which can be written approximately as

$$B \simeq \frac{\eta}{\pi} \left(\frac{b^2}{aw} \right), \tag{1.8}$$

where η is the local viscosity coefficient. This has the fortunate consequence of considerably reducing the number of important mechanisms. Furthermore, if dislocation-line motion is decomposed into a series of kink motions, then the possible sources of viscosity become even more localized.

1.4 Sources of Viscosity

In order to smooth out differences in velocities within a medium, momentum must be transferred from regions that are moving fast to slower ones. The means for this to occur were analyzed long ago by Maxwell (1867); modern developments began with the work of Frenkel (1926) and Andrade (1934). There are two general categories of viscous mechanisms: "gas-like" and "solid-like" (Green 1952). In the gas-like mode particles (or quasiparticles) are free to traverse relatively long distances between collisions. As the particles cross an imaginary plane perpendicular to a velocity gradient, they carry more momentum down the gradient (on the average) than they carry up it. Thus the velocity of the slower material tends to increase, while that of the faster tends to decrease. The net velocity relative to some fixed reference tends toward zero.

In the case of solids, the main "gases" are formed by free electrons, and phonons. Other excitations that create mobile quasiparticles may also act in this fashion, but their densities may be too small to cause significant viscosity.

In the solid-like mode, direct interactions between sliding molecules tend to smooth out velocity differences. The molecules may be constrained to remain in their own layers, but a faster moving layer sliding over a slower one exerts a dragging force that tends to speed up the latter. At the same time the faster layer tends to slow down. On the average one can think of the sliding molecules as being temporarily coupled together by a force, or per unit area by a coupling shear stress, σ_c. If the mean time that the coupling lasts is called τ, then the viscosity coefficient is (Maxwell 1867)

$$\eta = \sigma_c \tau, \tag{1.9}$$

and if a single process causes most of the loss, the damping constant becomes

$$B \simeq \left(\frac{b^2}{aw}\right) \frac{\sigma_c \tau}{\pi}. \tag{1.10}$$

The coupling relaxation time may be a function of such factors as the applied stress and the temperature, depending on what particular loss mechanism operates.

For dislocations moving through otherwise perfect crystals, the Peierls stress couples the molecules across the glide plane. If it is small, as for close-packed glide planes in pure metals, then since τ is roughly the reciprocal Debye frequency (say 10^{-12} sec), and σ_c is certainly less than about 10^5 dyn/cm^2, the local viscosity coefficient may be as small as $\sim 10^{-7}$ P, which is very small compared with the viscosity of a typical liquid metal ($\sim 10^{-2}$ P).

On the other hand, if the Peierls stress is large as in covalent crystals such as Ge and Si (say $\sim 10^{10}$ dyn/cm^2), and the coupling force is localized so that the velocity is directly related to the coupling time, which is given by $b/V \simeq 5 \times 10^{-4}$ sec, for a velocity of one micron per second; then the local viscosity coefficient is $\sim 5 \times 10^6$ P, which is moderately large.

In imperfect crystals the viscosity is heterogeneous. Dislocations may move quite freely over glide-plane areas that are free of imperfection, impeded only by electron, or phonon, gas viscosity. At imperfections, strong local bonding may create a strong coupling force across the glide plane, or weak bonding may destroy the local periodicity and thereby raise the effective Peierls stress. Nonviscous drag will also result if the moving dislocation intersects another one and acquires a jog so that it leaves a dipole in its wake. This constitutes a net change in the internal structure and leads to strain-hardening, but is qualitatively different from a viscous loss mechanism.

1.5 Viscosity at High Velocities

A linear dependence of dislocation velocity on applied stress has been observed in several pure crystals: copper (Greenman, Vreeland, and Wood, 1967); zinc (Pope, Vreeland, and Wood, 1967); and germanium (Schafer, 1967). It is apparent, however, that the velocity cannot continue to be proportional to stress indefinitely because it would soon exceed sonic velocities. Therefore, it is legitimate to wonder how the damping constant depends on velocity. Taylor (1968) has suggested that relativistic effects cause the damping constant to take the form (screw dislocation)

$$B = B_0 (1 - v^2/c_t^2)^{-1}, \tag{1.11}$$

where B_0 is the constant at low velocities, v is the instantaneous velocity, and c_t is the transverse elastic-wave velocity. According to this expression, the drag force increases without limit as v approaches c_t. Therefore, for any feasible applied force, v cannot exceed c_t (in a linear system).

The present author offers the following simple justification of Taylor's suggestion. The damping constant has the form $B = p_x/A_y$ where p is momentum and A is area. That is, it measures momentum transfer per unit area. It is well known that the Lorentz transformation for a momentum component is $p_x = p_x^0/\beta$ where $\beta = (1 - v^2/c^2)^{1/2}$; and since sliding occurs in one direction, the Fitzgerald contraction of the area is given by $A_y = \beta A_y^0$. Thus Equation 1.11 follows.

For steady-state motion, the applied force $\sigma_s b$ on a dislocation equals the drag force $B_0 V/\beta^2$ so that the velocity is explicitly related to the applied stress

$$v = \frac{c}{x}[(1 + x^2)^{1/2} - 1],\tag{1.12}$$

with $x =$ reduced stress $= \sigma_s(2b/B_0 c)$. This equation is plotted in Figure 1.3 to display its linear and saturation limits.

1.6. Nonlinear Viscous Drag

When localized coupling forces exist across glide planes, and the temperature is low (that is, low compared with the Debye temperature associated with the local coupling force) the flow velocity is not proportional to the applied stress. Instead, it is observed that the flow rate is very small until some critical stress is reached, and then "yielding" occurs. In other words the flow is stress-activated. In terms of Equation 1.9, the mean coupling time is a function of stress and it rather suddenly decreases when a critical applied stress is reached. A simple analytic form that relates velocity and stress in this case has been proposed by the author:

$$v = v^* e^{-D/\sigma_s},\tag{1.13}$$

where v^* is a terminal velocity and D is a drag stress. This form is followed by several sets of data (Gilman 1965), and is consistent with the idea that stress can activate dislocation motion via quantum-mechanical tunneling (Gilman 1968a). It is not the purpose here to discuss these matters, however, so Equation 1.13 will simply be asserted as a reasonable form.

At steady state the work done on a moving dislocation equals the power dissipated in the form of both heat and structural defects. Assuming that heat production dominates, the effective viscous damping constant is the ratio of the driving force per unit length to the velocity

$$B_{\text{eff}} = \frac{\sigma_s b}{v} = \left(\frac{b}{v^*}\right)\sigma_s e^{D/\sigma_s}\tag{1.14}$$

Thus the effective damping constant is infinite when the stress is zero (the dislocation is "pinned"). Then the damping decreases rapidly with increasing stress until a minimum value is reached when $\sigma_s = D/2$; and then it increases nearly linearly with further decreases of the stress. The corresponding velocity-stress curve for this "solid-like" viscosity is compared with the curve for the "gas-like" viscosity in Figure 1.3.

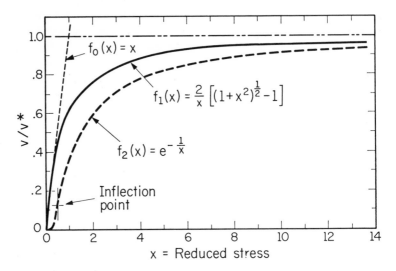

FIGURE 1.3. Comparison of velocity-stress functions. $f_0 =$ linear viscosity (Newton); $f_1 =$ relativistic viscosity (Taylor); $f_2 =$ yielding function(Gilman).

1.7. Amplification of Viscous Resistance by Internal Stresses

In addition to the complications caused by the heterogeneous nature of real materials, the internal stress fluctuations that are usually present in them can have important effects, especially for nonlinear flow.

For linear viscosity if the internal stresses fluctuate about some mean value, σ, with an average amplitude, $\Delta\sigma$, and an average wavelength, λ, it can be readily shown that the effective damping constant is

$$B'_{\text{eff}} = B_0 \left[1 - \left(\frac{\Delta\sigma}{\sigma} \right)^2 \right]^{-1}, \tag{1.15}$$

so that for small fluctuations there is a negligible effect.

For the nonlinear case, since $e^{1/1+\delta}$ is small compared with $e^{1/1-\delta}$ for moderate values of δ, the effective damping constant has the approximate value

$$B''_{\text{eff}} \simeq B_{\text{eff}}\, e^{(\sigma/\sigma - \Delta\sigma)}, \tag{1.16}$$

so that the effective viscosity increases very rapidly with increasing stress fluctuation amplitude. A more detailed discussion of this effect was first given by Chen, Gilman, and Head (1964); more recently it has been considered by Li (1968) and by Argon (1968).

1.8. Summary

It has been shown that by relating dislocation motion to the concept of fluid viscosity, it is possible to unify the description of mobility, and to take advantage of the existing theories of fluids in interpreting dislocation behavior. Also, this viewpoint places plastic solids between elastic solids and fluids in a well-defined way. This is their natural position.

REFERENCES

Andrade, E. N. da C., 1934, *Phil. Mag.*, **17**, 497, 698.

Argon, A. S., 1968, *Mat. Sci. Eng.*, **3**, 24.

Bethge, H., 1962, *Phys. Stat. Sol.*, **2**, 3.

Chen, H. S., J. J. Gilman, and A. K. Head, 1964, *J. Appl. Phys.*, **35**, 2502.

Dey, B. N., 1967, *J. Appl. Phys.*, **38**, 4144.

Eyring, H., D. Henderson, B. Stover, and E. Eyring, 1964, *Statistical Mechanics and Dynamics*, New York: Wiley Chapter 16.

Frenkel, J., 1926, *Z. Physik*, **35**, 652.

Gilman, J. J., 1960, *Australian J. Phys.*, **13**, 327.

Gilman, J. J., 1965, *J. Appl. Phys.*, **36**, 3195.

Gilman, J. J., 1968a, in A. R. Rosenfield et al. (eds.), *Dislocation Dynamics*, New York; McGraw-Hill, p. 3.

Gilman, J. J., 1968b, *Phys. Rev. Letters*, **20**, 157.

Green, H. S., 1952, *The Molecular Theory of Fluids*, New York: Interscience.

Greenman, W. F., T. Vreeland, and D. S. Wood, 1967, *J. Appl. Phys.*, **38**, 3595.

Li, J. C. M., 1968, in A. R. Rosenfield et al. (eds.), *Dislocation Dynamics*, New York: McGraw-Hill, p. 87.

Mason, W. P., 1960, *J. Acoust. Soc. Amer.*, **32**, 458.

Maxwell, J. C., 1867, *Phil. Trans. Roy. Soc.*, **A157**, 49.

Pope, D. P., T. Vreeland, and D. S. Wood, 1967, *J. Appl. Phys.*, **38**, 4011.

Schafer, S., 1967, *Phys. Stat. Sol.*, **19**, 297.

Schoeck, G., and A. Seeger, 1959 *Acta Met.*, **7**, 469.

Taylor, J. W., 1968, private communication.

ABSTRACT. The present paper presents a novel approach for treating dislocations within the framework of finite elasticity. It makes use of the idea that approximate solutions for materials of small or moderate compressibility may be obtained by a perturbation treatment starting from the so-called universal solutions for incompressible isotropic materials. The general formalism of this approach is developed and is then applied to a screw dislocation in an isotropic body. The specialization to Murnaghan's elastic energy function is compared with the treatment of Seeger and Mann (1959) based on the so-called quadratic elasticity theory; complete agreement is found. A number of possible further applications of the theory are discussed.

2. A Treatment of Screw Dislocations by Finite Elasticity

A. SEEGER AND Z. WESOŁOWSKI

2.1. Introduction

The linearized theory of elasticity has proved to be a most powerful tool in dislocation theory. Nevertheless, there are quite a number of important problems in the theory of dislocations in continuous media which lie outside the realm of that approximation. Various approaches have been devised to handle such problems. The oldest one is that suggested in 1939 by E. Orowan to R. Peierls (Peierls 1940),[1] which has since been developed further by a number of workers (see Seeger 1955, Seeger and Schiller 1966, Nabarro 1967). It is essentially a nonlinear treatment of the elastic properties in the glide plane of the dislocation, the usual linearized theory being employed everywhere else. For a screw dislocation such an approach may appear somewhat unnatural, since in an isotropic elastic medium the strain field of a screw dislocation has rotational symmetry. No single distinguished glide plane exists.

This drawback may be avoided by employing the theory of finite elastic deformations, as suggested by one of the present writers in 1955 (Seeger 1955, p. 563). In a series of papers the so-called quadratic approximation

---

[1] For outlines of the historical development the reader is referred to Orowan (1965) and Peierls (1968).

to the theory of finite deformations was developed in considerable detail and applied successfully to a number of problems not tractable by the linearized theory, e.g., the volume expansion caused by a dislocation and the scattering of elastic waves by static dislocations. (For brief summaries see Seeger 1964 and Seeger, Bross, and Gruner 1964). In its practical application this approach is essentially a perturbation treatment starting from the linearized theory and retaining in a systematic way the lowest-order terms that yield the physical effects under investigation. Such a perturbation approach is bound to fail near the dislocation core, and is useful only for treating physical effects (such as the two examples mentioned above) whose main contributions come from the more distant environment of the dislocation center.

The present paper constitutes an attempt to avoid the perturbation treatment at least partially and to obtain elastic solutions for screw dislocations that remain useful closer to the dislocation core than those obtained so far. (Inside the core any treatment based on the notion of a continuous medium will break down and will have to be replaced by an atomistic treatment. This region is outside the scope of the present paper.) The present approach is based on the concept of universal solutions of the equations of (finite) static elasticity, i.e., deformations that are possible in any elastic medium irrespective of its stress-strain relation. Such solutions permit an analytical treatment of finite elasticity without having to specify the stress-strain relation. This constitutes a considerable advantage, since at present for materials of interest to dislocation theory not much is known beyond the quadratic terms in the stress-strain relation.

The only universal solution valid for all homogeneous *compressible* elastic materials is the homogeneous deformation (Truesdell and Noll 1965). In the case of *incompressible* isotropic bodies universal solutions for five families of deformations have been found, among them the deformation corresponding to a screw dislocation. (Another one of these universal solutions in incompressible isotropic media of interest in solid-state physics is that of a so-called wedge disclination, which has recently been observed in the so-called Abrikosov lattice of type II superconductors [Anthony, Essmann, Seeger, and Träuble 1968]). The existence of this universal solution for a screw dislocation suggests that the solution for a screw dislocation in a moderately compressible elastic material may be found by improving the universal solution for the incompressible bodies by means of a perturbation technique. The essential point is that the strain components that become large near the dislocation center (and that are responsible for the multi-valuedness of the displacement) are already taken into account in the universal solution. For elastic materials which are not too compressible, the additional strains may indeed be expected to be small. This is further supported by the fact that in the linearized theory the universal solution 2.26, which forms the starting point for our treatment, holds also for compressible bodies. We may therefore expect that a moderate compressibility will not change the character of the solution radically and that a perturbation treatment for the additional strains will be adequate. In the present paper we shall use the perturbation theory of Green, Rivlin, and Shield (1952) for the superposition of a small deformation on a finite deformation in an isotropic medium (see also Green and Zerna 1954). An extension of this theory to curvilinearly orthotropic bodies has been given by Urbanowski (1959) and can be used for treating screw dislocations in, say, cubic and hexagonal crystals, but space does not permit the inclusion of this generalization here. A brief outline of the present work has been presented at a recent symposium (Wesołowski and Seeger 1968).

2.2. Basic Equations

Consider an elastic body in the natural state $B°$. Introduce in this state the in general curvilinear coordinate system θ^i, $i = 1, 2, 3$. The Cartesian coordinates $x_i°$ of an arbitrary point $P°$ are uniquely determined by the coordinates θ^i

$$x_i° = x_i°(\theta^r). \tag{2.1}$$

The coordinate system θ^i introduces in $B°$ a geometric structure. The position vector of the point $P°$ is denoted by

$$\mathbf{r}° = x_r° \mathbf{k}_r° \tag{2.2}$$

where $\mathbf{k}_r°$ are unit vectors along the axes $x_r°$.

The partial derivatives of $\mathbf{r}°$ with respect to θ^i are called the covariant base vectors $\mathbf{g}_i°$. The reciprocals of the covariant base vectors are the contravariant base vectors:

$$\mathbf{g}_i° = \frac{\partial \mathbf{r}°}{\partial \theta^i} = \frac{\partial x_r°}{\partial \theta^i} \mathbf{k}_r, \quad \mathbf{g}_i° \cdot \mathbf{g}^{°j} = \delta_i^j. \tag{2.3}$$

The base vectors are connected with the covariant and contravariant metric tensors $g_{ij}°$, $g^{°ij}$ by the relations

$$g_{ij}° = \mathbf{g}_i° \cdot \mathbf{g}_j° = \frac{\partial x_r°}{\partial \theta^i} \frac{\partial x_r°}{\partial \theta^j}, \quad g^{°ij} = \mathbf{g}^{°i} \cdot \mathbf{g}^{°j}. \tag{2.4}$$

The distance $ds°$ between two neighboring points θ^i and $\theta^i + d\theta^i$ in the body $B°$ is determined by the metric tensor $g_{ij}°$ according to

$$ds°^2 = g_{ij}° \, d\theta^i \, d\theta^j. \tag{2.5}$$

It is evident that $g_{ij}°$ determines also the angle between two arbitrary linear elements.

Let us now subject the body $B°$ to a finite deformation and denote the resulting body by B. The typical point $P°$ of the body $B°$ changes into the typical point P of the body B. Let the coordinate system θ^i deform with the body, so that the curvilinear coordinates θ^i of the point P° in B° and the point P in B remain the same (convected coordinates). In static problems such coordinates are especially convenient. Denote the Cartesian coordinates of the point P in B by

$$x_i = x_i(\theta^r) \qquad (2.6)$$

and the position vector of the point P by

$$\mathbf{r} = x_r \mathbf{k}_r, \qquad (2.7)$$

where $\mathbf{k}_r$ are the unit vectors along the axes x_r.

The coordinate system θ^i induces a geometric structure in B. Denoting by $\mathbf{g}_i$, $\mathbf{g}^i$ the base vectors and by g_{ij} and g^{ij} the covariant and contravariant components of the metric tensor in B, we have

$$\mathbf{g}_i = \frac{\partial x_r}{\partial \theta^i}\mathbf{k}_r, \quad \mathbf{g}_i \cdot \mathbf{g}^j = \delta_i^j,$$

$$g_{ij} = \mathbf{g}_i \cdot \mathbf{g}_j = \frac{\partial x_r}{\partial \theta^i}\frac{\partial x_r}{\partial \theta^j}, \quad g^{ij} = \mathbf{g}^i \cdot \mathbf{g}^j. \qquad (2.8)$$

The distance ds between two neighboring points θ^i and $\theta^i + d\theta^i$ in B is determined by g_{ij} according to

$$ds^2 = g_{ij}\, d\theta^i\, d\theta^j. \qquad (2.9)$$

The relations (2.5) and (2.9) allow us to introduce the strain tensor

$$\gamma_{ij} = \tfrac{1}{2}(g_{ij} - g^\circ_{ij}). \qquad (2.10)$$

The three invariants of the strain tensor γ_{ij} are defined in the following way:

$$I_1 = g^\circ_{rs}g^{rs}, \quad I_2 = \tfrac{1}{2}(I_1^2 - g^{\circ rm}g^{\circ sn}g_{rs}g_{mn}),$$

$$I_3 = \det g_{ij}/\det g^\circ_{ij}. \qquad (2.11)$$

The metric tensor components g_{ij}, g^{ij} (2.8) determine the Christoffel symbols of the second kind,

$$\Gamma^i_{jk} = \frac{1}{2}g^{is}\left(\frac{\partial g_{js}}{\partial \theta^k} + \frac{\partial g_{ks}}{\partial \theta^j} - \frac{\partial g_{jk}}{\partial \theta^s}\right). \qquad (2.12)$$

For an isotropic body the stored energy W is a function of the invariants I_k only. The stress tensor τ^{ij} is the first derivative of W with respect to γ_{ij} and can be represented by the formula

$$\tau^{ij} = \Phi g^{\circ ij} + \Psi b^{ij} + p g^{ij}, \qquad (2.13)$$

where

$$\Phi = \frac{2}{\sqrt{I_3}} \frac{\partial W}{\partial I_1}, \qquad \Psi = \frac{2}{\sqrt{I_3}} \frac{\partial W}{\partial I_2}, \qquad p = 2\sqrt{I_3}\,\frac{\partial W}{\partial I_3}, \qquad (2.14)$$

$$b^{ij} = I_1 g^{\circ ij} - g^{\circ ir} g^{\circ js} g_{rs}. \qquad (2.15)$$

The equilibrium equations and boundary conditions are

$$\nabla_i \tau^{ij} + \rho f^j = \partial_i \tau^{ij} + \Gamma_{ir}^j \tau^{ir} + \Gamma_{rs}^r \tau^{sj} + \rho f^j = 0 \qquad (2.16)$$

$$\tau^{ij} n_i = l^j, \qquad (2.17)$$

where n_i is the unit normal to the surface of the body B, l^j the surface traction, f^j the body force and ρ the mass density in the state B.

We shall now consider small additional deformations and assume that the points of the body B are subjected to displacement $\varepsilon w = \varepsilon w_i \mathbf{g}^i$, where ε is a small parameter to be neglected in comparison with unity. As a result of this additional deformation the body B passes to another state (denoted by $B + \varepsilon B'$) and all quantities Q characteristic for the state B change into $Q + \varepsilon Q'$. We have (Green and Zerna 1954)

$$g'_{ij} = \nabla_i w_j + \nabla_j w_i, \qquad g'^{ij} = -g^{ir} g^{js} g'_{rs}, \qquad (2.18)$$

$$\Gamma'^r_{ij} = \frac{1}{2}\, g^{rs}\!\left(\frac{\partial g'_{si}}{\partial \theta^j} + \frac{\partial g'_{sj}}{\partial \theta^i} - \frac{\partial g'_{ij}}{\partial \theta^s}\right)$$

$$+ \frac{1}{2}\, g'^{rs}\!\left(\frac{\partial g_{si}}{\partial \theta^j} + \frac{\partial g_{sj}}{\partial \theta^i} - \frac{\partial g_{ij}}{\partial \theta^s}\right), \qquad (2.19)$$

$$I'_1 = g^{\circ rs} g'_{rs}, \quad I'_2 = g_{rs}(g'^{rs} I_3 + g^{rs} I'_3), \quad I'_3 = I_3\, g^{rs} g'_{rs}, \qquad (2.20)$$

$$b'^{ij} = (g^{\circ ij} g^{\circ rs} - g^{\circ ir} g^{\circ js}) g'_{rs}, \qquad (2.21)$$

$$\left.\begin{array}{l}
\Phi' = AI'_1 + FI'_2 + EI'_3 - \dfrac{\Phi}{2I_3}\, I'_3 \\[2ex]
\Psi' = FI'_1 + BI'_2 + DI'_3 - \dfrac{\Psi}{2I_3}\, I'_3 \\[2ex]
p' = I_3(EI'_1 + DI'_2 + CI'_3) + \dfrac{p}{2I_3}\, I'_3
\end{array}\right\} \qquad (2.22)$$

$$A = \frac{2}{\sqrt{I_3}} \frac{\partial^2 W}{\partial I_1^2}, \; B = \frac{2}{\sqrt{I_3}} \frac{\partial^2 W}{\partial I_2^2}, \; C = \frac{2}{\sqrt{I_3}} \frac{\partial^2 W}{\partial I_3^2}$$

$$D = \frac{2}{\sqrt{I_3}} \frac{\partial^2 W}{\partial I_2 \partial I_3}, \; E = \frac{2}{\sqrt{I_3}} \frac{\partial^2 W}{\partial I_3 \partial I_1}, \; F = \frac{2}{\sqrt{I_3}} \frac{\partial^2 W}{\partial I_1 \partial I_2} \qquad (2.23)$$

$$\tau'^{ij} = g^{ij}\Phi' + B^{ij}\Psi' + B'^{ij}\Psi + G'^{ij}p + G^{ij}p'. \qquad (2.24)$$

The equilibrium equations (2.16) change into

$$\nabla_i \tau^{ij} + \rho f^j + \varepsilon(\nabla_i \tau'^{ij} + \Gamma'^j_{ir}\tau^{ir} + \Gamma''^r_{rs}\tau^{sj} + \rho' f^j + \rho f'^j) = 0.$$

$$(2.25)$$

2.3 The Screw Dislocation

Denoting by r, θ, z the cylindrical coordinate system in B and by r°, θ°, z° the cylindrical coordinate system in B°, the universal solution for a screw dislocation in an isotropic incompressible body (found by J. L. Ericksen) reads

$$r = r^\circ, \quad \theta = \theta^\circ, \quad z = z^\circ - c\theta^\circ. \tag{2.26}$$

The constant c is related to the Burgers vector $\mathbf{b}$ by $b = 2\pi c$. Without loss of generality we may assume that the system θ^i coincides in B with the system (r, θ, z). On the basis of the formulae given in Section 2.2 we can now calculate the metric tensors and the stress tensor. We obtain

$$g^\circ_{ij} = \begin{bmatrix} 1 & 0 & 0 \\ 0 & r^2+c^2 & c \\ 0 & c & 1 \end{bmatrix}, g^{\circ ij} = \begin{bmatrix} 1 & 0 & 0 \\ 0 & 1/r^2 & -c/r^2 \\ 0 & -c/r^2 & 1+c^2/r^2 \end{bmatrix}, \det g^\circ_{ij} = r^2$$

$$(2.27)$$

$$g_{ij} = \begin{bmatrix} 1 & 0 & 0 \\ 0 & r^2 & 0 \\ 0 & 0 & 1 \end{bmatrix}, g^{ij} = \begin{bmatrix} 1 & 0 & 0 \\ 0 & 1/r^2 & 0 \\ 0 & 0 & 1 \end{bmatrix}, \det g_{ij} = r^2 \tag{2.28}$$

$$I_1 = I_2 = 3 + c^2/r^2, \qquad I_3 = 1 \tag{2.29}$$

$$b^{ij} = \begin{bmatrix} 2+c^2/r^2 & 0 & 0 \\ 0 & 2 & -c/r \\ 0 & -c/r & 2+c^2/r^2 \end{bmatrix} \tag{2.30}$$

$$\Phi = 2\frac{\partial W}{\partial I_1}, \qquad \Psi = 2\frac{\partial W}{\partial I_2}, \qquad p = 2\frac{\partial W}{\partial I_3} \tag{2.31}$$

$$\left. \begin{aligned} \tau^{11} &= \Phi + (2+c^2/r^2)\Psi + p \\ r^2\tau^{22} &= \Phi + 2\Psi + p \\ \tau^{33} &= (1+c^2/r^2)\Phi + (2+c^2/r^2)\Psi + p \\ r\tau^{23} &= -\frac{c}{r}(\Phi + \Psi) \\ \tau^{31} &= \tau^{12} = 0. \end{aligned} \right\} \tag{2.32}$$

In the case of an incompressible body the arbitrary hydrostatic pressure can be chosen such that the stress tensor (2.32) satisfies the equilibrium equations for zero body force $f_i = 0$. However, in the present case the material is compressible and the stress is uniquely determined by the deformation (which is isochoric). Therefore, equilibrium with zero body forces is in general not possible. The equilibrating body forces can be calculated from Equation 2.16. Since the present problem is axially symmetric, only the radial component differs from zero. We do not give the corresponding relations, since they are not required for further investigations.

Our aim is to find a state of strain where the equilibrating body force equals zero. Such a state can be reached when the body B is subjected to additional radial displacement εu. According to the formulae given above the displacement εu produces increments in all the quantities characteristic for the state B. After some calculations one obtains

$$g'_{ij} = 2\begin{bmatrix} \dfrac{du}{dr} & 0 & 0 \\ 0 & ru & 0 \\ 0 & 0 & 0 \end{bmatrix}, \quad g'^{ij} = -2\begin{bmatrix} \dfrac{du}{dr} & 0 & 0 \\ 0 & u/r^3 & 0 \\ 0 & 0 & 0 \end{bmatrix} \tag{2.33}$$

$$I'_1 = I'_3 = 2\left(\frac{du}{dr} + \frac{u}{r}\right), \quad I'_2 = 2\left[\left(2 + \frac{c^2}{r^2}\right)\left(\frac{du}{dr} + \frac{u}{r}\right) - \frac{c^2}{r^2}\frac{u}{r}\right] \tag{2.34}$$

$$\Gamma'^1_{11} = \frac{d^2 u}{dr^2}, \quad \Gamma'^1_{22} = r\frac{du}{dr} - u, \quad \Gamma'^2_{12} = \Gamma'^2_{21} = \frac{1}{r^2}\left(r\frac{du}{dr} - u\right), \tag{2.35}$$

all other $\Gamma'^i_{jk} = 0$,

$$B'^{ij} = 2\begin{bmatrix} \dfrac{u}{r} & 0 & 0 \\ 0 & \dfrac{1}{r^2}\dfrac{du}{dr} & -\dfrac{c}{r^2}\dfrac{du}{dr} \\ 0 & -\dfrac{c}{r^2}\dfrac{du}{dr} & \dfrac{du}{dr} + \dfrac{u}{r} + \dfrac{c^2}{r^2}\dfrac{du}{dr} \end{bmatrix} \tag{2.36}$$

$$A = 2\frac{\partial^2 W}{\partial I_1^2}, \quad B = 2\frac{\partial^2 W}{\partial I_2^2}, \quad C = 2\frac{\partial^2 W}{\partial I_3^2},$$

$$D = 2\frac{\partial^2 W}{\partial I_2 \partial I_3}, \quad E = 2\frac{\partial^2 W}{\partial I_1 \partial I_3}, \quad F = 2\frac{\partial^2 W}{\partial I_1 \partial I_2}, \tag{2.37}$$

$$\Phi' = (2A + 2E - \Phi)\left(\frac{du}{dr} + \frac{u}{r}\right) + 2F\left[\left(2 + \frac{c^2}{r^2}\right)\left(\frac{du}{dr} + \frac{u}{r}\right) - \frac{c^2}{r^2}\frac{u}{r}\right]$$

$$\Psi' = (2F + 2D - \Psi)\left(\frac{du}{dr} + \frac{u}{r}\right) + 2B\left[\left(2 + \frac{c^2}{r^2}\right)\left(\frac{du}{dr} + \frac{u}{r}\right) - \frac{c^2}{r^2}\frac{u}{r}\right]$$

$$p' = (2E + 2C + p)\left(\frac{du}{dr} + \frac{u}{r}\right) + 2D\left[\left(2 + \frac{c^2}{r^2}\right)\left(\frac{du}{dr} + \frac{u}{r}\right) - \frac{c^2}{r^2}\frac{u}{r}\right]$$

$$\tag{2.38}$$

$$\tau'^{11} = \Phi' + \left(2 + \frac{c^2}{r^2}\right)\Psi' + p' + 2\Psi\frac{u}{r} - 2p\frac{du}{dr}$$

$$r^2\tau'^{22} = \Phi' + 2\Psi' + p' + 2\Psi\frac{du}{dr} - 2p\frac{u}{r}$$

$$\tau'^{33} = \left(1 + \frac{c^2}{r^2}\right)\Phi' + \left(2 + \frac{c^2}{r^2}\right)\Psi' + p' + 2\Psi\left(\frac{du}{dr} + \frac{u}{r} + \frac{c^2}{r^2}\frac{du}{dr}\right)$$

$$r\tau'^{23} = -\frac{c}{r}\Phi' - \frac{c}{r}\Psi' - 2\frac{c}{r}\Psi\frac{du}{dr}$$

$$\tau'^{31} = \tau'^{12} = 0.$$

$$(2.39)$$

From now on we take into account that the additionally deformed body $B + \varepsilon B'$ is in equilibrium with zero body forces. In Equation 2.25 we may then put

$$\rho f_i + \varepsilon(\rho' f_i + \rho f_i') = 0. \qquad (2.40)$$

After substituting Equations 2.32, 2.35, and 2.39 into 2.25, two of those equations are satisfied automatically and the third one gives

$$\frac{d}{dr}\tau^{11} + \frac{c^2}{r^3}\Psi + \varepsilon\Bigg\{\frac{d}{dr}\tau'^{11} + 2\Bigg[\Phi\left(\frac{d^2u}{dr^2} + \frac{1}{r}\frac{du}{dr} - \frac{u}{r^2}\right)$$

$$+ \Psi\left(2\frac{d^2u}{dr^2} + \frac{1}{r}\frac{du}{dr} - \frac{u}{r^2}\right) + p\frac{d^2u}{dr^2}\Bigg]$$

$$+ 2\frac{c^2}{r^2}\Bigg[(F + 2B + D)\left(\frac{1}{r}\frac{du}{dr} + \frac{u}{r^2}\right)$$

$$+ \left(\frac{d^2u}{dr^2} - \frac{u}{r^2}\right)\Bigg] + 2\frac{c^4}{r^4}B\frac{1}{r}\frac{du}{dr}\Bigg\} = 0. \qquad (2.41)$$

So far the calculations have been kept at the highest level of generality with respect to the function W. Equations 2.41 is valid for every energy function $W(I_1, I_2, I_3)$. As long as the solution εu of the equation 2.41 is small compared to c, 2.41 gives the correct additional displacement. Equation 2.41 can always be solved with the aid of numerical computations. In the next section it will be considered for the important case of the material proposed by Murnaghan (1951).

2.4. Solution for Murnaghan's Material

We confine ourselves here to Murnaghan's material defined by the relation (Murnaghan 1951)

$$W = \frac{l + 2m}{3}J_1^3 + \frac{\lambda + 2\mu}{2}J_1^2 - 2mJ_1J_2 - 2\mu J_2 + nJ_3, \qquad (2.42)$$

where λ, μ, l, m, and n are material constants, and J_1, J_2, J_3 the strain invariants. Denoting by X_i the initial and by x_i the final Cartesian coordinates, we have

$$J_1 = \gamma_{ii}, \; J_2 = \tfrac{1}{2}(\gamma_{ii}\gamma_{jj} - \gamma_{ji}\gamma_{ji}), \; J_3 = \det \gamma_{ij}$$

$$\gamma_{ij} = \frac{1}{2}\left(\frac{\partial x_r}{\partial X^i}\frac{\partial x_r}{\partial X^j} - \delta_{ij}\right). \tag{2.43}$$

It is possible to express the invariants J_k as a function of the invariants I_k according to

$$J_1 = \tfrac{1}{2}(I_1 - 3), \; J_2 = \tfrac{1}{4}[(I_2 - 3) - 2(I_1 - 3)],$$

$$J_3 = \tfrac{1}{8}[(I_3 - 1) - (I_2 - 3) + (I_1 - 3)]. \tag{2.44}$$

Substituting Equation 2.44 into 2.42, one obtains

$$W = \frac{l + 2m}{24}(I_1 - 3)^3 + \frac{\lambda + 2\mu + 4m}{8}(I_1 - 3)^2 + \frac{8\mu + n}{8}(I_1 - 3)$$

$$- \frac{m}{4}(I_1 - 3)(I_2 - 3) - \frac{4\mu + n}{8}(I_2 - 3) + \frac{n}{8}(I_3 - 1). \tag{2.45}$$

Here λ and μ are the Lamé constants, and l, m, and n are constants of the second-order elasticity (sometimes denoted as third-order elastic constants). An extensive literature devoted to the experimental determination of these elastic constants exists (see, for example, Seeger and Buck, 1960).

Specializing now the results of the preceding sections to the material in Equation 2.45, we obtain in accordance with 2.31 and 2.37 for the first and second derivatives of W

$$\left.\begin{aligned}
\Phi &= \frac{l + 2m}{4}\frac{c^4}{r^4} + \frac{\lambda + 2\mu + 3m}{2}\frac{c^2}{r^2} + \frac{8\mu + n}{4}, \\[2mm]
\Psi &= -\frac{m}{2}\frac{c^2}{r^2} - \frac{4\mu + n}{4}, \\[2mm]
p &= \frac{n}{4}, \\[2mm]
A &= \frac{l + 2m}{2}\frac{c^2}{r^2} + \frac{\lambda + 2\mu + 4m}{2}, \quad F = -\frac{m}{2}, \\[2mm]
B &= C = D = E = 0.
\end{aligned}\right\} \tag{2.46}$$

Therefore the stresses τ^{ij} and the stress increments τ'^{ij} according to Equations 2.32 and 2.39 are

$$\tau^{11} = \frac{l}{4}\frac{c^4}{r^4} + \frac{2\lambda + 2m - n}{4}\frac{c^2}{r^2},$$

$$r^2\tau^{22} = \frac{l + 2m}{4}\frac{c^4}{r^4} + \frac{\lambda + 2\mu + m}{2}\frac{c^2}{r^2},$$

$$\tau^{33} = \frac{l + 2m}{4}\frac{c^6}{r^6} + \frac{2\lambda + 4\mu + 6m + l}{4}\frac{c^4}{r^4} + \frac{\lambda + 4\mu + m}{2}\frac{c^2}{r^2},$$

$$r\tau^{23} = -\frac{l + 2m}{4}\frac{c^5}{r^5} - \frac{\lambda + 2\mu + 2m}{2}\frac{c^3}{r^3} - \mu\frac{c}{r},$$

$$\tau^{31} = \tau^{21} = 0,$$

$$(2.47)$$

and

$$\tau'^{11} = -\frac{l}{4}\left(\frac{du}{dr} + \frac{u}{r}\right)\frac{c^4}{r^4} + \frac{-2\lambda - 2m + 4l + n}{4}\left(\frac{du}{dr} + \frac{u}{r}\right)\frac{c^2}{r^2}$$

$$+ \frac{2\lambda + 4\mu + n}{2}\left(\frac{du}{dr} + \frac{u}{r}\right) - \frac{4\mu + n}{2}\frac{u}{r} - \frac{n}{2}\frac{du}{dr}. \qquad (2.48)$$

We do not give the corresponding expressions for the other components τ'^{ij} (easy to calculate from Equations 2.38, 2.39, and 2.46) since they are not required for the further calculations.

Substituting Equations 2.46 to 2.48 into 2.41, one obtains the ordinary differential equation

$$L(r) \equiv -\left(l + \frac{m}{2}\right)\frac{c^4}{r^5} - \left(\lambda + \mu + m - \frac{n}{4}\right)\frac{c^2}{r^3}$$

$$+ \frac{c^4}{r^4}\left[\frac{l}{4}\frac{d^2v}{dr^2} + \left(m + \frac{5}{4}l\right)\frac{1}{r}\frac{dv}{dr} + \frac{3}{4}l\frac{v}{r^2}\right]$$

$$+ \frac{c^2}{r^2}\left[\frac{2\lambda + 2m + 4l - n}{4}\frac{d^2v}{dr^2}\right.$$

$$+ \frac{6\lambda + 6m - 4l + 8\mu - 3n}{4}\frac{1}{r}\frac{dv}{dr}$$

$$+ \left.\frac{2\lambda - 12l - 6m - n}{4}\frac{v}{r^2}\right]$$

$$+ (\lambda + 2\mu)\left(\frac{d^2v}{dr^2} + \frac{1}{r}\frac{dv}{dr} - \frac{v}{r^2}\right) = 0, \qquad (2.49)$$

where v denotes the displacement increment

$$v = \varepsilon u. \tag{2.50}$$

If we consider Equation 2.45 as an exact expression for the energy W valid for every deformation, 2.49 is the exact equation for the additional displacement v provided the solution is small compared to the constant c. In the general case 2.49 cannot be solved analytically. A special case of 2.49 was obtained and solved by Seeger and Mann (1959). Their equation may be obtained from 2.49 by assuming that c is small compared with r. In this case we have

$$\frac{d^2v}{dr^2} + \frac{1}{r}\frac{dv}{dr} - \frac{v}{r^2} = \frac{\lambda + \mu + m - n/4}{\lambda + 2\mu}\frac{c^2}{r^3} \tag{2.51}$$

with the solution

$$v = -\frac{\lambda + \mu + m - n/4}{2(\lambda + 2\mu)}\frac{c^2}{r}\ln\frac{r}{c} + C_1 r + C_2/r, \tag{2.52}$$

where C_1 and C_2 are integration constants. These constants can be determined in such a manner that the deformation field satisfies *two* boundary conditions, e.g., zero radial displacements on two given radii (characterizing an outer and an inner cut-off radius).

Equation 2.49 can be solved analytically in another limiting case, namely for small r/c. In this case it reduces to

$$\frac{l}{4}\frac{d^2v}{dr^2} + \left(m + \frac{5}{4}l\right)\frac{1}{r}\frac{dv}{dr} + \frac{3}{4}\frac{v}{r^2} = \frac{1}{r}\left(l + \frac{m}{2}\right). \tag{2.53}$$

With the aid of the substitution $t = \ln r$ the solution

$$v = \frac{2l + m}{2(m + 2l)}r + C_1 r^{\nu_1} + C_2 r^{\nu_2} \tag{2.54}$$

is found, where

$$\nu_{1,2} = \frac{1}{2}\left(-m - l \pm \sqrt{m^2 + \frac{l^2}{4} + 2ml}\right). \tag{2.55}$$

It should be mentioned that in the case of Murnaghan's material the method proposed by Spencer (1964) for treating almost incompressible bodies is not applicable.

2.5. Further Possible Applications

So far we have considered Equation 2.42 as an exact expression for the elastic energy. It is more realistic to consider it as an approximation for small (not necessarily infinitesimal) strains.

We have emphasized that the treatment in Sections 2.2 and 2.3 holds for arbitrary functions W. One may choose a functional form of W which coincides for small deformations with Equation 2.42 and which takes into account some features of crystals. Due to the periodicity of the arrangement of atoms in crystals, W is not a monotonously increasing function for increasing shear but returns to zero for homogeneous deformations that restore the atomic arrangement of the undeformed crystal. An appropriate choice of W should enable us to give a continuum description of this feature of crystals and to permit an appropriate treatment of screw dislocations in such a material. However, since the periodicity of W is closely linked to the anisotropy of the crystal, it appears advisable to base such a treatment not on the theory of isotropic bodies but on that of curvilinearly orthotropic bodies mentioned in Section 2.1.

Another possible application (and slight extension) of the present theory is that to the scattering of elastic waves from dislocations, which is an important process in the theory of low-temperature heat conductivity. The calculations so far carried out (see Seeger, Bross, and Gruner 1964) were based entirely on Murnaghan's energy expression (2.42). A treatment based on a more general energy function W would permit estimates of the effects of the large strains near the dislocation center on the scattering process. In a sense this would be similar to replacing Born's approximation in the theory of electron scattering by a more exact solution of the Schrödinger equation.

The authors acknowledge gratefully the support by the Deutscher Akademischer Austauschdienst (DAAD), which has made the present collaboration possible. They are also grateful to Drs. C. Teodosiu and E. Mann for critical readings of the final manuscript.

REFERENCES

Anthony, K. H., U. Essmann, A. Seeger, and H. Träuble, 1968, in E. Kröner (ed.), *Mechanics of Generalized Continua* (*Symposium on the Generalized Cosserat Continuum and the Continuum Theory of Dislocations with Applications, at Freudenstadt/Stuttgart*), Berlin-Heidelberg-New York; Springer, p. 355.

Green, A. E., R. S. Rivlin, and R. T. Shield, 1952, *Proc. Roy. Soc.*, London, **A211**, 128.

Green, A. E., and W. Zerna, 1954, *Theoretical Elasticity*, Oxford: Clarendon Press.

Murnaghan, F. D., 1951, *Finite Deformations of an Elastic Solid*, New York: Wiley.

Nabarro, F. R. N., 1967, *Theory of Crystal Dislocations*, Oxford: Clarendon Press.

Orowan, E., 1965, in C. S. Smith (ed.), *The Sorby Centennial Symposium on the History of Metallurgy*, New York: Gordon and Breach, p. 359.

Peierls, R., 1940, *Proc. Phys. Soc.*, London, **52**, 34.

Peierls, R., 1968, in A. R. Rosenfield, et al. (eds.), *Dislocation Dynamics*, New York: McGraw-Hill, pp. XIII, XVII.

Seeger, A., 1955, in S. Flügge (ed.), *Encyclopedia of Physics*, Berlin: Springer, vol. VII/1, p. 383.

Seeger, A., 1964, in M. Reiner and D. Abir (eds.), *Second-Order Effects in Elasticity, Plasticity, and Fluid Dynamics*, New York: Macmillan, p. 129.

Seeger, A., and E. Mann, 1959, *Z. Naturforsch.*, **14a**, 154.

Seeger, A., and O. Buck, 1960, *Z. Naturforsch.*, **15a**, 1056.

Seeger, A., H. Bross, and P. Gruner, 1964, *Disc. Faraday Soc.*, **38**, 69.

Seeger, A., and P. Schiller, 1966, in W. P. Mason (ed.), *Physical Acoustics*, New York and London: Academic, vol. III A, p. 361.

Spencer, A. J. M., 1964, in M. Reiner and D. Abir (eds.), *Second-Order Effects in Elasticity, Plasticity, and Fluid Dynamics*, New York: Macmillan, p. 200.

Truesdell, C., and W. Noll, 1965, in S. Flügge (ed.), *Encyclopedia of Physics*, Berlin-Heidelberg-New York: Springer, vol. III/3.

Urbanowski, W., 1959, *Arch. Mech. Stos.*, **11**, 223.

Wesołowski, Z., and A. Seeger, 1968, in E. Kröner (ed.), *Mechanics of Generalized Continua (Symposium on the Generalized Cosserat Continuum and the Continuum Theory of Dislocations with Applications, at Freudenstadt/Stuttgart)*, Berlin-Heidelberg-New York: Springer, p. 295.

ABSTRACT. The forces between parallel edge dislocations in elastically isotropic specimens are calculated, taking into account climb as well as glide forces. Diagrams are given to show the directions in the resulting force fields for several important cases. For given relative orientation, the magnitude of the force is always inversely proportional to the distance of separation.

3. Forces Between Parallel Edge Dislocations in Elastically Isotropic Crystals

D. KUHLMANN-WILSDORF

3.1. Introduction

The forces of interaction between parallel dislocations are well known, certainly insofar as the relevant mathematical equations are concerned (see for example Nabarro 1952, Cottrell 1953, and Read 1953). Moreover, specific cases have been discussed repeatedly. In particular, discussions of the glide forces between parallel edge dislocations on parallel slip planes have come close to being a standard item in relevant textbooks. Correspondingly, the present paper is not meant to add any new fundamental knowledge. It is written to point out some unexpected and/or useful aspects of the problem which may have hitherto escaped attention.

3.2. General Considerations

Two different forces, both normal to the dislocation axis, act on edge dislocations in stress fields, namely glide forces, acting in the slip plane, and climb forces, acting normal to the slip plane. If the magnitudes of these, per unit length of dislocation line, are denoted by the symbols $_gF$ and $_cF$, respectively, then we know that

$$_gF = b\tau_{gb} \tag{3.1}$$

29

and

$$_cF = -b\tau_{bb} \tag{3.2}$$

where τ_{gb} is the shear stress parallel to the glide plane (indicated by the suffix g) in the direction of the Burgers vector (indicated by the suffix b), evaluated at the position of the dislocation axis, while τ_{bb} is the normal stress component parallel to the direction of the Burgers vector, at the position of the dislocation axis. Here, in order to avoid possible complications due to image forces, due to Bardeen-Herring climb forces (Bardeen and Herring 1952), and due to externally imposed hydrostatic pressures (Weertman 1965, Lothe and Hirth 1967, and de Wit 1968,) we restrict ourselves to infinitely large, isotropic crystals, free of vacancy supersaturations, and free of externally imposed stresses.

The forces of interaction between two parallel edge dislocations (or any dislocations, for that matter) arise because of the stress fields associated with them. In order to find $_cF$ and $_gF$ for various cases it is thus necessary and sufficient to determine the values of τ_{gb} and τ_{bb} due to one dislocation at the position of the other dislocation.

One may easily convince oneself that the law of action equals reation applies to the total forces between dislocations but not, in general, to the glide forces alone, nor to the climb forces alone. For example, an edge dislocation situated on the glide plane of another edge dislocation will not be subject to a climb force because, for an edge dislocation, all normal stress components vanish in the slip plane. However, if the Burgers vectors of the two dislocations are rotated with respect to each other, the first dislocation will not lie on the slip plane of the second dislocation and thus must suffer a climb force. On the other hand, we may see immediately that action equals reaction by considering that the forces of interaction can be found from the gradient of the interaction energy between any two dislocations.

Since, usually, dislocations are much more mobile in glide than in climb, it is very tempting to concentrate on glide forces and neglect climb forces, which to a large extent has been done in the past. In the present paper, both glide and climb forces are considered, and the total force fields for four basic cases are derived. As will be seen, these have simple characteristics which are not apparent from consideration of the glide forces alone.

Throughout, we shall consider the forces which a mobile edge dislocation parallel to the z axis and with a Burgers vector of magnitude b, suffers in the stress field of a positive edge dislocation whose axis coincides with the z axis and whose Burgers vector, $\mathbf{b}'$, is parallel to the x axis. The stress field of the latter dislocation is given by

$$\tau_{xy} = A \frac{x(x^2 - y^2)}{(x^2 + y^2)^2}$$

$$\tau_{yz} = \tau_{zx} = 0$$

$$\tau_{xx} = -A \frac{y(3x^2 + y^2)}{(x^2 + y^2)^2} \tag{3.3}$$

$$\tau_{yy} = A \frac{y(x^2 - y^2)}{(x^2 + y^2)^2}$$

$$\tau_{zz} = -2vA \frac{y}{x^2 + y^2}$$

in Cartesian coordinates, and by

$$\tau_{r\theta} = A \frac{\cos \theta}{r}$$

$$\tau_{\theta z} = \tau_{zr} = 0$$

$$\tau_{rr} = \tau_{\theta\theta} = -A \frac{\sin \theta}{r} \tag{3.4}$$

$$\tau_{zz} = -2vA \frac{\sin \theta}{r}$$

in cylindrical coordinates where

$$A = \frac{Gb'}{2\pi(1 - v)}, \tag{3.5}$$

with G the modulus of rigidity and v Poisson's ratio. In Equations 3.4 above, $\theta = 0 \pm n\pi$ is the slip plane. Four basic cases of relative orientation between the dislocations will be considered as sketched in Figure 3.1.

3.3. Dislocations on Parallel Slip Planes

In this case, $\tau_{gb} = \tau_{xy}$ and $\tau_{bb} = \tau_{xx}$ of Equations 3.3, so that Equations 3.1 and 3.2 can be written

$$_gF_{\perp\perp} = bA \frac{x(x^2 - y^2)}{(x^2 + y^2)^2} \tag{3.6a}$$

$$_cF_{\perp\perp} = bA \frac{y(3x^2 + y^2)}{(x^2 + y^2)^2}, \tag{3.7}$$

choosing the mobile dislocation to be of positive type, as also indicated by the suffices $\perp\perp$.

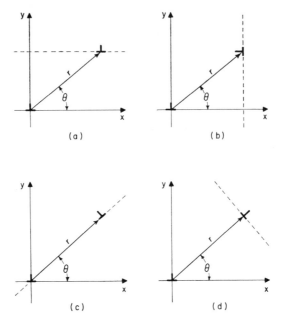

FIGURE 3.1. The four basic cases of relative orientation between two parallel edge dislocations that are considered in the present paper. These are: (a) two parallel edge dislocations on parallel slip planes; (b) two parallel edge dislocations on mutually perpendicular slip planes; (c) two parallel edge dislocations one of which lies on the slip plane of the other; (d) two parallel edge dislocations arranged such that the slip plane of one of them is normal to the closest distance between the dislocations. In all cases the dislocation axes are assumed to be parallel to the z axis, normal to the plane of the figure.

As stated in the Introduction, Equation 3.6a has been exhaustively discussed in quite a few places. Its more important properties are that $_gF_{\perp\perp}$ vanishes for $x = y$ and for $x = 0$, that for $y = $ const (i.e., on any one possible slip plane of the mobile dislocation), $_gF_{\perp\perp}$ has extrema of magnitude $bA/4y$ at $x = 0.414y$ and $x = 2.414y$, and that $_gF_{\perp\perp}$ is repulsive for $|x| > |y|$ but is attractive for $|x| < |y|$. Rewriting Equation 3.6a into cylindrical coordinates one obtains (Nabarro 1952)

$$_gF_{\perp\perp} = bA \frac{\sin 4\theta}{4y}. \tag{3.6b}$$

In this form the important properties of Equation 3.6 are more clearly revealed. In particular, it is now quite obvious that, for any given θ-value, the glide force falls off inversely with y, i.e., the distance between the two glide planes concerned, and that the extrema lie at $\theta = \pi/8 + n\pi/4$, $n = 1, 2, \ldots$.

The climb force according to Equation 3.7 is positive for all positive values of y, i.e., above the slip plane of the fixed dislocation, and is negative below it. Thus, in response to climb forces, parallel edge dislocations with parallel Burgers vectors will always climb apart, while those with anti-parallel Burgers vectors will climb together. Interestingly, for $y = \text{const}$, $_cF_{\perp\perp}$ has a relative minimum of magnitude bA/y at $\theta = 90$ deg, the position in which the dislocations are arranged in the configuration of an incipient tilt boundary, while it attains its maximum at $\theta = \pm 60$ deg with a value $_cF_{\perp\perp} = 1.125\, bA/y$. The radial and tangential components of the inter-action force are (Cottrell 1953)

$$_rF_{\perp\perp} = {_g}F_{\perp\perp} \cos\theta + {_c}F_{\perp\perp} \sin\theta = bA/r \qquad (3.8)$$

and

$$_\theta F_{\perp\perp} = {_c}F_{\perp\perp} \cos\theta - {_g}F_{\perp\perp} \sin\theta = bA\, \frac{\sin 2\theta}{r}. \qquad (3.9)$$

Combining the force components to obtain the force vector, we obtain for the absolute value of $F_{\perp\perp}$

$$F_{\perp\perp} = ({_g}F_{\perp\perp}^2 + {_c}F_{\perp\perp}^2)^{1/2} = ({_r}F_{\perp\perp}^2 + {_\theta}F_{\perp\perp}^2)^{1/2} = \frac{bA}{r}\,(1 + \sin^2 2\theta)^{1/2}, \qquad (3.10)$$

i.e., for any fixed value of θ, the force of interaction falls off simply as $1/r$. The angle α, at which $\mathbf{F}_{\perp\perp}$ is inclined against the x axis is found from

$$\tan\alpha = \frac{_cF_{\perp\perp}}{_gF_{\perp\perp}} = \frac{\sin^3\theta + 3\sin\theta\cos^2\theta}{\cos^3\theta - \sin^2\theta\cos\theta} = \frac{1 + 3\cot^2\theta}{\cot\theta(\cot^2\theta - 1)} \qquad (3.11)$$

"Lines of force" which represent the force of interaction between the two dislocations in accordance with Equations 3.9 and 3.10 are drawn, semiquantitatively, in Figure 3.2. In this form, the forces of interaction are committed to memory much more easily than from diagrams showing the glide forces and climb forces singly.

By the rule of the squares of the Burgers vectors one expects a net repulsion between parallel dislocations if their Burgers vectors include an angle of less than 90 deg between them, expects an attraction if that same angle is larger than 90 deg, and expects neither attraction nor repulsion if the two Burgers vectors involved are at right angles. In this light, the glide attraction between parallel edge dislocations on parallel slip planes for $\theta > 45$ deg, which is so well known and which was briefly mentioned above, is quite unexpected. However, as revealed in Equation 3.8 and Figure 3.2, it is *only* the glide component of the force of interaction which can change sign, in an apparent violation of the rule of the squares of the Burgers vectors, while the total force of interaction between the dislocations is repulsive everywhere, and thus is in accord with the said rule.

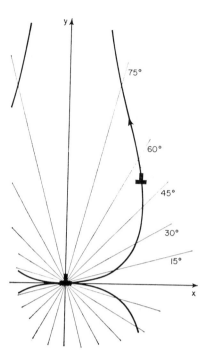

FIGURE 3.2. Lines of force giving the direction of the total force exerted on a mobile positive edge dislocation normal to the plane of the drawing by the stress field due to a fixed positive edge dislocation, also normal to the plane of the drawing, situated at the origin of the coordinate system (see Figure 3.1.a) The forces are generally repulsive if the dislocations are of same sign, attractive if they are of opposite sign.

3.4. Dislocations on Mutually Perpendicular Slip Planes

The appropriate expressions for τ_{gb} and τ_{bb} of Equations 3.1 and 3.2 for this case, with the dislocation orientations as indicated in Figure 3.1b, are $-\tau_{xy}$ and $-\tau_{yy}$ (Equation 3.3), respectively, to yield

$$_gF_{\perp\dashv} = -bA\frac{x(x^2 - y^2)}{(x^2 + y^2)^2} \tag{3.12}$$

$$_cF_{\perp\dashv} = bA\frac{y(x^2 - y^2)}{(x^2 + y^2)^2}. \tag{3.13a}$$

If one wishes to examine the forces of a mobile dislocation on its own slip plane, a constant value of x must be chosen. The function $_gF_{\perp\dashv}$ for constant x, as dependent on $y/x = \tan\theta$, was given Nabarro (1952). For a positive value of x it is negative (i.e., downward in Figure 3.1b) for $|y| < |x|$, and positive for $|y| > |x|$, vanishing at $x = y$. For large distances it approaches $(_gF_{\perp\dashv})_{y \gg x} = bAx/y^2$ and has an intermediate maximum at $y/x = \pm\sqrt{3}$, at which point $_gF_{\perp\dashv}$ assumes the value of $bA/8x$. For $y = 0$ it has a minimum with the value of $-bA/x$.

The climb component as a function of y for constant x (Equation 3.13a) has a form symmetrical to Equation 3.6a and may similarly be rewritten as

$$_cF_{\perp\dashv} = bA\,\frac{\sin 4\theta}{4x} \tag{3.13b}$$

Thus $_cF_{\perp\dashv}$ for $x = $ const. depends on y exactly as $_gF_{\perp\perp}$ for $y = $ const depends on x.

Combining the glide and climb force to find the radial and tangential components of the total force yields

$$_rF_{\perp\dashv} = {_cF_{\perp\dashv}}\cos\theta + {_gF_{\perp\dashv}}\sin\theta = 0 \tag{3.14}$$

and

$$_\theta F_{\perp\dashv} = {_gF_{\perp\dashv}}\cos\theta - {_cF_{\perp\dashv}}\sin\theta = -bA\,\frac{\cos 2\theta}{r}. \tag{3.15}$$

The corresponding lines of force are given in Figure 3.3. As seen from Equation 3.14 and Figure 3.3, the rule of the squares of the Burgers vectors

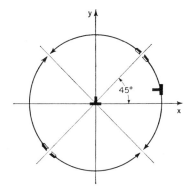

FIGURE 3.3. As Figure 3.2, but for the case that the Burgers vector of the mobile dislocation is rotated through 90° (see Figure 3.1b).

holds true also in this case, inasmuch as no net force of attraction or repulsion exists between the two dislocations with mutually perpendicular Burgers vectors. Even so, the glide as well as climb forces, singly, are about as strong as for dislocations with parallel Burgers vectors.

3.5. One Dislocation Situated on the Other's Slip Plane

These are configurations of the type indicated in Figure 3.1c. In this case, τ_{gb} of Equation 3.1 is given by $\tau_{r\theta}$, and τ_{bb} by τ_{rr} (Equation 3.4) so that

$$_gF_{\perp\gamma} = {_rF_{\perp\gamma}} = b\tau_{r\theta} = bA\,\frac{\cos\theta}{r} \tag{3.16}$$

$$_cF_{\perp\gamma} = {_\theta F_{\perp\gamma}} = -b\tau_{rr} = bA\,\frac{\sin\theta}{r} \tag{3.17}$$

The magnitude of the resulting total force is

$$F_{\perp\gamma} = \frac{bA}{r} \tag{3.18}$$

independent of θ, but its direction depends on θ. Namely, the angle α^*, which $F_{\perp\gamma}$ subtends on the direction connecting the two dislocations, i.e., subtends on the slip plane of the mobile dislocation, is found as

$$\alpha^* = \text{arc tan}(_cF_{\perp\gamma}/_gF_{\perp\gamma}) = \theta. \tag{3.19}$$

The angle between the x axis and the slip plane of the mobile dislocation is thus equal to the angle between the force $F_{\perp\gamma}$ and that same slip plane, so that the angle α, included between the total force and the x axis, is simply

$$\alpha = 2\theta \tag{3.20}$$

as indicated in Figure 3.4a.

As the roles of the two dislocations involved were entirely symmetrical in the cases considered up to this point, the forces on the dislocation considered to be stationary, coincident with the z axis, needed not to be discussed. Now, however, the same symmetry does not apply, and for clarification Figure 3.4b indicates the forces when the roles of the "mobile" and "stationary" dislocations are reversed. As in all preceding cases, the figures do not indicate all possible arrangements. In Figures 3.1, 3.3, and 3.4a, the remaining arrangements can be found by reversing the Burgers vector of the "stationary" dislocation, which will reverse the direction of the force, while Figure 3.4b may be mirrored in the plane normal to the x axis.

3.6. The Line Connecting the Dislocations is Normal to One of the Slip Planes

The geometry of this last case is depicted in Figure 3.1d. The appropriate expressions for τ_{gb} and τ_{bb} of Equation 3.1 are the same as in the preceding section, namely $\tau_{r\theta}$ and τ_{rr} (Equation 3.4), respectively.

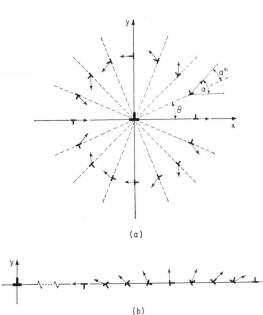

(a)

(b)

FIGURE 3.4. The direction of the forces between two edge dislocations arranged such that one of the dislocations is situated on the other's slip plane. The magnitude of the interaction forces is bA/r, independent of orientation.

Correspondingly, the total force is again given by Equation 3.18. The orientation of the force, however, is different — and much simpler. We find

$$_gF_{\perp\curlyvee} = {}_\theta F_{\perp\curlyvee} = b\tau_{r\theta} = bA\,\frac{\cos\theta}{r} \tag{3.21}$$

$$_cF_{\perp\curlyvee} = {}_rF_{\perp\curlyvee} = -b\tau_{rr} = bA\,\frac{\sin\theta}{r}, \tag{3.22}$$

whence

$$_xF_{\perp\curlyvee} = {}_rF_{\perp\curlyvee}\cos\theta - {}_\theta F_{\perp\curlyvee}\sin\theta = 0 \tag{3.23}$$

$$_yF_{\perp\curlyvee} = {}_rF_{\perp\curlyvee}\sin\theta + {}_\theta F_{\perp\curlyvee}\cos\theta = \frac{bA}{r}, \tag{3.24}$$

i.e., the force is always of magnitude bA/r, normal to the slip plane of the "stationary" dislocation in Figure 3.1d, directed upward everywhere under the conditions indicated in that figure, and downward if the sign of the stationary dislocation is reversed.

   As the forces on the "stationary" dislocations are equal and opposite to the forces on the "mobile" dislocation, it is subject only to climb forces of magnitude bA/r no matter where the mobile dislocation should happen to be situated. This is, of course, not surprising since all shear stresses in the stress field of an edge dislocation vanish in the plane, containing the dislocation axis oriented normal to the slip plane, and that is the plane in which the stationary dislocation is situated with respect to the mobile dislocation.

3.7. Parallel Dislocations with Arbitrary Burgers Vectors

Even while only some special cases have been considered so far, the results of the present investigation are helpful to estimate, at a glance, the direction and magnitude of the forces between any two parallel dislocations: The Burgers vectors of the two dislocations may always be resolved into their edge and screw components, and the interaction be found as the sum of the interactions between the screw components alone and the edge components alone since, in isotropic elasticity theory, parallel edge and screw dislocations do not interact. The two edge components must necessarily always be close to one of the four basic cases considered above, while the interactions between parallel screw dislocations are most simple, always acting in the direction of closest approach between the dislocations and falling off as r^{-1}.

3.8. Summary

The forces between parallel edge dislocations have been investigated for four basic orientation relationships. The results are to some extent unexpected and, when committed to memory, are most useful in estimating the forces between parallel dislocations with arbitrary Burgers vectors.

REFERENCES

Bardeen, J., and C. Herring, 1952, in W. Shockley et al. (eds.), *Imperfections in Nearly Perfect Crystals*, New York: Wiley, p. 261.

Cottrell, A. H., 1953, *Dislocations and Plastic Flow in Crystals*, Oxford: Clarendon Press, p. 45.

de Wit, R., 1968, *J. Appl. Phys.*, **39**, 138.

Lothe, J., and J. P. Hirth, 1967, *J. Appl. Phys.*, **38**, 845.

Nabarro, F. R. N., 1952, *Advances in Physics*, **1**, 269.

Read, W. T., 1953, *Dislocations in Crystals*, New York, McGraw-Hill, Chapter 5.

Weertman, J., 1965, *Phil. Mag.*, **11**, 1217.

ABSTRACT. The line integral expressions of Burgers for dislocation displacements, of Peach and Koehler for stresses and forces, and of Blin for energy are considered. Explicit expressions for these integrals are presented for a finite, straight dislocation segment. It is shown that the elastic properties of complex dislocation arrays can be determined in an easy way by replacing the exact array by an approximate one composed of such segments.

4. Stress Fields of Dislocation Segments and Forces on Them

J. P. HIRTH AND J. LOTHE

4.1. Introduction

In the continuum elastic theory of dislocations, four basic formulas are available which, in principle, enable one to derive the elastic properties of any dislocation array. The first is the Burgers displacement equation. Burgers (1939) imagined a cut made on a surface A, terminating at a line C'. The materials on opposite sides of the cut are displaced by $\mathbf{b}'$, creating a closed dislocation loop along C', as shown in Figure 4.1. The loop is generated by a line segment $d\mathbf{l}' = dl'\xi$, where ξ is a unit vector tangent to the line, dl' is at a point $\mathbf{r}' = (x_1', x_2', x_3')$ and has components dx_1', dx_2', and dx_3'. The convention is used of $\mathbf{b}'$ coincident with ξ for a right-handed screw dislocation. Then at a point $\mathbf{r} = (x_1, x_2, x_3)$, the displacements produced by the loop are

$$\mathbf{u}(\mathbf{r}) = \frac{\mathbf{b}'}{4\pi} \int_A \frac{\mathbf{R} \cdot d\mathbf{A}}{R^3} - \frac{1}{4\pi} \oint_C \frac{\mathbf{b}' \times d\mathbf{l}'}{R}$$

$$+ \frac{1}{8\pi(1-v)} \nabla \oint_C \frac{(\mathbf{b}' \times \mathbf{R}) \cdot d\mathbf{l}'}{R} \tag{4.1}$$

where $\mathbf{R} = \mathbf{r}' - \mathbf{r}$, $d\mathbf{A} = \mathbf{n} dA$ with $\mathbf{n}$ a unit vector normal to the surface and v is Poisson's ratio.

39

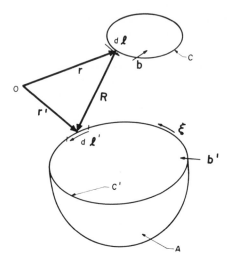

FIGURE 4.1.

From Equation 4.1, Peach and Koehler (1950) derived the equation for the stress field at **r** (see also deWit 1960 for a detailed derivation).

$$\sigma_{\alpha\beta} = -\frac{\mu}{8\pi}\oint_C b'_j \varepsilon_{ij\alpha}\frac{\partial}{\partial x'_i}\nabla^2 R\,dx'_\beta - \frac{\mu}{8\pi}\oint_C b'_j \varepsilon_{ij\beta}\frac{\partial}{\partial x'_i}\nabla^2 R\,dx'_\alpha$$

$$-\frac{\mu}{4\pi(1-\nu)}\oint_C b'_j\varepsilon_{ijk}\left(\frac{\partial^3 R}{\partial x'_i\,\partial x'_\alpha\,\partial x'_\beta} - \delta_{\alpha\beta}\frac{\partial}{\partial x'_i}\nabla^2 R\right)dx'_k \quad (4.2)$$

where μ is the shear modulus, ε_{ijk} is the Einstein permutation operator, $\delta_{\alpha\beta}$ is the Kronecker delta and the subscripts assume the values, 1, 2, 3.

Now consider that an additional dislocation loop is at **r** and is generated by a line segment $d\mathbf{l}$ with components dx_1, dx_2, dx_3, and with Burgers vector **b**. The force on the element $d\mathbf{l}$, caused by the first loop is given by the Peach-Koehler force formula.

$$d\mathbf{F} = b_\alpha \sigma_{\alpha\beta}(\mathbf{e}_\beta \times d\mathbf{l}), \quad (4.3)$$

where the $\mathbf{e}_\beta$ are unit vectors. Finally, from Equation 4.2, Blin (1955) derived the expression for the interaction energy between the two loops:

$$W_{\text{int}} = -\frac{\mu}{2\pi}\oint_C\oint_{C'}\frac{(\mathbf{b}\times\mathbf{b}')\cdot(d\mathbf{l}\times d\mathbf{l}')}{R} + \frac{\mu}{4\pi}\oint_C\oint_{C'}\frac{(\mathbf{b}\cdot d\mathbf{l})(\mathbf{b}'\cdot d\mathbf{l}')}{R}$$

$$+\frac{\mu}{4\pi(1-\nu)}\oint_C\oint_{C'}(\mathbf{b}\times d\mathbf{l})\cdot\mathbf{T}\cdot(\mathbf{b}'\times d\mathbf{l}') \quad (4.4)$$

where **T** is a tensor with components $T_{ij} = \partial^2 R/\partial x_i\,\partial x_j$.

For generally curved dislocations, the above integrals are quite complex, and usually can be solved only numerically. Hence a number of methods have been developed for approximating a curved configuration by a sequence of finite, straight segments and then carrying out exact integrations for these approximate configurations to estimate the elastic properties. For the interaction energy this was accomplished by Jøssang et al. (1965), with later variants by Gottlieb (1967), Bacon and Crocker (1966), and deWit (1967). Some misprints and errors in these treatments are clarified by Jøssang (1968) and by Hirth and Lothe (1968). Our own position on the interaction energy is well documented in these references and hence is not discussed here.

For the stresses, Yoffe (1960, 1961) started with Equation 4.1 and derived the stress field of an angular dislocation with its two arms extended infinitely. Hokanson (1963) extended her treatment. The stress field of a closed loop can be determined by superposing a set of such angular dislocations. Li (1964) noted that the stress fields of the two arms were independent of one another and pointed out that stress fields of loops composed of straight segments could be generated by superposing single semi-infinite dislocations, corresponding to one of the arms of the angular dislocation.

Here, we show that the result for the stress field of a single finite dislocation segment can be derived quite straightforwardly from Equation 4.2. In one form, the result reduces to Li's result. Some other forms of the result, useful in some cases, are also presented. The results can be extended to give the result for the angular dislocation, for the infinitesimal loop, or for any configuration composed of straight segments.

4.2. Stress Field of a Dislocation Segment

Let us first develop an extended form of Equation 4.2. We define the relative coordinates $x = x'_1 - x_1$, $y = x'_2 - x_2$, and $z = x'_3 - x_3$. Then, the stress field of Equation 4.2 is developed specifically for the component dx'_3 of $d\mathbf{l}'$, yielding

$$\frac{d\sigma_{11}}{\sigma_0} = b'_1 \left[\frac{y}{R^3} + \frac{3x^2 y}{R^5} \right] dx'_3 + b'_2 \left[\frac{x}{R^3} - \frac{3x^3}{R^5} \right] dx'_3$$

$$\frac{d\sigma_{22}}{\sigma_0} = b'_1 \left[-\frac{y}{R^3} + \frac{3y^3}{R^5} \right] dx'_3 + b'_2 \left[-\frac{x}{R^3} - \frac{3xy^2}{R^5} \right] dx'_3$$

$$\frac{d\sigma_{33}}{\sigma_0} = b'_1 \left[\frac{2vy}{R^3} - \frac{y(x^2 + y^2)}{R^5} \right] dx'_3 + b'_2 \left[-\frac{2vx}{R^3} + \frac{x(x^2 + y^2)}{R^5} \right] dx'_3$$

$$\frac{d\sigma_{12}}{\sigma_0} = b'_1 \left[-\frac{x}{R^3} + \frac{3xy^2}{R^5} \right] dx'_3 + b'_2 \left[\frac{y}{R^3} - \frac{3x^2 y}{R^5} \right] dx'_3 \qquad (4.5)$$

$$\frac{d\sigma_{23}}{\sigma_0} = b_1'\left[\frac{vz}{R^3} - \frac{3y^2z}{R^5}\right] dx_3' + b_2'\left[\frac{3xyz}{R^5}\right] dx_3' \qquad \text{(4.5 cont.)}$$

$$+ b_3'\left[-\frac{x(1-v)}{R^3}\right] dx_3'$$

$$\frac{d\sigma_{31}}{\sigma_0} = b_1'\left[-\frac{3xyz}{R^5}\right] dx_3' + b_2'\left[-\frac{vz}{R^3} + \frac{3x^2z}{R^5}\right] dx_3'$$

$$+ b_3'\left[\frac{y(1-v)}{R^3}\right] dx_3',$$

where $\sigma_0 = \mu b/4\pi(1-v)$.

The stress components produced by the components dx_1' and dx_2' of dl' can be obtained from Equation 4.5 by cyclic permutation $1 \to 2$, $2 \to 3$, $3 \to 1$, etc.

Of course, the stress tensor of Equation 4.5 satisfies neither the equilibrium equations nor the compatibility equations of elasticity. However, when Equation 4.5 is integrated over a closed loop of dislocation, over a continuous network of dislocations, or for a dislocation terminating at a free surface, with the inclusion of image terms, these equations are satisfied. Hence, no problems arise when applying Equation 4.5 to a physically realistic situation. Furthermore, the stresses of Equation 4.5 transform like a tensor for each segment of dislocation line, so that the stresses can all be transformed to a common coordinate system and then summed.

Now let us extend Equation 4.5 to describe a finite straight segment. We choose the x_3' axis to lie along the dislocation line (Figure 4.2), so that

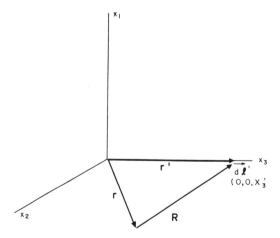

FIGURE 4.2.

$x'_1 = x'_2 = 0$. Then in Equation 4.5 $x = -x_1$, $y = -x'_2$, $z = x'_3 - x_3$. The stresses associated with a finite segment AB are then given by

$$\sigma_{ij} = \sigma_{ij}[x'_3(B)] - \sigma_{ij}[x'_3(A)] = \int_{x'_3(A)}^{x'_3(B)} d\sigma_{ij}. \tag{4.6}$$

All of the integrals involved in Equation 4.6, with Equation 4.5 substituted for $d\sigma_{ij}$, are elementary. Furthermore, all terms not containing x'_3 cancel when the definite limits are taken. Thus, it suffices simply to present $\sigma_{ij}[x'_3]$. The solution for σ_{ij} can then be easily determined from Equation 4.6 by inserting the appropriate values $x'_3 = x'_3(B)$ and $x'_3 = x'_3(A)$ and subtracting. The result for the stress tensor $\sigma_{ij}[x'_3]$ is

$$\frac{\sigma_{11}}{\sigma_0} = b'_1\left[-\frac{x_2}{R(R-z)}\left(1 + \frac{x_1^2}{R^2} + \frac{x_1^2}{R(R-z)}\right)\right]$$

$$+ b'_2\left[\frac{-x_1}{R(R-z)}\left(1 - \frac{x_1^2}{R^2} - \frac{x_1^2}{R(R-z)}\right)\right]$$

$$\frac{\sigma_{22}}{\sigma_0} = b'_1\left[\frac{x_2}{R(R-z)}\left(1 - \frac{x_2^2}{R^2} - \frac{x_2^2}{R(R-z)}\right)\right]$$

$$+ b'_2\left[\frac{x_1}{R(R-z)}\left(1 + \frac{x_2^2}{R^2} + \frac{x_2^2}{R(R-z)}\right)\right]$$

$$\frac{\sigma_{33}}{\sigma_0} = b'_1\left[\frac{x_2 z}{R^3} - \frac{2vx_2}{R(R-z)}\right] + b'_2\left[-\frac{x_1 z}{R^3} + \frac{2vx_1}{R(R-z)}\right] \tag{4.7}$$

$$\frac{\sigma_{12}}{\sigma_0} = b'_1\left[\frac{x_1}{R(R-z)}\left(1 - \frac{x_2^2}{R^3} - \frac{x_2^2}{R(R-z)}\right)\right]$$

$$+ b'_2\left[\frac{-x_2}{R(R-z)}\left(1 - \frac{x_1^2}{R^2} - \frac{x_1^2}{R(R-z)}\right)\right]$$

$$\frac{\sigma_{23}}{\sigma_0} = b'_1\left[\frac{v}{R} - \frac{x_2^2}{R^3}\right] + b'_2\left[\frac{x_1 x_2}{R^3}\right] + b'_3\left[\frac{x_1(1-v)}{R(R-z)}\right]$$

$$\frac{\sigma_{13}}{\sigma_0} = b'_1\left[-\frac{x_1 x_2}{R^3}\right] + b'_2\left[-\frac{v}{R} + \frac{x_1^2}{R^3}\right] + b'_3\left[-\frac{x_2(1-v)}{R(R-z)}\right],$$

where as before $z = x'_3 - x_3$, but now $R^2 = x_1^2 + x_2^2 + z^2$.

The above form follows directly from Equation 4.2 by integration. Hence, as stated at the outset, the Peach-Koehler equation for $\sigma_{\alpha\beta}$ leads directly to the stress field for a finite straight segment. However, because terms independent of x'_3 can be added to Equation 4.7 without changing

σ_{ij} in Equation 4.6, other forms of $\sigma_{ij}(x'_3)$ are possible. One form is (see Hirth and Lothe 1968)

$$\frac{\sigma_{11}}{\sigma_0} = b'_1\left[\frac{x_2}{R(R+z)}\left(1 + \frac{x_1^2}{R^2} + \frac{x_1^2}{R(R+z)}\right)\right]$$

$$+ b'_2\left[\frac{x_1}{R(R+z)}\left(1 - \frac{x_1^2}{R^2} - \frac{x_1^2}{R(R+z)}\right)\right] \qquad (4.8)$$

$$\frac{\sigma_{23}}{\sigma_0} = b'_1\left[\frac{v}{R} - \frac{x_2^2}{R^3}\right] + b'_2\left[\frac{x_1 x_2}{R^3}\right] + b'_3\left[-\frac{x_1(1-v)}{R(R+z)}\right]$$

It is obvious from a comparison of Equations 4.7 and 4.8 how the remaining terms in Equation 4.8 can be written out: change all factors $(R - z)$ to $(R + z)$ and change the sign of all terms multiplied by $(R - z)^{-1}$ or $(R - z)^{-2}$. Equation 4.8 is the form that specifically yields Li's (1964) result. His case is given by Equation 4.6 with $x'_3(A) = 0$, $x'_3(B) = \infty$. For this case Equation 4.8 yields

$$\frac{\sigma_{11}}{\sigma_0} = b'_1\left[-\frac{x_2}{R(R-x_3)}\left(1 + \frac{x_1^2}{R^2} + \frac{x_1^2}{R(R-x_3)}\right)\right]$$

$$+ b'_2\left[\frac{-x_1}{R(R-x_3)}\left(1 - \frac{x_1^2}{R^2} - \frac{x_1^2}{R(R-x_3)}\right)\right] \qquad (4.9)$$

etc., with $R^2 = x_1^2 + x_2^2 + x_3^2$. Equation 4.9 is identical to Li's expression.

A third form, convenient for use in limiting cases where $x'_3 \to 0$ or $x'_3 \to \infty$, is obtained from Equation 4.7 by replacing the factor $(R - z)^{-1}$ by z/ρ^2 and the factor $R^{-1}(R - z)^{-2}$ by $2z/\rho^4$, where $\rho^2 = x_1^2 + x_2^2$. Thus

$$\frac{\sigma_{11}}{\sigma_0} = b'_1\left[-\frac{x_2 z}{R\rho^2}\left(1 + \frac{2x_1^2}{\rho^2} + \frac{x_1^2}{R^2}\right)\right] + b'_2\left[\frac{-x_1 z}{R\rho^2}\left(1 - \frac{2x_1^2}{\rho^2} - \frac{x_1^2}{R^2}\right)\right]$$

$$(4.10)$$

Li (1964) already has noted that the stress fields of two semi-infinite arms can be superposed to give the angular dislocation result. Since Equation 4.8 reduces to the Li result, evidently the angular dislocation can be generated in a straightforward manner from Equations 4.7, 4.8, or 4.10. Similarly, as demonstrated by Hirth and Lothe (1968), Kroupa's (1960, 1962a, 1962b) result for the stress field of an infinitesimal closed dislocation loop can be generated from the above equations. Thus, as was asserted at the outset, all stress-field equations for configurations composed of straight-line segments can be developed directly from the Peach-Koehler Equation 4.2. In all such equations, a portion of the integration of Equation 4.2 already has been performed, reducing the labor in the determination of a final solution. Equations 4.7–4.10 involve the least amount of algebra of the various methods for achieving such solutions.

4.3. Interaction Forces

With the above results for the stress tensor σ_{ij}, the interaction forces between a configuration composed of a set of straight segments l'_k (or approximated by such a set) and an element dl can be developed from Equation 4.3. However, in general, this procedure involves the transformation of components, of the set of stress tensors, which produce no force on dl. Thus, some effort is wasted. An alternative method is to combine Equations 4.1 and 4.3. After some analysis, one finds the result (Hirth and Lothe 1968)

$$dF = -\frac{\mu}{8\pi} \int [(b' \times b) \cdot \nabla(\nabla^2 R)](dl \times dl')$$

$$-\frac{\mu}{8\pi} \int \{[b' \times \nabla(\nabla^2 R)] \times dl\}(b \cdot dl')$$

$$-\frac{\mu}{4\pi(1-v)} \int [(b' \times dl') \cdot \nabla](dl \times b \cdot T) \qquad (4.11)$$

$$+\frac{\mu}{4\pi(1-v)} \int [(b' \times dl') \cdot \nabla(\nabla^2 R)](dl \times b)$$

where T is a tensor with the components $T_{\alpha\beta} = \partial^2 R/\partial x_\alpha\, \partial x_\beta$. Equation 4.11 is the interaction force analogue of Equation 4.2. Again the total force dF on dl can be found by integrating Equation 4.11 over each segment l'_k contributing to the interaction. Since the explicit integrals of Equation 4.11 are similar to those discussed above in connection with Equation 4.2, and since the detailed result is presented elsewhere (Hirth and Lothe 1968), the solution to Equation 4.11 is not given here. Equation 4.11 is quite useful in image force problems, in problems involving dislocation bends, and in the justification for the line tension approximation (Lothe 1967a, 1967b, 1967c; Brown 1967; Indenbom and Dubnova 1967; Indenbom and Orlov 1967).

One can also use Equation 4.11 to determine the *self-force* on a portion dl of a segmented, closed loop caused by the other segments l'_k of the loop. This is the analogue of the self-stress introduced by Brown (1964). However, in such a procedure one must include the self-force on dl caused by the segment to which dl belongs (Figure 4.3). The latter term is given by

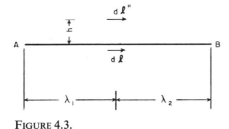

FIGURE 4.3.

the force on an element $d\mathbf{l}''$, a distance h removed from $d\mathbf{l}$ but with the same $\mathbf{b}$, caused by the segment AB minus the force on $d\mathbf{l}''$ caused by an infinite straight dislocation containing AB, in the limit that $h \to 0$,

$$d\mathbf{F} = \frac{\mu v}{4\pi(1-v)}\left(\frac{1}{\lambda_1} + \frac{1}{\lambda_2}\right)(\mathbf{b} \cdot d\mathbf{l})[\xi \times (\xi \times \mathbf{b})]. \tag{4.12}$$

Here λ_1 and λ_2 are defined in Figure 4.3.

This completes the treatment of stresses and interaction forces associated with straight dislocation segments. The analogous development for the interaction energy is presented elsewhere (Jøssang et al. 1965; Jøssang 1968).

This work was supported by the Directorate of Materials and Processes of the U.S. Air Force Systems Command.

REFERENCES

Bacon, D. J., and A. G. Crocker, 1966, *Phil. Mag.*, **13**, 217.

Blin, J., 1955, *Acta Met.*, **3**, 199.

Brown, L. M., 1964, *Phil. Mag.*, **10**, 441.

Brown, L. M., 1957, *Phil. Mag.*, **15**, 363.

Burgers, J. M., 1939, *Proc. Kon. Ned. Akad. Wetenschap.*, **42**, 293.

Gottlieb, H. P. W., 1967, *Phil. Mag.*, **15**, 290.

Hirth, J. P., and J. Lothe, 1968, *Theory of Dislocations*, New York: McGraw-Hill.

Hokanson, J. L., 1963, *J. Appl. Phys.*, **34**, 2337.

Indenbom, V. L., and G. N. Dubnova, 1967, *Sov. Phys. Solid State*, **9**, 915.

Indenbom, V. L., and C. C. Orlov, *Sov. Phys. Cryst.*, **12** 971.

Jøssang, T., 1968, *Phys. Stat. Sol.*, **27**, 579.

Jøssang, T., J. Lothe, and K. Skylstad, 1965, *Acta Met.*, **13**, 271.

Kroupa, F., 1960, *Czech. J. Phys.*, **10B**, 384.

Kroupa, F., 1962a, *Czech. J. Phys.*, **12B**, 191.

Kroupa, F., 1962b, *Phil. Mag.*, **7**, 783.

Li, J. C. M., 1964, *Phil. Mag.*, **10**, 1097.

Lothe, J., 1967a, *Phil. Mag.*, **15**, 9.

Lothe, J., 1967b, *Phil Mag.*, **15**, 353.

Lothe, J., 1967c, *Physica Norvegica*, **2**, 153.

Peach, M. O., and J. S. Koehler, 1950, *Phys. Rev.*, **80**, 436.

de Wit, R., 1960, in F. Seitz and D. Turnbull (eds.), *Solid State Physics*, New York: Academic, vol. 10, p. 249.

de Wit, R., 1967, *Phys. Stat. Sol.*, **20**, 575.

Yoffe, E., 1960, *Phil. Mag.*, **5**, 161.

Yoffe, E., 1961, *Phil. Mag.*, **6**, 1147.

ABSTRACT. A new formalism is presented for treating the nonlinear aspects of the elastic theory of continuous distributions of dislocations and point defects. The concept of deformations of tensors and connections associated with an imperfect continuum is introduced and used to obtain relevant geometric field equations together with expressions for the change of defect content without invoking either "non-Euclidean spaces or anholonomic coordinates." Attention is drawn to the fact that the ordinarily used Cauchy strain is a generally invalid strain measure and the theory is therefore developed using both the correct Green strain and other appropriate strain measures. The correct defect measure is the Cauchy torsion tensor shown to be, in general, independent of the analogous Green torsion tensor which occurs in the geometric field equation for the Green strain. The consequent difficulties are discussed and procedures for solving two new types of nonlinear defect problems are given

5. On the Deformation of an Imperfect Solid

R. BULLOUGH AND J. A. SIMMONS

5.1. Introduction

Professor Orowan (1934) was, together with Polanyi (1934) and Taylor (1934), the first person to recognize the significant role played by defects in the deformation of an imperfect crystalline medium. In particular he conceived that the contribution made by defects to the total deformation of such a medium was specifically due to the motion of dislocations. Since that time dislocation theory has developed into a major branch of material science, ranging in scope from atomic calculations in crystalline lattices to field theoretic studies of distributions of dislocations. This paper will be devoted to the latter aspects of dislocation theory together with related topics concerning point defects.

The differential geometrical description of continuous distributions of dislocations and the important relationship between the Burgers vector and the torsion tensor of a material affine connection have been developed in a series of papers by Kondo, Bilby, and Kröner (see for example Kondo 1955; Bilby et al. 1955; Kröner and Seeger 1959). These treatments have provided descriptions of the defects *in situ* in an imperfect body and have freely utilized the concepts of non-Euclidean spaces and anholonomic coordinates. In addition to the intrinsic mathematical difficulties associated with such concepts, these treatments have used only the Cauchy (or Eulerian) strain as a strain measure and have thereby excluded the possibility of dealing with the nonlinear aspects of an anisotropic body. Furthermore these formalisms are not well adapted to the discussion of the actual deformation of a given solid.[1]

In this paper we wish to present a new geometric formalism, specifically avoiding the use of non-holonomic coordinates, which permits both the characterization of an existing imperfect body as well as its deformations. The presentation herein will be restricted to displacement type defects, such as dislocations and point defects, which can be treated in a formalism with zero curvature tensor. A more general treatment including the appropriate stress analysis requires more extensive discussion and will be presented elsewhere.

5.2. Initial Concepts

We consider a solid body said to be in a given state α in ordinary space, which we designate laboratory space and for which we select a particular fixed Cartesian coordinate system x^1, x^2, x^2. We refer to this system as the laboratory coordinate system. It is assumed that the solid body has a local crystalline structure, that is, (1) at each material point of the body a reference configuration of the triad of vectors $\mathbf{x}_i$ is selected which are sufficient to characterize any fixed Bravais reference lattice; (2) an experimental procedure exists for identifying the three local crystallographic vectors $\mathbf{a}_i$ which correspond to the three reference vectors $\mathbf{x}_i$. (See the theory of oriented media, Truesdell and Toupin 1960). This local crystalline structure characterizes the defect content of the body and is considered to determine its elastic energy density.

It is convenient to introduce a mathematical reference state $r\alpha$ whose material points coincide with those of the state α but in which the local crystallographic vectors $_{r\alpha}\mathbf{a}_i$ are in the reference configuration:

$$_{r\alpha}\mathbf{a}_i = \mathbf{x}_i. \tag{5.1}$$

In keeping with the later discussion of deformations the vectors $_r\mathbf{a}_i$ may be considered to be produced from the corresponding lattice triad $_{r\alpha}\mathbf{a}_i$ in the reference state by a lattice deformation $(\alpha E r\alpha)$; using a notation similar to Bilby (1960), the letter E is introduced to indicate that it is this lattice deformation upon which the *elastic* energy density depends. Then any lattice vector $\mathbf{a}$ with components a^i in the reference state may be deformed to a corresponding vector designated

$$\langle \mathbf{a}; r\alpha E\alpha \rangle = \mathbf{x}_i (\alpha E r\alpha)^i_j a^j \tag{5.2}$$

---

[1] For some attempts in this direction see Bilby, Gardner, and Stroh (1957) and Stojanovich (1962).

in the state α. On the other hand a covariant lattice form **b** with components b_i (compare reciprocal lattice vector) in the state α may be deformed under the same lattice deformation to the corresponding form

$$\langle \mathbf{b}; \alpha Er\alpha \rangle = b_i(\alpha Er\alpha)^i_j \mathbf{x}^j \tag{5.3}$$

in the state $r\alpha$, where $\mathbf{x}^i$ are the covariant laboratory base. We shall assume that this lattice deformation $(\alpha Er\alpha)$ is nonsingular so that its inverse $(r\alpha E\alpha)$ may also be introduced. Analogous expressions to Equations 5.2 and 5.3 for the deformations of vectors and forms involving $(r\alpha E\alpha)$ follow immediately. In fact we may now write the structural notation for the deformation of a lattice tensor T in $r\alpha$ from $r\alpha$ to α as $\langle T; r\alpha E\alpha \rangle$ where the covariant components of T are deformed by $(r\alpha E\alpha)$ and the contravariant components by $(\alpha Er\alpha)$. Clearly $(r\alpha E\alpha)$ and $(\alpha Er\alpha)$ are two state tensors connecting local tensors at the same material points in the states α and $r\alpha$; these tensors and all other geometric quantities will, in this work for ease of presentation, be expressed in terms of the underlying coordinates. The methods for converting such quantities to general curvilinear coordinate systems are well known.

Let us now introduce pertinent strain measures. We have already assumed that the elastic energy density of the imperfect body is specified by a knowledge of the local crystallographic base $_\alpha\mathbf{a}_i$. This in turn is specified by a knowledge of the lattice deformation $(\alpha Er\alpha)$ with respect to the reference configuration in the state $r\alpha$; i.e., the elastic energy density may be considered a tensor function of the tensor $(\alpha Er\alpha)$. Several other entirely equivalent ways of expressing the energy dependence on the lattice deformation also exist. In particular it may equally well be considered a function of the tensor $(r\alpha E\alpha)$, or, because the energy density does not depend on the particular orientation of the crystallographic triads $_\alpha\mathbf{a}_i$, as a function of the Green strain metric

$$\langle c; \alpha Er\alpha \rangle = (\alpha Er\alpha)^r_i(\alpha Er\alpha)^r_j \mathbf{x}^i\mathbf{x}^j, \tag{5.4}$$

where c is the Euclidean metric such that[2]

$$c_{ij} \overset{n}{=} \delta_{ij}. \tag{5.5}$$

The Green strain itself, denoted by $(\alpha er\alpha)$, is related to the Green strain metric by

$$(\alpha er\alpha) = \tfrac{1}{2}(\langle c; \alpha Er\alpha \rangle - c). \tag{5.6}$$

On the other hand the Cauchy strain metric

$$\langle c; r\alpha E\alpha \rangle = (r\alpha E\alpha)^r_i(r\alpha E\alpha)^r_j \mathbf{x}^i\mathbf{x}^j \tag{5.7}$$

is not, in general, a satisfactory strain measure because its value is insensitive to the local orientation of the reference configuration $_{r\alpha}\mathbf{a}_i$ — it is thus

---

[2] The symbol $\overset{n}{=}$ indicates numerical rather than tensor equality

inappropriate for an anisotropic body — and sensitive to the local orientation of the actual lattice configuration $_\alpha\mathbf{a}_i$. Its only use is then for isotropic or linear anisotropic bodies. The relevant Cauchy strain $(r\alpha e\alpha)$ has the form:

$$(r\alpha e\alpha) = \tfrac{1}{2}(c - \langle c; r\alpha E\alpha\rangle). \tag{5.8}$$

Let us now consider the deformation of a body from an initial state o to a final state f. In order to specify the body in the final state we must be given the material point positions and the local crystallographic configurations and their relation to the corresponding initial point positions and local crystallographic configurations. Let the position of a typical point P in the initial state be given by $_ox^i$ and its position in the final state by $_fx^i$. The relation of the point positions between the initial and final states determines the macroscopic shape change of the body under deformation. If $_fx^i(o)$ denotes the final state coordinates of a point parameterized with respect to its initial positions, then we may introduce the local macroscopic deformation tensor (fJo) as the Jacobian:

$$(fJo)^i_j = \frac{\partial_f x^i(o)}{\partial_o x^j}. \tag{5.9}$$

This tensor is clearly a two-state tensor and it relates "macroscopic vectors" of the initial state with those in the final state. Note that from the definition of the reference state it is possible to write the analogous expression for the identity local shape change between the state α and its reference state, i.e.,

$$(\alpha J r\alpha)^i_j \overset{n}{=} \delta^i_j \tag{5.10}$$

Let the final state lattice configuration at the point P be given by the triad of vectors $_f\mathbf{a}_i$ and let these be represented in terms of the deformation $(fErf)$ from the reference state as

$$_f\mathbf{a}_i = \mathbf{x}_j(fErf)^j_i \tag{5.11}$$

If we now make use of the fact that the reference configuration coincides with the laboratory coordinate bases in all reference states, we may write

$$(rfEro)^i_j \overset{n}{=} \delta^i_j \tag{5.12}$$

so that simple transitivity of deformations enables us to introduce the relative lattice deformation

$$(fEo) = (fErf)(rfEro)(roEo). \tag{5.13}$$

In terms of this relative deformation we may write the relative Green strain metric and relative Green strain as

$$\langle c; fEo\rangle = (fEo)^r_i(fEo)^r_j \mathbf{x}^i\mathbf{x}^j \tag{5.14}$$

and

$$(feo) = \tfrac{1}{2}(\langle c; fEo \rangle - c), \tag{5.15}$$

respectively. The corresponding relative Cauchy strain metric and relative Cauchy strain are

$$\langle c; oEf \rangle = (oEf)_i^r(oEf)_j^r \mathbf{x}^i \mathbf{x}^j \tag{5.16}$$

and

$$(oef) = \tfrac{1}{2}(c - \langle c; oEf \rangle), \tag{5.17}$$

respectively.

5.3. Material Connections and their Deformation

We have seen in the previous section that the elastic energy density of a body in the state α is determined by the lattice deformation $(\alpha Er\alpha)$ or equivalently $(r\alpha E\alpha)$. In geometrical terms the energy is thus determined by a comparison of a deformed Cartesian triad with an undeformed Cartesian triad at every point of the body. In the same way the metric strain measures stem from the comparison of a deformed Euclidean metric with an undeformed Euclidean metric. It is thus suggestive that further geometrical properties of a deformed body may be deduced by a comparison of its deformed Euclidean properties with its undeformed properties. We shall in particular study the Euclidean connection and its deformation and show both that it leads directly to the geometric field equations for the strain measures and that its invariants provide useful defect measures.

In order to know how to deform a connection associated to a given state it is necessary to understand the principle by which ordinary tensors and functions deform. The feature which these have in common with affine connections is that they may all be regarded as linear operators on a linear space producing values in another linear space. Thus if we are given an operator O_α on the state α, we mean by the deformed operator O_β on the state β, that operator which leaves invariant the relations existing in the state α; more explicitly we define O_β as that operator which does to deformed objects the equivalent of what the undeformed operator O_α does to undeformed objects. In equation form this "golden rule" for deformation is:

$$O_\beta[u_\beta] = C_V(O_\alpha[\tilde{C}_U(u_\beta)]), \tag{5.18}$$

where the operator O_α maps the linear space $U_\alpha \to V_\alpha$ and the operator O_β maps the linear space $U_\beta \to V_\beta$. $\tilde{C}_U$ and C_V define the correspondences between the spaces U_β, U_α and V_α, V_β, respectively and are determined by the deformation; the u_β in Equation 5.18 is any element of U_β. Equation 5.18 is a familiar one from many branches of physics and mathematics, expressing, for example, a similarity transformation if all the mappings in Equation 5.18 are considered elements of a transformation group.

Equation 5.18 may now be applied to obtain the deformation of an affine connection. Let Γ be an affine connection in the state α with connection coefficients Γ_{qr}^p, i.e., the covariant derivative of the vector field $\mathbf{a} = \mathbf{x}_i a^i$ in the $\mathbf{x}_k$ direction has the form

$$\overset{\Gamma}{\nabla}_{\mathbf{x}_k}[\mathbf{x}_i a^i] = \mathbf{x}_i\left(\frac{\partial a^i}{\partial x^k} + \Gamma_{rk}^i a^r\right). \tag{5.19}$$

Then if Γ is thought of as an operator on lattice vector fields which deform via the lattice deformation $(\beta E\alpha)$ to the state β, Equation 5.18 yields[3] the deformed connection $\langle\Gamma; \alpha E\beta\rangle$ in the state β:

$$\langle\Gamma; \alpha E\beta\rangle_{qr}^p = (\beta E\alpha)_j^p \frac{\partial(\alpha E\beta)_q^j(\beta)}{\partial_\beta x^r}$$

$$+ (\beta E\alpha)_j^p \Gamma_{sk}^j(\alpha E\beta)_q^s(\alpha J\beta)_r^k. \tag{5.20}$$

In this expression $(\alpha J\beta)$ is the macroscopic deformation from the state β to the state α, and $(\alpha E\beta)_q^j(\beta)$ means that the tensor $(\alpha E\beta)$ is parameterized with respect to the point positions in the state β. If Γ is thought of as an operator on macroscopic vector fields, then the deformed connection $\langle\Gamma; \alpha J\beta\rangle$ has the identical form to Equation 5.20 except that all E's are replaced by J's. It is the deformed connection in Equation 5.20, however, based as it is on the lattice vector fields, that yields information on the lattice defect content of the body.

In particular when $\Gamma = C$, where C is the Euclidean connection defined by

$$\overset{n}{C_{jk}^i} = 0, \tag{5.21}$$

we obtain, from Equation 5.20, the deformed Euclidean connection

$$\langle C; \alpha E\beta\rangle_{qr}^p = (\beta E\alpha)_j^p \frac{\partial(\alpha E\beta)_q^j(\beta)}{\partial_\beta x^r}. \tag{5.22}$$

Furthermore the "golden rule" for deformation implies that since $\mathbf{x}_i$ and c are invariants of the Euclidean connection (i.e., all their covariant derivatives vanish) then $\langle\mathbf{x}_i; \alpha E\beta\rangle$ and $\langle c; \alpha E\beta\rangle$ are invariants of the deformed connection $\langle C; \alpha E\beta\rangle$. We shall call $\langle C; \alpha E\beta\rangle$ the material connection of the state β relative to the state α and, more simply, when $\alpha = r\beta$ the quantity $\langle C; r\beta E\beta\rangle$ the material connection of the state β. It is a simple matter to show that these connections are curvature free.

---

[3] The detailed derivation of this expression will be given elsewhere.

5.4. Geometric Field Equations and Defect Content

The geometric field equations are those relations satisfied by either the lattice deformations $(\alpha E\beta)$ or their associated strains $(\alpha e\beta)$. For certain types of problems in either perfect or imperfect solids they provide some of the basic field equations for determining the internal stress. The expressions for the two basic invariants of the material connection of the state β relative to the state α are (see for example Schouten 1954) (i) the torsion tensor

$$
\begin{aligned}
\overset{\langle C;\,\alpha E\beta\rangle}{T^p_{qr}} &= \tfrac{1}{2}\{\langle C;\,\alpha E\beta\rangle^p_{qr}\}_{[qr]_A} \\
&= \tfrac{1}{2}(\beta E\alpha)^p_l\left\{\frac{\partial(\alpha E\beta)^l_q\,(\beta)}{\partial_\beta x^r}\right\}_{[qr]_A}
\end{aligned}
\tag{5.23}
$$

and (ii) the curvature tensor

$$
\begin{aligned}
\overset{\langle C;\,\alpha E\beta\rangle}{B^*_{ijkl}} &= \left\{\frac{\partial\langle C;\,\alpha E\beta\rangle^*_{ijl}(\beta)}{\partial_\beta x^k}\right. \\
&\left. + \langle c;\,\alpha E\beta\rangle^{rs}_{**}\langle C;\,\alpha E\beta\rangle^*_{sjl}\langle C;\,\alpha E\beta\rangle^*_{irk}\right\}_{[kl]_A} \\
&= \left\{\frac{\partial \overset{\langle C;\,\alpha E\beta\rangle}{M_{ijl}(\beta)}}{\partial_\beta x^k} + \frac{\partial \overset{\langle C;\,\alpha E\beta\rangle}{\theta_{ijl}(\beta)}}{\partial_\beta x^k}\right. \\
&\quad + \langle c;\,\alpha E\beta\rangle^{rs}_{**}\left(\overset{\langle C;\,\alpha E\beta\rangle}{M_{sjl}} + \overset{\langle C;\,\alpha E\beta\rangle}{\theta_{sjl}}\right) \\
&\quad \left. \times \left(\overset{\langle C;\,\alpha E\beta\rangle}{M_{irk}} + \overset{\langle C;\,\alpha E\beta\rangle}{\theta_{irk}}\right)\right\}_{[kl]_A}
\end{aligned}
\tag{5.24}
$$

where $M\,(\langle C;\,\alpha E\beta\rangle)^4$ has Cartesian components given by

$$
\overset{\langle C;\,\alpha E\beta\rangle}{M_{ijk}} = \frac{1}{2}\left(\frac{\partial\langle c;\,\alpha E\beta\rangle_{ij}(\beta)}{\partial_\beta x^k} + \frac{\partial\langle c;\,\alpha E\beta\rangle_{ki}(\beta)}{\partial_\beta x^j} - \frac{\partial\langle c;\,\alpha E\beta\rangle_{jk}(\beta)}{\partial_\beta x^i}\right)
\tag{5.25}
$$

and where the tensor $\theta\,(\langle C;\,\alpha E\beta\rangle)$ has Cartesian components:

$$
\overset{\langle C;\,\alpha E\beta\rangle}{\theta_{ijk}} = \overset{\langle C;\,\alpha E\beta\rangle}{T^*_{ijk}} + \overset{\langle C;\,\alpha E\beta\rangle}{T^*_{jki}} + \overset{\langle C;\,\alpha E\beta\rangle}{T^*_{kji}}
\tag{5.26}
$$

In these expressions the asterisk $(*)$ indicates that the index has been raised or lowered with the *deformed* metric $\langle c;\,\alpha E\beta\rangle$. It is seen by substitution

---

[4] This is a contracted notation of the tensor M. Other such contractions will be used in the text below.

of the covariant form of the connection $\langle C, \alpha E\beta \rangle$ given in Equation 5.22 into Equation 5.24 or by geometrical considerations that the curvature $B(\langle C; \alpha E\beta \rangle)$ vanishes:

$$\underset{B}{\langle C; \alpha E\beta \rangle} = 0. \tag{5.27}$$

Equation 5.27 is an inherent property of the formalism presented in this paper and is a direct consequence of the assumption that there is a unique lattice correspondence $(\alpha E\beta)$ between the two states α and β. The consequences of modifying this assumption will be presented elsewhere.

5.4.1. *The Single State*

Let us now consider the geometric field equations associated to a given state which we will formulate relative to its reference state $r\alpha$. The first of these geometric field equations which involves $(\alpha Er\alpha)$ is

$$\begin{aligned} \underset{\tau^i_{jk}}{\langle C; r\alpha E\alpha \rangle} &\equiv (r\alpha E\alpha)^i_p \quad \underset{T^p_{jk}}{\langle C; r\alpha E\alpha \rangle} \\ &= \frac{1}{2} \left\{ \frac{\partial (r\alpha E\alpha)^i_j(\alpha)}{\partial_\alpha x^k} \right\}_{[jk]A} \\ &= \tfrac{1}{2}\varepsilon_{jkl}\, \overset{\alpha}{A^{il}} \end{aligned} \tag{5.28}$$

where $A^{il}(\alpha)$ are a set of nine quantities assumed to be experimentally given and may be formally identified with the true dislocation tensor as introduced by Bilby et al. (1955). The tensor $A(\alpha)$ has the interpretation that for any area element Γ with area ΔS and covariant unit normal components n_l there is a resultant true Burgers vector

$$\mathbf{b}(\Gamma) = \mathbf{x}_i\, b^i = \mathbf{x}_i\, \overset{\alpha}{A^{il}} n_l\, \Delta S. \tag{5.29}$$

In the context of this presentation this Burgers vector not only arises from the continuum description of the actual dislocations that thread Γ but also includes the effective dislocations threading Γ defined by the distribution of infinitesimal dislocation loops (Kroupa 1963). These loops may be either slip or prismatic in kind and may be used to synthesize the continuum description of a distribution of point defects (extra matter) or thermal distortions, etc.

As will be discussed below, the analogous quantity $\tau\,(\langle C; \alpha Er\alpha \rangle)$ does not have any consistent physical interpretation in terms of defects, nor indeed does there appear to be any useful geometric field equation directly in terms of the quantity $(r\alpha\, E\alpha)$; thus $(\alpha E\, r\alpha)$ appears to be the only valid distortion strain measure for use in defect problems.

The geometric field equations which involve the quadratic strain measures $(\alpha er\alpha)$ and $(r\alpha e\alpha)$ derive formally from the vanishing of the curvatures $B(\langle C; \alpha Er\alpha\rangle)$ and $B(\langle C; r\alpha E\alpha\rangle)$. Their explicit form follows from Equation 5.24 with the appropriate substitutions for α and β and with the deformed metric $\langle c; \alpha E\beta\rangle$ replaced by its appropriate expression in terms of the strain measures given by Equations 5.6 and 5.8. They are (i) On the Green strain $(\alpha er\alpha)$:

$$\left\{ \frac{\partial \overset{\langle C;\,\alpha Er\alpha\rangle}{M_{ijl}}(r\alpha)}{\partial_{r\alpha} x^k} + \frac{\partial \overset{\langle C;\,\alpha Er\alpha\rangle}{\theta_{ijl}}(r\alpha)}{\partial_{r\alpha} x^k} + (2(\alpha er\alpha) + c)^{rs}_{**} \right.$$

$$\times (\ \overset{\langle C;\,\alpha Er\alpha\rangle}{M_{sjl}} + \overset{\langle C;\,\alpha Er\alpha\rangle}{\theta_{sjl}}\ )(\ \overset{\langle C;\,\alpha Er\alpha\rangle}{M_{irk}} + \overset{\langle C;\,\alpha Er\alpha\rangle}{\theta_{irk}}\ )\left. \right\}_{[kl]_A}$$

$$= 0 \tag{5.30}$$

where

$$\overset{\langle C;\,\alpha Er\alpha\rangle}{M_{ijk}} = \frac{\partial(\alpha er\alpha)_{ij}(r\alpha)}{\partial_{r\alpha} x^k} + \frac{\partial(\alpha er\alpha)_{ki}(r\alpha)}{\partial_{r\alpha} x^j} - \frac{\partial(\alpha er\alpha)_{jk}(r\alpha)}{\partial_{r\alpha} x^i} \tag{5.31}$$

and $\theta(\langle C; \alpha Er\alpha\rangle)$ is given by Equation 5.26. The asterisk refers to shifting the index with the deformed metric, which in this case is $\langle c; \alpha Er\alpha\rangle = 2(\alpha er\alpha) + c$.
(ii) On the Cauchy strain $(r\alpha e\alpha)$:

$$\left\{ \frac{\partial \overset{\langle C;\,r\alpha E\alpha\rangle}{M_{ijl}}(\alpha)}{\partial_\alpha x^k} + \frac{\partial \overset{\langle C;\,r\alpha E\alpha\rangle}{\theta_{ijl}}(\alpha)}{\partial_\alpha x^k} \right.$$

$$+ (c - 2(r\alpha e\alpha))^{rs}_{**}(\ \overset{\langle C;\,r\alpha E\alpha\rangle}{M_{sjl}} + \overset{\langle C;\,r\alpha E\alpha\rangle}{\theta_{sjl}}\ )$$

$$\times (\ \overset{\langle C;\,r\alpha E\alpha\rangle}{M_{irk}} + \overset{\langle C;\,r\alpha E\alpha\rangle}{\theta_{irk}}\ )\left. \right\}_{[kl]_A}$$

$$= 0 \tag{5.32}$$

where

$$\overset{\langle C;\,r\alpha E\alpha\rangle}{M_{ijk}} = - \left(\frac{\partial(r\alpha e\alpha)_{ij}(\alpha)}{\partial_\alpha x^k} + \frac{\partial(r\alpha e\alpha)_{ki}(\alpha)}{\partial_\alpha x^j} - \frac{\partial(r\alpha e\alpha)_{jk}(\alpha)}{\partial_\alpha x^i} \right) \tag{5.33}$$

and $\theta(\langle C; r\alpha E\alpha\rangle)$ is given by Equation 5.26. Again the asterisk refers to shifting the index with the deformed metric, which in this case is $\langle c; r\alpha E\alpha\rangle = c - 2(r\alpha e\alpha)$.

Equation 5.32 for the Cauchy strain ($r\alpha e\alpha$) is identical with that given by Kröner and Seeger (1959) for a single-state material with defects *in situ* who derived it by using a non-Euclidean natural state. This geometric field equation involves, together with the sought-after Cauchy strain, the Cauchy torsion $T(\langle C; r\alpha E\alpha\rangle)$ which has an almost direct relation to the measure A (α) of the defect content (see Equation 5.28); an iteration procedure can easily be constructed in which successive approximations for $T(\langle C; r\alpha E\alpha\rangle)$, derived from the effective dislocation density, are used. Thus when the Cauchy strain is a valid elastic energy measure, Equation 5.32 provides a tractable geometric field equation for the Cauchy strain.

In the general situation, however, the correct quadratic strain measure is the Green strain ($\alpha e r\alpha$) whose geometric field Equation 5.30 involves the quantity $T(\langle C; \alpha E r\alpha\rangle)$, which we may call the Green torsion. Unfortunately the Green torsion has no simple physical meaning. An analysis of closure failure circuits for the Green torsion, analogous to that which demonstrates the physical validity of the Cauchy torsion, requires the ability to allow free translations of lattice vectors in the generally inhomogeneously strained state α. Such freedom is physically unreasonable. As an example let us define a perfect state α as one in which the effective dislocation tensor vanishes, so that from Equation 5.28 the Cauchy torsion also vanishes. It is then possible to construct a perfect natural state (in the sense used in nonlinear elasticity by Doyle and Ericksen 1956, rather than that used by Kröner 1960); that is a state whose local crystal configuration is that of the reference lattice and which may be macroscopically deformed to the state α by a point deformation for which the relation

$$(\alpha J n) = (\alpha E n) \tag{5.34}$$

is valid. This elementary construction, which is not given here, is valid so long as the state α is simply connected. The relation of the state n to the state α can be fully discussed within the confines of conventional nonlinear elasticity for which the geometric field equations

$$\overset{\langle C; nE\alpha\rangle}{T} = \overset{\langle C; \alpha En\rangle}{T} = 0 \tag{5.35}$$

and

$$\overset{\langle C; nE\alpha\rangle}{B} = \overset{\langle C; \alpha En\rangle}{B} = 0 \tag{5.36}$$

hold. These equations are merely the usual nonlinear compatibility equations on both the Green and Cauchy strains. If we calculate, however, the torsions with respect to the reference state, the Cauchy torsion $T(\langle C; r\alpha E\alpha\rangle)$ remains unaffected by the change from the perfect natural to the reference state, whereas the Green torsion $T(\langle C; \alpha E r\alpha\rangle)$ becomes, in general, a nonvanishing quantity depending on the point positions of the state α. That is, even a perfect continuum has, in general, a nonvanishing Green torsion with respect to its reference state. Since a general *imperfect* continuum does not have a perfect natural state there is no preferred reference state with respect to which the Green torsion should be computed. Thus although the Green strain is quite independent of the choice of reference state, the Green torsion depends explicitly on the point positions in the reference state; the latter therefore contains information extraneous to that characterizing the defect content of the physical material. In addition the Green torsion, which is effectively computed of the curl of ($\alpha Er\alpha$) and the Cauchy torsion, which is effectively computed of the curl of ($r\alpha E\alpha$) appear to be mathematically unrelated without specific knowledge of the tensor ($\alpha Er\alpha$).

There is thus an essential dilemma in that the geometric field Equation 5.30 on the physically relevant Green strain involves the nonphysical Green torsion, while the geometric field Equation 5.32 involving the physically relevent defect measure (the Cauchy torsion) is on a strain measure of restricted applicability. Because of this dilemma it is not clear how to set a physically well-posed problem for a general nonlinear anisotropic medium involving a knowledge of only the Cauchy torsion defect measure when the quadratic Green strain measure is used.

5.4.2. *Relative Deformations*

Let us consider an initial state o which is, in general, an imperfect continuum, and let this state be deformed to the final state f. The geometric field equation on the lattice deformation (fEo) follows formally from Equation 5.23 by substituting o for α and f for β and vice-versa for the geometric field equation on (oEf). The geometric field equations on the relative Green and Cauchy strains (feo) and (oef) respectively follow analogously from Equations 5.23 and 5.27. The torsion tensors involved in these equations are the relative Green torsion $T(\langle C; fEo \rangle)$ and the relative Cauchy torsion $T(\langle C; oEf \rangle)$. As in the previous section, in the case of a single state, we may only expect that the relative Cauchy torsion will be directly relatable to the defect content. It is reasonable to expect that this torsion will depend on both the Cauchy torsion of the final state and certain information about the initial state, together with both relative deformations (fEo) and (fJo). That is so is seen directly by application of the "golden rule" (Equation 5.18) for deformation. It is first convenient to establish the equality

$$\langle \langle C; \alpha E\beta \rangle; \beta E\gamma \rangle = \langle C; \alpha E\gamma \rangle. \tag{5.37}$$

The Euclidean connection C has the complete set of invariant vectors $\mathbf{x}_i$ so that according to Equation 5.18 the connection $\langle C; \alpha E\beta \rangle$ has the set of invariant vectors

$$\langle \mathbf{x}_i ; \alpha E\beta \rangle = \mathbf{x}_j (\beta E\alpha)_i^j \tag{5.38}$$

and similarly the connection $\langle\langle C; \alpha E\beta\rangle; \beta E\gamma\rangle$ has

$$\langle\langle \mathbf{x}_i; \alpha E\beta\rangle; \beta E\gamma\rangle = \langle \mathbf{x}_i; \alpha E\gamma\rangle \tag{5.39}$$

as a set of invariant vectors, which coincide with the complete set of invariants for the connection $\langle C; \alpha E\gamma\rangle$. Since the two connections in Equation 5.37 have the same complete set of vector invariants they are equal, which establishes Equation 5.37.

Applying Equation 5.37 twice, one has

$$\langle C; rfEf\rangle = \langle\langle\langle C; rfEro\rangle; roEo\rangle; oEf\rangle. \tag{5.40}$$

Clearly, from Equation 5.12, we have

$$\langle C; rfEro\rangle = C \tag{5.41}$$

so that

$$\langle C; rfEf\rangle = \langle\langle C; roEo\rangle; oEf\rangle, \tag{5.42}$$

which by Equation 5.20 may be written

$$\langle C; rfEf\rangle^p_{qr} = (fEo)^p_j \frac{\partial(oEf)^j_q(f)}{\partial_f x^r}$$

$$+ (fEo)^p_j\langle C; roEo\rangle^j_{sk}(oEf)^s_q(oJf)^k_r. \tag{5.43}$$

From this follows the required relation

$$\begin{matrix} \langle C; rfEf\rangle & & \langle C; oEf\rangle \\ T^p_{qr} & = & T^p_{qr} \end{matrix} \quad + \tfrac{1}{2}(fEo)^p_j\langle C; roEo\rangle^j_{sk}\{(oEf)^s_q(oJf)^k_r\}_{[qr]_A} \tag{5.44}$$

It is thus apparent that the relative Cauchy torsion is directly related to the final-state Cauchy torsion, the relative deformations (fEo) and (fJo) and the material connection of the initial state. Equation 5.44 thus permits the introduction of the physically accessible quantities $\langle C; roEo\rangle$ and $\tau(\langle C; rfEf\rangle)$ into the geometric field equations for the Cauchy-type strain measures (oEf) and (oef). In this case one must expect further information relating (oEf) and (oJf) to be given; an iterative procedure using successive approximations for (oEf) and therefore for $T(\langle C; oEf\rangle)$ should then be possible. The relation analogous to Equation 5.44 on the relative Green torsion, which is involved in the geometric field equations of Green type, can be similarly deduced but it appears at the present time to be of dubious value for other than the deformation of a perfect distortion-free initial state. In fact an example of a suitable iterative procedure for solving a problem involving the Green strain associated with the imperfect deformation of such an initial state will be outlined in the next section.

Finally, Equation 5.44 may be thought of as simply describing the change of defect content under deformation. Then if the initial state is perfect and distortion free and subjected to an arbitrary deformation, the defect content of the final state is described by the relative torsion $T(\langle C; oEf \rangle)$ and is independent of the point positions of the initial state, as should be the case for a valid intrinsic defect measure. Alternatively if the initial state is imperfect but is subjected to a perfect deformation, i.e., $(fJo) = (fEo)$, then the final-state Cauchy torsion is merely the macroscopically deformed initial-state Cauchy torsion. Thus, in this case,

$$\underset{T}{\langle C; rfEf \rangle} = \langle \underset{T}{\langle C; roEo \rangle} ; \; oJf \rangle. \tag{5.45}$$

Equation 5.45 implies that the true effective dislocation density is conserved under a perfect deformation; for by Equation 5.28

$$\underset{\tau^i_{jk}}{\langle C; rfEf \rangle} = \underset{\tau^i_{pq}}{\langle C; roEo \rangle} (oJf)^p_j(oJf)^q_k. \tag{5.46}$$

If now Γ_0 is an oriented surface in the initial state, then Equation 5.29 gives the resultant Burgers vector of all the effective dislocations that thread Γ_0. Equation 5.46 then states that the resultant Burgers vector of the effective dislocations threading the macroscopically deformed surface Γ_f is the same as that for Γ_0 in the initial state. Thus we see that a perfect deformation of an imperfect medium creates no new effective dislocations; it merely shifts their positions macroscopically.

5.5. Applications

In this section we wish to outline two types of potential application of the formalism developed above. Both of these applications emphasize what we believe to be aspects of nonlinear defect theory that have not been previously treated.

The two examples we shall consider involve the relative deformation from a distortion-free perfect initial state; however both may be generalized to the situation when the initial state is imperfect. Let the two states be o and f connected by the two deformations (fEo) and (fJo). We introduce the "defect deformation" tensor (oDo) and its inverse $(o\tilde{D}o)$ defined by

$$(oDo) \equiv (oEf)(fJo) \tag{5.47}$$

$$(o\tilde{D}o) \equiv (oJf)(fEo), \tag{5.48}$$

from which follow the relations

$$(fJo) = (fEo)(oDo) \tag{5.49}$$

and

$$(oJf) = (o\tilde{D}o)(oEf). \tag{5.50}$$

We further introduce the quantities $\Delta(fJo)$, $\Delta(fEo)$, $\Delta(oDo)$, and $\Delta(o\tilde{D}o)$, defined by the relations

$$
\left.
\begin{aligned}
I + \Delta(fJo) &= (fJo) \\
I + \Delta(fEo) &= (fEo) \\
I + \Delta(oDo) &= (oDo) \\
I + \Delta(o\tilde{D}o) &= (o\tilde{D}o)
\end{aligned}
\right\},
\tag{5.51}
$$

where the tensor I is given in Cartesian form by

$$
I^i_j \overset{n}{=} \delta^i_j.
\tag{5.52}
$$

Combining Equations 5.49 and 5.51 with the above definitions we obtain the relations

$$
\begin{aligned}
(fJo) &= (fEo) + (fEo)\Delta(oDo) \\
&\equiv (fEo) + \Delta GD
\end{aligned}
\tag{5.53}
$$

and

$$
\begin{aligned}
(oJf) &= (oEf) + \Delta(o\tilde{D}o)(oEf) \\
&\equiv (oEf) + \Delta CD,
\end{aligned}
\tag{5.54}
$$

the quantities ΔGD and ΔCD in the last two equations being directly related to the geometric field equation on the Green and Cauchy strains, respectively, as will be seen. From the above equations one also has

$$
\Delta GD = -(fEo)\Delta CD(fEo)(I + \Delta CD(fEo))^{-1}.
\tag{5.55}
$$

5.5.1. Initial Defect Deformation Given

The first type of problem we shall consider is a lattice-invariant defect deformation (oDo) imposed within the initial state as an internal boundary condition. The material is now allowed to undergo any imaginable lattice deformation (fEo) such that the product $(fEo)(oDo)$ is compatible, i.e. defines a final state via (fJo) as demanded by Equation 5.49. The system, however, chooses that particular (fEo) which both minimizes its total elastic-free energy and is consistent with any other imposed boundary conditions, (these latter boundary conditions must, of course, be consistent with the geometrical constraints, and we here assume that they are sufficient to determine (fEo) uniquely). By referring to (oDo) as a *lattice-invariant* defect deformation we mean merely that the local lattice configuration is not changed by (oDo). Since the "lattice parameter" in this continuum-field theory is necessarily infinitesimal, any vector may be interpreted as a slip vector and any plane as a lattice slip plane. It is thus possible to conceive of (oDo) as a distribution of infinitesimal cuts in the initial state across which lattice slip is imposed so that the local lattice structure is unaffected, but a local shape change does take place. For example an infinitesimal internally ruled vector $\mathbf{l} = \mathbf{x}_i l^i$ at the point P acquires the shape

$$\langle \mathbf{l}; oDo \rangle \equiv \mathbf{x}_i (oDo)^i_j \, l^j \tag{5.56}$$

as a result of the slip. The important features of this shape change are that it is imposed and that it does not in itself contribute to the elastic energy; it does not even effect the form of the elastic energy density function. The subsequent response of the material is then to adapt elastically to any defects which are produced. It is characteristic of a nonlinear problem that while the initial defect deformation has been imposed a priori the resulting final defect distribution can only be determined when the positions of the material points in the final state have been determined; since only when one knows (fJo) can one determine the precise Cauchy torsion.

The relations (Equation 5.53) already suggest on iterative procedure for finding the deformations (fJo) and (fEo) arising from an imposed (oDo) when appropriate additional boundary conditions are provided. Suppose we have obtained the kth iteration to (fEo), i.e., $(fEo)_k$ and $(oEf)_k$ are known. From Equation 5.53 we define

$$\Delta GD_k \equiv (fEo)_k \, \Delta(oDo) \tag{5.57}$$

and then, also from Equation 5.53, we have

$$\mathrm{curl}\,(fEo)_{k+1} = -\mathrm{curl}\,\Delta GD_k \tag{5.58}$$

where the appropriate curl operation is performed in an initial-state coordinate system. Hence we obtain the Green torsion

$$\begin{aligned}\langle C; fEo \rangle \\ T_{k+1} &= -\tfrac{1}{2}(oEf)_k\,\mathrm{curl}\,\Delta GD_k. \end{aligned} \tag{5.59}$$

Equation 5.58 is a geometric field equation for $(fEo)_{k+1}$ and Equation 5.59 together with the stipulation of zero curvature forms a geometric field equation for $(feo)_{k+1}$ (see Equation 5.30 for the reference-state geometric field equation). Once the strain $(feo)_{k+1}$ is known, $(fEo)_{k+1}$ follows by integration. In either case we now assume that $(fEo)_{k+1}$ can be obtained by elasticity considerations, thus completing the iteration cycle. This process is expected to converge if the initial guess, $(fEo)_1$, is sufficiently good; for instance the linear elastic solution, using

$$\Delta GD = \Delta(oDo) \tag{5.60}$$

would seem a sensible choice. When the iteration scheme has produced (fEo) of sufficient accuracy (say after k' iterations) one can employ Equation 5.53 to obtain an approximate (fJo)

$$(fJo)_k = (fEo)_{k'+1} + \Delta GD_{k'}, \tag{5.61}$$

which is always exactly integrable. The resulting defect content of the final state is then given by the Cauchy torsion:

$$\begin{array}{c}\langle C;\,oEf\rangle\\ T^p_{qr}\end{array} = \tfrac{1}{2}(fEo)^p_l\left\{\frac{\partial(oEf)^1_q(0)}{\partial_0 x^s}\,(oJf)^s_r\right\}_{[qr]_A}, \tag{5.62}$$

where, as would appear convenient, we have used initial point positions for carrying out the calculation.

5.5.2. *Final Defect Deformation Given*

The second type of defect problem also takes the form of an internal boundary condition, but this time it is imposed in the final state. In this case let us consider the quantity ΔCD as given. We see from Equation 5.54 that knowledge of ΔCD in the final state completely determines the effective dislocation tensor $A\,(f)$ defined by Equation 5.28. If we now use the quantity (oEf) as the strain measure, then

$$\mathrm{curl}\,(oEf) = \overset{f}{A} \tag{5.63}$$

provides an appropriate geometric field equation for (oEf). However, as noted previously, it does not appear possible to usefully relate ΔCD with the Green torsion when only the final-state positions are known. It thus appears most unlikely that the Green strain (feo) can be directly obtained. In the special circumstance that the material is isotropic, so that the Cauchy strain is a valid strain measure, the geometric field equation for the Cauchy strain (cf. Equation 5.32 for the reference state form) becomes applicable. In this situation, since the Cauchy strain geometric field equation involves the Cauchy torsion and we are only given the effective dislocation density, an iterative method of solution analogous to that outlined in the previous example, should be appropriate (c.f. Kröner and Seeger 1959, where the Cauchy torsion is assumed given). In any event we assume that appropriate boundary conditions have been specified so that (oEf) can be found by the methods of elasticity. Having determined (oEf) we may now return to Equation 5.54 to determine (oJf) and thus obtain the initial perfect state from which this final state has arisen. In addition from ΔCD we can determine (oDo) and thus finally discover the precise defect deformation in the initial state that produced the final state.

REFERENCES

Bilby, B. A., 1960, in *Prog. Solid. Mech.*, **1**, 329.

Bilby, B. A., R. Bullough, and E. Smith, 1955, *Proc. Roy. Soc.*, **A231**, 263.

Bilby, B. A., L. R. T. Gardner, and A. N. Stroh, 1957, *Extrait des Actes, 9 Congr. Int. Mec. Appl.*, **8**, 35.

Doyle, T. C., and J. L. Ericksen, 1956, *Adv. Appl. Mech.*, **4**, 53.

Kondo, K., 1955, *R.A.A.G. Mem.*, **1**, 458.

Kröner, E., 1960, *Archs. Ration. Mech. Analysis*, **4**, 273.

Kröner, E., and A. Seeger, 1959, *Archs. Ration. Mech. Analysis*, **3**, 97.

Kroupa, F., 1963, *Czech, J. Phys.*, **A13**, 301.

Orowan, E., 1934, *Z. Physik.*, **89**, 634.

Polanyi, M., 1934, *Z. Physik.*, **89**, 660.

Schouten, J. A., 1954, *Ricci-Calculus*, Heidelberg: Springer, pp. 131 ff.

Stojanovitch, R., 1962, *Phys. Stat. Sol.*, **2**, 566.

Taylor, G. I., 1934, *Proc. Roy. Soc.*, **A145**, 362.

Truesdell, C., and R. Toupin, 1960, in S. Flügge (ed.), *Encyclopedia of Physics*, Berlin-Heidelberg-New York: Springer, vol. III/1.

ABSTRACT. The conditions for the invisibility of dislocations in the electron microscope are considered when the approximation of elastic isotropy is replaced by the real elastic anisotropy of most crystals. Some modifications to the classical $\mathbf{g} \cdot \mathbf{b} = 0$ rule are necessary and the possibility of new types of invisibility is shown by a dislocation which is completely invisible when $\mathbf{g} \cdot \mathbf{b} = 6$.

6. The Invisibility of Dislocations

A. K. HEAD

6.1. Introduction

The postulate by Orowan (1934), Polanyi (1934), and Taylor (1934) that dislocations were the agency by which crystalline solids deformed plastically was followed by two decades during which the existence of dislocations was a theoretical necessity but experimental evidence was sparse and indirect. The discovery that dislocations could be directly observed by transmission electron microscopy through thin metal foils (Hirsch, Horne, and Whelan 1956; Bollman 1956) was both a confirmation of the essential correctness of the theoretical postulate and the start of what has become the most powerful single method for investigating dislocations (e.g. Hirsch et al. 1965).

The visibility of dislocations in the electron microscope is not by resolution of the positions of the atoms around the dislocation, as this is usually beyond the capability of the instrument, but by the mechanism of diffraction contrast. The incident electron beam in passing through the crystal will be diffracted by any crystallographic plane which is at, or near, the correct orientation for Bragg reflection of the electron waves. If there are regions of the crystal where the atoms are displaced from their regular crystallographic positions, as in the neighborhood of a dislocation, then the diffracting power of these regions will differ from that of perfect crystal. If an image is formed from a diffracted beam or from the undeviated main beam, then this electron micrograph shows, in a general sense, the variations in the diffracting power of the crystal from region to region. The interpretation of such diffraction contrast images is simplest if the crystal is oriented so that only one set of planes diffracts strongly (Hirsch, Howie, and Whelan 1960) and the theory of this two beam case (the main beam and one diffracted beam) developed by Howie and Whelan (1961, 1962) enables the calculation of the image to be expected from any given atomic displacements.

This theory showed that it is only the component of the atomic displacements in the direction of the diffracting vector $\mathbf{g}$ (which is the normal to the reflecting crystallographic plane) which contributes to the diffraction contrast in the micrograph and that the displacement components in the reflecting plane give no contrast. This means that electron micrographs of the same dislocation taken on different reflections will generally show a number of different images, corresponding to different components of the displacement field being selected. In particular, the observation that a dislocation could be invisible for a particular reflection (Bradley and Phillips 1957) has the interpretation that the reflecting plane has remained flat.

The question of what atomic planes are left flat in the presence of a dislocation is a problem in elasticity and Hirsch, Howie, and Whelan (1960) showed that both pure screw and pure edge dislocations had this property. For a pure screw dislocation, all planes which contain the direction of the Burgers vector $\mathbf{b}$ remain flat so that for any diffracting vector $\mathbf{g}$ which is perpendicular to $\mathbf{b}$, the dislocation will be invisible. For a pure edge dislocation the planes which are cut normally by the dislocation line remain flat and as the Burgers vector lies in this plane, $\mathbf{b}$ and $\mathbf{g}$ are perpendicular, i.e., $\mathbf{g} \cdot \mathbf{b} = 0$. Hirsch et al. suggested that whenever $\mathbf{g} \cdot \mathbf{b} = 0$ it could be expected that the image of any general dislocation would be, if not invisible as in the above cases, at least faint, and that it should be possible to determine the Burgers vector of any dislocation by examining the contrast in several reflections and finding two or more reflections for which the contrast vanishes. This method for the direct determination of Burgers vectors has indeed become widely used.

A limitation on this method was found during an investigation of dislocations in beta-brass (Head, Loretto, and Humble 1967 a and b). It was found that for most dislocations it was impossible to find any reflecting vector $\mathbf{g}$ which would make the dislocation image even faint let alone invisible. A reconsideration of the rule of Hirsch et al. showed that it was only true in general if the material was elastically isotropic and that in an anisotropic crystal (such as beta-brass) pure screw or pure edge dislocations would only obey the strict invisibility rule if the direction of the dislocation line, $\mathbf{u}$, were perpendicular to a symmetry plane of the crystal. For an isotropic crystal every plane is an elastic symmetry plane so the rule of Hirsch et al. is a special case of the general rule. In a cubic crystal there are two sets of symmetry planes, {100} and {110}. So in a f.c.c. crystal the screw dislocation with $\mathbf{b} = \mathbf{u} = \langle 110 \rangle$ will be invisible if $\mathbf{g} \cdot \mathbf{b} = 0$ whatever the elastic anisotropy but the screw dislocation in a b.c.c. crystal with $\mathbf{b} = \mathbf{u} = \langle 111 \rangle$ will be invisible only if the crystal is elastically isotropic.

The examination of the dislocations in beta-brass also showed that the diffraction image of dislocations is very rarely just a featureless black line but that in general there is a lot of characteristic detail in the image. The variation of this character for different Burgers vectors and for different diffracting conditions was sufficiently great to enable the identification of several different types of dislocations in beta-brass by comparison with the corresponding theoretical predictions. This detailed examination of the character of dislocation images has become a routine procedure with the development of computer-generated electron micrographs (Head 1967b) by which theoretical predictions are produced by the computer as pictures for direct comparison with experiment.

However, it is still true that the most striking characteristic of a dislocation image is if it can be made invisible. We have seen that when the real elastic anisotropy of most crystals is taken into account, then the invisibility rule of Hirsch et al. is restricted. The question which will be examined here is whether elastic anistropy introduces any new categories of invisibility. The discussion will be confined to cubic crystals and also to exact invisibility since this is an absolute criterion and is independent of experimental sensitivity.

6.2. The Flat Plane

In this section we will consider the natural extension of the invisibility criterion of Hirsch et al. and inquire if it is possible that a dislocation in an elastically anisotropic crystal can leave a set of atomic planes flat. If this is so, then of course it would be invisible in an electron micrograph taken using that set of planes as the operative reflection.

One possibility was stated above, that if the dislocation line is perpendicular to a symmetry plane of the crystal, then pure screw and pure edge dislocations have the required property, and this is true generally for any crystal symmetry, not just cubic.

A necessary condition, also for crystals of any symmetry, can be derived by considering a Burgers circuit around the dislocation with the circuit lying in the plane which remains flat. The closure failure of this circuit is, by definition, the Burgers vector $\mathbf{b}$ of the dislocation but since the plane remains flat, $\mathbf{b}$ can have no component out of the plane. Hence $\mathbf{g} \cdot \mathbf{b} = 0$ is a necessary condition that the set of planes with normal $\mathbf{g}$ are left flat by a dislocation of Burgers vector $\mathbf{b}$.

A further necessary condition comes from the detailed elastic analysis of the displacements of atoms around a dislocation in an anisotropic crystal. If $\mathbf{R}(\mathbf{r})$ is the displacement vector of the atom at $\mathbf{r}$ and $\mathbf{g}$ is the operative reflection, then the image theory of Howie and Whelan depends on the combination

$$\frac{d}{dz}(\mathbf{g} \cdot \mathbf{R}), \tag{6.1}$$

where z is the direction of the electron beam. The general form of this is known to be (Eshelby, Read, and Shockley 1953; Stroh 1958)

$$\sum_{\alpha=1}^{3} \frac{P_\alpha x_1 + Q_\alpha x_2}{(x_1 - R_\alpha)^2 + (S_\alpha x_2)^2}, \tag{6.2}$$

where x_1 and x_2 are coordinates perpendicular to the dislocation line.

The 12 constants P_α, Q_α, R_α, and S_α however can be derived only numerically. In particular $R_\alpha + iS_\alpha$ are three roots of the sextic polynomial, the coefficients of which depend on the elastic constants and the crystallographic direction of the dislocation line. P_α and Q_α depend on R_α and S_α and also on $\mathbf{g}$ and $\mathbf{b}$.

The dislocation is invisible if the expression 6.2 is identically zero. For this to be possible it is first necessary that two of the terms in 6.2 have identical denominators; then it might be possible to choose $\mathbf{g}$ and $\mathbf{b}$ such that the two numerators are equal and opposite and also that the numerator of the third term is zero.

Since identical denominators means that the sextic polynomial has repeated roots a large quantity of existing calculations (Head 1967a, Atkinson and Head 1966) could be surveyed since it had been recorded if the sextic had equal or nearly equal roots. These covered cubic crystals with a wide range of elastic anisotropy and for directions of dislocation lines in {100}, {110}, and {111} planes and a few other directions in the interior of the stereographic triangle. This information suggested, and further computations have confirmed, that the following statement is true:

"For each anisotropic cubic crystal there is a direction of dislocation line in the {110} plane for which the sextic polynomial has double roots. For this direction there are two different Burgers vectors which have the property of leaving sets of planes flat. The Burgers vectors and normals to the flat planes (which must of course be perpendicular to the Burgers vectors) lie in the same {110} plane as the dislocation line."

For a cubic crystal which has three different elastic constants C_{11}, C_{12}, and C_{44}, the atomic displacements depend only on two ratios of these constants. It is convenient to take the ratios

$$A = \frac{2C_{44}}{C_{11} - C_{12}} \qquad B = \frac{C_{11} + 2C_{12}}{C_{44}}.$$

Humble (1967) has given a diagram of the location of various cubic crystals on an (A, B) plot.

It has been found that this invisiblity of dislocations depends mainly on A and only to a small extent on B. For $B = 4$, Figure 6.1 shows how $\mathbf{u}$,

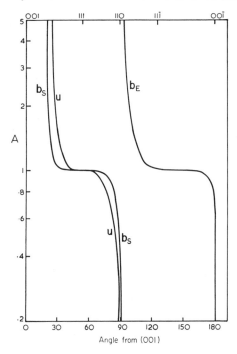

FIGURE 6.1. Invisible dislocations on the $(1\bar{1}0)$ plane of a cubic crystal. Direction of dislocation line, u, and Burgers vectors b_S, b_E, as a function of elastic anisotropy A for $B = 4$.

the direction of the invisible dislocation line, varies with A on both sides of isotropy ($A = 1$). The corresponding two invisible Burgers vectors are shown as $\mathbf{b}_S$ and $\mathbf{b}_E$ since they are approximately screw and edge dislocations. Such dislocations will be possible in a crystal only if the Burgers vector is in a simple crystallographic direction. However, it will be seen that the curves cut $\langle 100 \rangle$, $\langle 110 \rangle$, and $\langle 111 \rangle$ directions, which give a number of real possibilities of dislocation which are invisible in an anisotropic crystal, although, since they are neither pure screw nor pure edge, they would not be invisible in an isotropic crystal.

6.3. The Spiral Staircase

The above discussion has been based on the original observation of Hirsch et al. that a dislocation is invisible when the reflecting plane is left flat in the presence of the dislocation and that a necessary condition for this is $g \cdot b = 0$. In this section we show that this is not the only possibility in anisotropic crystals.

In a cubic crystal, dislocations which run in $\langle 100 \rangle$ or $\langle 110 \rangle$ directions have elastic properties which are very similar to the isotropic case and explicit expressions can be derived for stresses, strains, displacements, etc. But for an arbitrary direction of dislocation line, the sextic polynomial cannot be explicitly factorized and the elastic properties can only be obtained numerically. The one exception is the $\langle 111 \rangle$ direction. This is not perpendicular to a symmetry plane so it is not just a modification of the isotropic case but because of the three-fold symmetry it is possible to factorize the sextic polynomial and obtain explicit solutions for stresses etc. (Head 1964, Hirth and Lothe 1966, Chou and Mitchell 1967).

For a screw dislocation running in a $\langle 111 \rangle$ direction, it can be shown that the component of the displacements in the direction of the dislocation line is given by

$$R_z = \frac{b}{6\pi} \arctan(C \tan 3\theta) \tag{6.3}$$

where

$$C = \sqrt{\frac{(2A + 1)(AB + A + 2B + 11)}{9(AB + 3A + 1)}}^{*} . \tag{6.4}$$

For isotropy ($A = 1$) this reduces to the well-known expression

$$R_z = \frac{b\theta}{2\pi} \tag{6.5}$$

which describes the fact that the planes which are perpendicular to the dislocation line have been deformed into a spiral ramp which rises by b for each revolution around the dislocation line.

Now suppose we consider the limit of Equation 6.3 as $A \to \infty$. C also becomes infinite and Equation 6.3 implies that the planes perpendicular to the dislocation line are deformed into a spiral staircase with 6 steps per revolution, the tread of each step being flat and the rise between steps being $b/6$. If an electron micrograph of this pure screw dislocation were taken with g along the dislocation line, then the flat treads of the steps contribute no contrast and also the rises will give no contrast if $g \cdot b/6$ is an integer. So this pure screw dislocation would be invisible for a g along the dislocation line (which in isotropy is the characteristic of an edge dislocation) and such that $g \cdot b = 6$ rather than $g \cdot b = 0$.

This dislocation becomes exactly invisible only for infinite anisotropy. The approach to invisibility as anisotropy increases is shown on the left side of Figure 6.2; for comparison, the right side shows the same dislocation

---

* This constant C was denoted by R in Head (1964).

FIGURE 6.2. Computer generated electron micrographs of a pure screw dislocation along ⟨111⟩ in a cubic crystal for increasing elastic anisotropy. Left: $\mathbf{g} \cdot \mathbf{b} = 6$; Right: $\mathbf{g} \cdot \mathbf{b} = 0$.

for $\mathbf{g} \cdot \mathbf{b} = 0$, which is invisible if isotropic, the contrast becoming increasingly strong and unusual with increasing anisotropy. These computer-generated electron micrographs follow the conventions described by Head (1967b), and the details are as follows:

> Burgers vector = $[111]$
> Dislocation line = $[111]$
> Foil thickness = $4\xi_g$
> Foil normal = $[001]$
> Beam direction = $[\bar{1}\bar{1}2]$
> Deviation from Bragg condition $w = 0.5$
> Anomalous absorption = 0.1
> Diffracting vector = Figure 6.2a $[222]$
> Figure 6.2b $[\bar{1}10]$
> and $A = 1$ to 10^6 in powers of 10 with $B = 4$.

6.4. Discussion

Because dislocation problems in elastic anisotropy can, in general, be solved only numerically, it is not possible to give an exhaustive survey of the conditions for which the electron-microscope image of a dislocation will be invisible. The two very different types of invisibility given here do illustrate that significant deviations from the isotropic rule do occur and that caution is needed, particularly if anisotropy is large.

In practice it is not exact invisibility which counts but experimental invisibility for which the image is fainter than some experimental threshold. A complete survey of this would be prohibitive but some indication of what can happen comes from an examination of a set of 68 computer-generated pictures of dislocations in beta-brass (Head 1967b). Of this 68, 19 were for $\mathbf{g} \cdot \mathbf{b} = 0$. If the set of 68 was arranged in order of faintness, then the faintest was a $\mathbf{g} \cdot \mathbf{b} = 0$ picture, but the next was for a $\mathbf{g} \cdot \mathbf{b} = 1$ picture that was fainter than the other 18 cases of $\mathbf{g} \cdot \mathbf{b} = 0$. In fact there was nothing to distinguish the 19 $\mathbf{g} \cdot \mathbf{b} = 0$ cases as a group from the other pictures since they ranged from faint to very intense. Other cases of the effect of elastic anisotropy on noninvisible images of dislocations have been considered by Humble (1967).

REFERENCES

Atkinson, C., and A. K. Head, 1966, *Int. J. Fracture Mech.*, **2**, 489.
Bollman, W., 1956, *Phys. Rev.*, **103**, 1588.
Bradley, D. E., and R. Phillips, 1957, *Proc. Phys. Soc.*, **B70**, 533.
Chou, Y. T., and T. E. Mitchell, 1967, *J. Appl. Phys.*, **38**, 1535.
Eshelby, J. D., W. T. Read, and W. Shockley, 1953, *Acta Met.* **1**, 251.
Head, A. K., 1964, *Phys. Stat. Sol.*, **6**, 461.
Head, A. K., 1967a, *Phys. Stat. Sol.*, **19**, 185.

Head, A. K., 1967b, *Australian J. Phys.*, **20**, 557.

Head, A. K., M. H. Loretto, and P. Humble, 1967a, *Phys. Stat. Sol.*, **20**, 505.

Head, A. K., M. H. Loretto, and P. Humble, 1967b, *Phys. Stat. Sol.*, **20**, 521.

Hirsch, P. B., R. W. Horne, and M. J. Whelan, 1956, *Phil. Mag.*, **1**, 677.

Hirsch, P. B., A. Howie, R. B. Nicholson, D. W. Pashley, and M. J. Whelan, 1965, *Electron Microscopy of Thin Crystals*, Butterworth, London.

Hirsch, P. B., A. Howie, and M. J. Whelan, 1960, *Phil. Trans. Roy. Soc.*, **A252**, 499.

Hirth, J. P., and J. Lothe, 1966, *Phys. Stat. Sol.*, **15**, 487.

Howie, A., and M. J. Whelan, 1961, *Proc. Roy. Soc.*, **A263**, 217.

Howie, A., and M. J. Whelan, 1962, *Proc. Roy. Soc.*, **A267**, 206.

Humble, P. 1967, *Phys. Stat. Sol.*, **21**, 733.

Orowan, E., 1934, *Z. Physik.*, **89**, 605 ff.

Polanyi, M., 1934, *Z. Physik*, **89**, 660.

Stroh, A. N., 1958, *Phil. Mag.*, **3**, 625.

Taylor, G. I., 1934, *Proc. Roy. Soc.*, **A145**, 362

ABSTRACT. The velocity-stress relationship is obtained for dislocations whose motion on the slip plane is controlled by a Newtonian viscous damping mechanism. In the subsonic velocity region the applied stress reaches a maximum at the shear wave velocity for screw dislocations and the Rayleigh wave velocity for climbing-edge or gliding-edge dislocations. In the case of gliding-edge dislocations moving in the transonic velocity region, the stress drops from a maximum value at the shear wave velocity as the velocity increases, reaches a minimum at $\sqrt{2}$ times the shear wave velocity, and then increases until a maximum is reached at the longitudinal wave velocity. Solutions for a climbing-edge dislocation between the Rayleigh and the longitudinal wave velocities and for a gliding-edge dislocation between the Rayleigh and the shear wave velocities are possible only if the dislocation moves on a slip plane which is in unstable equilibrium. In these velocity ranges it is more likely that the dislocation will split into partial supersonic dislocations. A Newtonian damping mechanism cannot prevent supersonic dislocations from moving at an arbitrary velocity on slip or climb planes that are in unstable equilibrium.

7. Stress Dependence on the Velocity of a Dislocation Moving on a Viscously Damped Slip Plane

J. WEERTMAN

7.1. Introduction

To our knowledge, Orowan (1940) wrote the first paper that considered the motion of damped dislocations. The present article adds another to the long list of papers that have since explored this subject originally illuminated by him.

Gilman (1968) has pointed out that the major portion of the dislocation damping arising from phonon and electron viscosity mechanisms occurs in or near the dislocation core. He has suggested that in consequence of this fact the damping force acting on a moving dislocation can be calculated, at least to a rough approximation, through use of the simplifying assumtion that all the viscous energy losses occur within the slip plane. In effect, according to his assumption the slip plane can be considered to contain a sheet of a Newtonian viscous fluid that is an atomic dimension in thickness. When the dislocation moves across the slip plane, this fluid is deformed in simple shear. The stress required to shear the fluid gives rise to a damping force on the dislocation.

Gilman made only an order-of-magnitude calculation of the damping force on a dislocation that moves at a slow velocity. In this paper we wish to develop an exact calculation, valid at any dislocation velocity, of the problem as posed by Gilman.

7.2. Theory

7.2.1. Basic Assumptions

We assume that a dislocation, either screw or edge in character, moves on a slip (or climb) plane which, in the absence of damping mechanisms, obeys a periodic stress versus displacement relationship. If D is the displacement across the slip (or climb) plane, and if σ_0 is the stress on this plane,

$$\sigma_0 = \sigma_1(D) = \sigma_1(D + b), \tag{7.1}$$

where b is the periodicity of the relationship and σ_1 is an arbitrary periodic function of b. (For subsonic dislocations $\int_0^b \sigma_1 \, dD$ must also equal zero.)

Assume next that the damping of the dislocation motion arises from the presence of a sheet of a Newtonian viscous fluid of viscosity η, whose thickness a is equal to the atomic spacing across the slip plane. When $\partial D/\partial t = \dot{D} \neq 0$ the viscous fluid gives rise to a stress $\eta \dot{D}/a$ across the slip or climb plane. (In the case of climb of an edge dislocation, the volume of the fluid must change as $\dot{D}$ changes.) For Gilman's problem Equation 7.1 is replaced by

$$\sigma_0 = \sigma_1(D) + \eta \dot{D}/a. \tag{7.2a}$$

In problems of uniform motion $\partial D/\partial t = -V \, \partial D/\partial x$, where V is the velocity of motion and x measures distance in the direction of dislocation motion in a coordinate system moving with velocity V. The function $B(x) = -\partial D/\partial x$ $(D(x) = \int_x^\infty B(x) \, dx)$ is equal to the density of infinitesimal dislocations on the slip (or climb) plane. Thus

$$\sigma_0 = \sigma_1(D) - \eta V B(Xx)/a. \tag{7.2b}$$

Assume further that the two half-spaces above and below the slip (or climb) plane are filled with isotropic elastic material containing no damping mechanisms. The stress σ_2 that acts on the surfaces of these two half-spaces which arises from a distribution of dislocations moving at subsonic velocities ($V \leq c$, where c is the shear wave sound velocity) is (Frank 1949, Eshelby 1949, Leibfried and Dietze 1949, Weertman 1961, 1967a)

$$\sigma_2 = (\mu A/\pi) \int_{-\infty}^{\infty} \{B(x')/(x - x')\} \, dx', \tag{7.3}$$

where μ is the shear modulus, $A = \beta/2$ for a screw dislocation, $A = (2c^2/V^2)$ $(\gamma - \alpha^4/\beta)$ for a gliding edge dislocation, and $A = (2c^2/V^2)(\beta - \alpha^4/\gamma)$ for a climbing edge dislocation. Here $\beta^2 = 1 - V^2/c^2$; $\gamma^2 = 1 - V^2/c_\lambda^2$, where c_λ is the longitudinal sound velocity; and $\alpha^2 = 1 - V^2/2c^2$.

For supersonic dislocations ($V > c$ for a screw dislocation and $V > c_\lambda$ for an edge dislocation) Equation 7.3 is replaced by (Eshelby 1956, Weertman 1967b)

$$\sigma_2 = -\mu S B(x), \tag{7.4}$$

where $S = \beta^*/2$ for a screw dislocation,

$$S = [4\gamma^* + (\beta^{*2} - 1)^2/\beta^*]/2(\beta^{*2} + 1)$$

for a gliding edge dislocation, and $S = [4\beta^* + (\beta^{*2} - 1)^2/\gamma^*]/2(\beta^{*2} + 1)$ for a climbing edge dislocation. Here $\beta^{*2} = -\beta^2$ and $\gamma^{*2} = -\gamma^2$.

In the transonic velocity range ($c \le V \le c_\lambda$) Equation 7.4 becomes (Weertman 1967b)

$$\sigma_2 = -\mu S B(x) + (\mu A/\pi) \int_{-\infty}^{\infty} \{B(x')/(x - x')\}\, dx', \qquad (7.5)$$

where $S = 2c^2\alpha^4/\beta^* V^2$ and $A = 2c^2\gamma/V^2$ for a gliding-edge dislocation, and $S = 2\beta^*c^2/V^2$ and $A = -2c^2\alpha^4/\gamma V^2$ for a climbing-edge dislocation.[1]

The solution of the problem requires that a dislocation density function $B(x)$ be found that gives rise to a stress $\sigma_2(x)$ equal to $-(\sigma_0 + \sigma_a)$, where σ_a is the applied stress. The density must also satisfy the condition that $\int_{-\infty}^{\infty} B(x)\, dx = b$.

7.2.2. A Subsonic Solution

A simple subsonic solution is found for Gilman's problem for the periodic stress-displacement relationship $\sigma_1(D)$ shown in Figure 7.1. This relationship is

$$-\infty < \sigma_1 < \infty \qquad (7.6a)$$

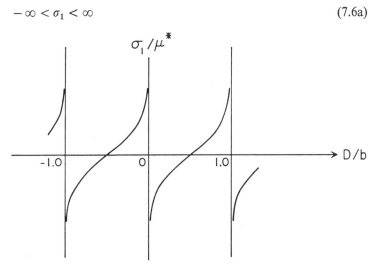

FIGURE 7.1. Periodic stress-displacement law across the slip plane.

[1] The terms A and S for a climbing edge dislocation in the transonic velocity range given above are different than those in Eq 46 of Weertman (1967b). Equations 46 through 49 in that reference are in error because the term α^2 should have been deleted in expressions for σ_{xy} in Eqs 43 and 45. Because of this error it was not realized in that reference that a climbing transonic edge dislocation can exist only on a climb plane that is in unstable equilibrium when $D = 0$ and $\sigma_a = 0$. In Equation 34 for the gliding edge dislocation γ^{*2} should have been γ^*.

for $D = 0$, $\pm b$, $\pm 2b$, etc., and the periodic repetitions of

$$\sigma_1 = -\mu^* \log |(D - b)/D| \tag{7.6b}$$

for noninteger values of D/b. The coefficient μ^* is an arbitrary positive constant. Equations 7.6 require the density function to be

$$
\begin{aligned}
B(x) &= b/2w \quad && \text{for } -w \le x \le w \\
&= 0 && |x| > w
\end{aligned} \tag{7.7}
$$

where

$$w = \mu A b / 2\pi\mu^* \tag{7.8}$$

and

$$V = 2w\sigma_a a/\eta b = \mu A a \sigma_a / \pi\eta\mu^* \tag{7.9}$$

Figures 7.2, 7.3, and 7.4 show plots of V versus applied stress obtained from this last equation. The velocity is proportional to stress at low stresses. It increases less than linearly with increasing stress near the shear wave velocity for a screw dislocation or the Rayleigh wave velocity c_r for an edge dislocation.

The behavior of edge dislocations in the velocity region $c_r \le V \le c$ requires special comment. No solution of an edge dislocation in uniform motion is possible in this velocity region for a periodic stress relationship $\sigma_1(D)$ such that the slip plane is in stable equilibrium when $D = 0$ and $\sigma_a = 0$. (In other words, when $d\sigma_1/dD < 0$ at $D = 0$.) However, if the periodic stress is such that the slip plane is in unstable equilibrium at $D = 0$ and $\sigma_a = 0$. (that is, $d\sigma_1/dD > 0$ at $D = 0$) solutions are possible.

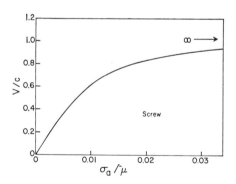

FIGURE 7.2. Normalized velocity versus normalized applied stress relationship for a subsonic screw dislocation. The curve at low velocities is made to fit the velocity data ($\sigma_a b/V = 6.5 \times 10^{-4}$ cgs) found in internal friction studies of copper crystals at room temperature (Granato and Lücke 1966). In this and the following figures, the following constants (all in cgs units) for copper were used (Kolsky 1953, p. 201): $c = 2.25 \times 10^5$; $c_\lambda = 4.56 \times 10^5$; $c_r = 2.12 \times 10^5$; $\mu = 4.5 \times 10^{11}$.

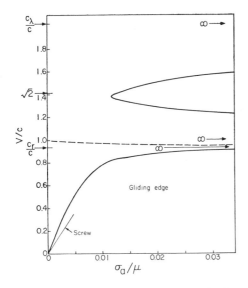

FIGURE 7.3. Normalized velocity versus normalized applied stress relationship for a subsonic and transonic gliding-edge dislocation. The dashed curve is calculated for a slip plane in unstable equilibrium.

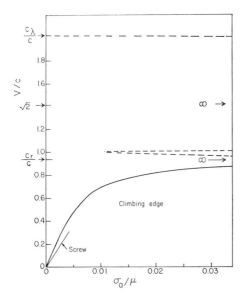

FIGURE 7.4. Normalized velocity versus normalized applied stress relationship for a subsonic and transonic climbing-edge dislocation. The dashed curves are calculated for a climb plane in unstable equilibrium.

(Solutions in this velocity region also are possible if the edge dislocation is discrete [Weertman 1961, 1967a] or, rather unlikely, if σ_1 is a double-valued function of D.) The dashed curves in Figures 7.3 and 7.4 in the region $c_r \leq V \leq c$ were calculated with a stress σ_1 given by Equation 7.6b but with a reversed sign ($\mu^* < 0$).

In the calculation of the curves in Figures 7.2 through 7.6, the term $\eta\mu^*$

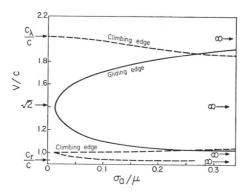

FIGURE 7.5. The transonic velocity region of Figures 7.3 and 7.4 shown with a different scale for the stress axis.

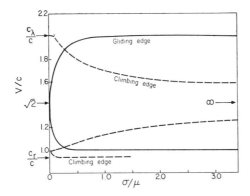

Figure 7.6 The transonic velocity region of Figures 7.3 and 7.4 shown with a different scale for the stress axis.

was chosen so that the theoretical values of the velocity of a screw dislocation at low stress levels agree with the velocities deduced from internal friction studies made on copper crystals at room temperature (Granato and Lücke 1966), p. 255.

7.2.3. *A Transonic Solution*

Solutions for edge dislocations in the transonic velocity range are found by combining Equations 7.2b, 7.5, and 7.6. The dislocation density function

$B(x)$ again is given by Equations 7.7 and 7.8. Equation 7.9 is replaced with (for $V > 0$)

$$\sigma_a = (\pi\mu^*/\mu A)(\mu S + \eta V/a). \tag{7.10}$$

Figures 7.3 through 7.6 show stress-velocity curves calculated from this equation using the same value of $\eta\mu^*$ employed in the subsonic region and assuming a value for μ^* that would give $w = b$ for a stationary screw dislocation ($\mu^* = 4\pi\mu$).

The term A is a negative quantity for a transonic climbing-edge dislocation. Therefore, as was the case in the velocity region $c_r \leq V \leq c$, a solution cannot exist when μ^* is a positive quantity. The curves shown in Figures 7.4, 7.5, and 7.6 for the climbing-edge dislocation were found after reversing the sign of Equation 7.6b and, in effect making μ^* a negative constant.

For a gliding-edge dislocation the applied stress required to satisfy Equation 7.10 is a minimum near $V = \sqrt{2}c$ and is approximately equal to

$$(\sigma_a) \rightarrow \approx (\pi\mu^*\eta V/\mu\gamma a)_{V=\sqrt{2}c}. \tag{7.11}$$

When no damping exists, the minimum stress is zero and occurs at exactly the velocity $V = \sqrt{2}c$. The stress of Equation 7.11 is approximately the same as that found by linearly extrapolating the subsonic curve found at low stresses and velocities. Obviously a gliding edge dislocation in the transonic velocity range will move at a velocity equal to or greater than $\sqrt{2}c$. At velocities lower than $\sqrt{2}c$ the stress-velocity relationship gives rise to unstable motion.

7.2.4. Supersonic Solutions

Supersonic solutions are found by combining Equations 7.2b and 7.4. A supersonic screw or edge dislocation must satisfy the equation (for $V > 0$)

$$B(x) = \{\sigma_1(D) + \sigma_a\}/\{\mu S + \eta V/a\}. \tag{7.12}$$

This equation can be satisfied only if $\int_0^b \{\sigma_1(D) + \sigma_a\}\, dD$ is greater than zero. Moreover, as was shown previously (Eshelby 1956, Weertman 1967b), a solution can exist only if the sum $\sigma_1(S) + \sigma_a$ never changes sign as D varies from 0 to b. If these two conditions are met the dislocation can move at any velocity in the supersonic velocity range. If the two conditions are met for $\sigma_a = 0$ the dislocation will move supersonically at any supersonic velocity despite the fact that a viscous damping force acts on the slip or climb plane.

7.3. Discussion

In the analysis presented in this paper a particular periodic force law was employed in calculating the stress-velocity relationship in the subsonic and transonic velocity regions. Other periodic force laws would lead to qualitatively similar curves. It is reasonable to expect that if the maximum stress of the periodic law were finite rather than infinite, the stress required to move dislocations at velocities $v = c_r$, c, $\sqrt{2}c$ or c_λ would equal or approximate this maximum stress rather than the infinite values shown in Figures 7.2 through 7.6.

The behavior of dislocations in the velocity region requiring a periodic force law that starts at $D = 0$ in unstable equilibrium is rather uncertain. Supersonic dislocations of Burgers vector equal to one-half the periodicity of the force law can propagate on a slip or climb plane that is in unstable equilibrium. Thus dislocations moving at velocities shown by dashed curves in Figures 7.3, 7.4, 7.5, and 7.6 can be eliminated from the lattice by the propagation of pairs of partial supersonic dislocations of Burgers vector $b/2$ that move in opposite directions away from the original dislocations.

If the applied stress is gradually increased from zero to a finite stress less than the maximum stress the slip or climb plane can support, it is clear that the limiting velocity of a screw dislocation is c, the shear wave velocity. The limiting velocity of a climbing-edge or gliding-edge dislocation is c_r, the Rayleigh wave velocity. Eshelby (1949) originally proposed this latter velocity as the limiting velocity for gliding edge dislocations. Solutions also exist for the gliding-edge dislocation in the transonic velocity range. Perhaps gliding-edge dislocations can be brought into this velocity region past the barrier that exists at $V = c_r$ through the sudden application of an external stress. The most serious obstacle to moving an edge dislocation at these velocities is that the minimum applied stress is rather large, as can be seen from Figure 7.3. This minimum stress is approximately the same as the stress found by extrapolating to the velocity $V = \sqrt{2}c$ the damping force on a slowly moving dislocation determined from internal friction studies.

The research upon which this paper was based was supported by the Advanced Research Projects Agency of the Department of Defense through the Northwestern University Materials Research Center under Contract SD–67.

REFERENCES

Eshelby, J. D., 1949, *Proc. Phys. Soc.*, **A62**, 307.

Eshelby, J. D., 1956, *Proc. Phys. Soc.*, **B69**, 1013.

Frank, F. C., 1949, *Proc. Phys. Soc.*, **A62**, 131.

Gilman, J. J., 1968, *Phys. Rev. Letters*, **22**, 157.

Granato, A. V., and K. Lücke, 1966, in W. P. Mason (ed.), *Physical Acoustics*, New York and London: Academic, vol. IV, p. 226.

Kolsky, H., 1953, *Stress Waves in Solids*, Oxford: Clarendon Press.

Leibfried, G., and H. D. Dietze, 1949, *Z. Physik*, **126**, 790.

Orowan, E., 1940, *Proc. Phys. Soc.*, **52**, 8.

Weertman, J., 1961, in P. Shewmon and V. Zackay (eds.), *Response of Metals to High Velocity Deformation*, New York: Interscience, p. 205.

Weertman, J., 1967a, *J. Appl. Phys.*, **38**, 2612.

Weertman, J., 1967b, *J. Appl. Phys.*, **38**, 5293.

ABSTRACT. The theory of homogeneous nucleation of mechanical twins and of martensite plates which was described by Orowan in 1954 is reexamined. There appear to be many difficulties in accepting this theory, even if the usual numerical estimates of relevant parameters are varied. The alternative "embryo" theory of martensite nucleation, which depends on special assumptions, has disadvantages comparable to those of the homogeneous theory. It seems probable that in most circumstances nucleation of both twins and martensite is defect-aided, but also requires high local driving forces.

8. The Nucleation of Mechanical Twins and of Martensite

J. W. CHRISTIAN

8.1. Introduction

In the published version of his famous contribution to the 1951 A.I.M.E. Seminar on Dislocations in Metals, Orowan (1954) developed the theory of homogeneous nucleation of mechanical twins and of martensite plates. His assumption that twins may form in defect-free regions of crystal has been rather widely accepted, especially since the publication of the experimental work of Price (1961) on zinc platelets, but the corresponding theory of homogeneous nucleation of martensite has encountered many difficulties and is generally regarded as discredited. It is my intention in this note briefly to re-examine the present status of Orowan's theory, and to relate it to recent developments.

The analogies between mechanical twinning and martensitic transformation are well known. The idealized or unconstrained shape change produced in a twinned region is a simple shear on the plane K_1 in the direction η_1, and the change of shape in a martensite plate is a combination of a simple shear on the habit plane with a uniaxial expansion or contraction normal to that plane. When a single crystal is twinned by stress, the operative variants are generally found to be those which most effectively reduce the potential of the applied stress, in accordance with Le Chatelier's principle.

Martensitic transformation is often induced solely by nondirectional chemical driving forces, but when transformation takes place in a stress field, the energy of interaction of this field with the shape change becomes important, and the preferred variants of the habit plane are then often those predicted by Le Chatelier's principle.

Attempts to express the conditions governing twinning or transformation in quantitative form have been less successful. The evidence for and against a critical resolved shear stress law for mechanical twinning is very confused, and there have been few attempts to test such a law for stress-induced transformation. However measurements of the chemical driving force for spontaneous formation of martensite at M_s are usually consistent, and in recent work it has also been shown that for many alloys the change in M_s caused by transformation under pressure or in a magnetic field is in agreement with the hypothesis that the M_s temperature is determined by a critical value of the total driving force (Radcliffe and Schatz 1962, Satyanarayan, Eliasz, and Miodownik 1968).

The assumption of homogeneous nucleation implies very high internal stress concentrations for twinning, and large driving forces for martensite. Many alternative theories of twinning have been developed by postulating that existing defect configurations are able to grow into twins, either under the influence of the average stress field applied externally, or at internal stress concentrations. The minimum nucleation stress for such a process is often expressed as α_f/b where α_f is the energy of the stacking fault which corresponds to a monolayer twin. Similar theories for martensite have generally not been developed in such detail as the dislocation models for twinning. The third possibility is that the nuclei are present initially, so that the process considered is entirely growth. Unless the preexisting nuclei are associated with defects, they must then exceed the critical size of classical nucleation theory before growth is possible; this assumption has generally been made in the "embryo" theory of martensite.

8.2. Theory of Homogeneous Nucleation

The description in this section is so arranged that twinning may be regarded as a special case of martensite formation. The treatment is generally similar to that originally given by Fisher, Holloman, and Turnbull (1949) and Orowan (1954) but incorporates later improvements. It will be assumed that the shape deformation is a simple shear γ on the habit plane and an expansion or contraction ε normal to this plane. The current crystallographic theories of martensite also allow for possible small deviations from this assumption of an exact invariant plane strain, and the simplest generalization is obtained by introducing a dilatation parameter δ. The necessary modifications to the nucleation theory when a dilatation is present have been considered by Christian (1958, 1959), but this complication will be ignored here.

Let the shear stress and normal stress on the habit plane be τ and σ, respectively, so that the mechanical work per unit volume of product (defined to be positive when these stresses aid transformation) is $\Delta g_{mech} = \tau\gamma + \sigma\varepsilon$. If the parent phase has a chemical free energy per unit volume which is higher than that of the product phase by Δg_{chem}, the total driving force is $\Delta g = \Delta g_{mech} + \Delta g_{chem}$. Opposing this driving force is the surface energy of the product and the strain energy attributable to the constraints of the matrix. For an ellipsoidal inclusion of semiaxes r, r, and y, the strain energy per unit volume may be written Ay/r where A has been calculated in closed form in the approximation of isotropic elasticity (Eshelby 1957).

The surface free energy is often written in the form $2\pi r^2 \alpha'$, but Orowan pointed out that it is more correct to express it as $2\pi r^2 \alpha + 4\pi r(y/d)\Gamma$. Here α is the specific surface energy of the flat interface, y/d is the number of atomic steps of height d needed to give the lenticular shape, and Γ is the energy per unit length of each step. Strictly the mean radius of the steps is less than r, but this small numerical factor ($\sim(2/3)^{1/2}$ for the oblate spheroidal shape assumed above) is unimportant. The steps are equivalent to dislocation lines of Burgers vector $\mathbf{b}_T$ where $b_T^2 = d^2(\gamma^2 + \varepsilon^2)$; they are often described as twinning or transformation dislocations, respectively. These dislocations glide in the interface, even though (in the martensite case) the Burgers vector has a component normal to the interface; usually, however, it is sufficient to assume $b_T = \gamma d$. It follows that Γ may be approximated as $C\mu b_T^2$ where μ is the shear modulus and $C \sim 1/4 - 1/2$, so that the total energy of the steps is $4\pi C\mu(b_T^2/d)ry$. This energy may also be written in the form $4\pi\beta ry$ where β is the equivalent incoherent surface free energy per unit area.

Some typical values of γ are $(2)^{-1/2}$ for twinning in cubic metals, $(32)^{-1/2} \simeq 0.18$ for martensite in steels, and ~ 0.14 for twinning in zinc. The true twinning stress is generally unknown, but has been measured for zinc as ~ 50 kg mm^{-2} (Price 1961), so that Δg_{mech} in this case is $\sim 7.10^8$ erg cm^{-3}. At the M_s temperature for martensite in steels, Δg_{chem} is $\sim 2.10^9$ erg cm^{-3}. The factor A is given by Eshelby as $\pi(2 - \nu)\mu\gamma^2/8(1 - \nu) \simeq \mu\gamma^2$ where ν is Poisson's ratio. Thus $A \sim 2.5.10^{10}$ erg cm^{-3} for steels and 6.10^9 erg cm^{-3} for zinc, and the elastic energy may be large even for small y/r. Orowan's suggestion that the elastic energy is only of order 10^7 erg cm^{-3} depends on the assumption of a rather low maximum stress in the matrix, and a high shear modulus, and in addition does not allow for the fact that the matrix volume of high strain energy is larger than the nucleus volume by a factor of order r/y. For a small nucleus, it seems likely that the matrix stresses could reach the theoretical defect-free strength, and the energy per unit volume of nucleus can then be of order 10^9 erg cm^{-3}.

The surface free energy terms are difficult to estimate. Measurements on face-centered cubic metals (Inman and Tipler 1963) suggest $\alpha = 10$–50 ergs cm^{-2} for coherent twins, but twin interfaces in body-centered cubic and hexagonal close-packed metals may well have higher energies. A dislocation model of the martensite interface in steels (Frank 1953) has led to estimates of $\alpha \sim 200$ ergs cm^{-2} (Knapp and Dehlinger 1956, Kaufman and Cohen 1958). The dislocations of this model match the structures at a planar interface, and are quite distinct from the steps discussed above. In practice, the matching of the two lattices is more frequently accomplished by internal twinning, but it will be assumed arbitrarily that the effective α for this substructure is also ~ 200 ergs cm^{-2}. Finally the value of $\beta = C\mu d\gamma^2$ is ~ 600 ergs cm^{-2} for $\{211\}$ twinning in iron if $C \sim 1/2$, and is much smaller for martensite and for twinning in zinc. It is perhaps preferable to regard β as an incoherent boundary energy, of order 500 ergs cm^{-2}, and this corresponds with Orowan's estimate, $\Gamma \sim 10^{-5}$ ergs cm^{-1}.

The volume of the new region at any stage is $v = 4\pi r^2 y/3$, and the change in free energy due to its formation may now be written

$$\frac{\Delta G}{v} = -\Delta g + \frac{3\alpha}{2y} + \frac{3\beta}{r} + A\frac{y}{r} \qquad (8.1)$$

The size and shape of the critical nucleus are obtained by setting $\partial\Delta G/\partial r = \partial\Delta G/\partial y = 0$ in order to obtain the saddle point in G. The shape of the nucleus is then defined by the equations

$$\Delta g \cdot r = 2Ay + 3\beta \qquad (8.2)$$

$$\alpha r = Ay^2 + \beta y \qquad (8.3)$$

The minimization procedure used by Orowan corresponds to $A = 0$, while most other published versions of the theory do not include β. The equations change slightly with different assumptions about the analytical shape of the nucleus; the geometry assumed by Orowan, for example, introduces additional factors of 4/3 and 3/4 into the second term on the right of Equation 8.2 and the first term on the right of Equation 8.3, respectively.

Equations 8.2 and 8.3 may be solved to give the critical thickness

$$y_c = -(p + q) \pm (p^2 + q^2 - 4pq)^{1/2}, \qquad (8.4)$$

where $p = -\alpha/\Delta g$ and $q = \beta/2A$.

Table 8.1 shows the results of numerical calculations using parameters (as above) thought to be appropriate to twinning in zinc and to martensite in steels. In order to examine the effects of wrong assumptions about the elastic energy and the incoherent surface energy, results are given both for $\beta = 0$ and $A = 0$ as well as for the general case. It will be seen that the computed energies for martensite in steels are all much too high for thermal activation to be feasible, and this failure of the homogeneous nucleation

TABLE 8.1. Typical calculated parameters for homogeneous nucleation. (Units of α, β are erg cm^{-2}, and A, Δg are erg cm^{-3}).

(a) Twinning in zinc ($\alpha = 20$, $\Delta g = 7 \times 10^8$)			
	$\beta = 0$ $A = 6.10^9$	$\beta = 100$ $A = 0$	$\beta = 100$ $A = 6.10^9$
$y_c(\text{Å})$	5.7	8.6	6.3
$r_c(\text{Å})$	98	43	151
$\Delta G_c(\text{eV})$	23	13	75

(b) Martensite in steels ($\alpha = 200$, $\Delta g = 2.10^9$)			
	$\beta = 0$ $A = 2.5.10^{10}$	$\beta = 10^3$ $A = 0$	$\beta = 10^3$ $A = 2.5.10^{10}$
$y_c(\text{Å})$	20	30	22
$r_c(\text{Å})$	500	150	700
$\Delta G_c(\text{eV})$	6,000	1,600	15,500

theory is now familiar. Reasonable values of ΔG_c can be obtained only by assuming much lower values of α and A.

The calculation for zinc is of especial interest, since Price's measurement of the nucleation stress has been widely held to be consistent with Orowan's calculation. Table 8.1 shows that the lowest estimate of ΔG_c is obtained by neglecting the strain energy, but even this most favorable estimate is still appreciably larger than the required value of ~ 1 eV claimed by Price to be predicted by the Orowan theory. The source of this discrepancy is not clear. If the experimental twinning stress is combined with the other parameters assumed by Orowan, his Equation 3.32 gives a value for ΔG_c that is 44 times larger than that given in the middle column of Table 8.1.[1] Equation 2 of Price's paper seems to be derived from Equation 3.32 of Orowan's paper by writing $\Gamma = (1/4)\mu b_T^2$ and arbitrarily equating the step height to b_T. The equation as published is, however, dimensionally incorrect; there is a factor $\tau \gamma^2$ (in present notation) omitted from the divisor.

The general conclusion is that it is very difficult to believe in homogeneous nucleation of twins or martensite unless there is a combination of very high-driving force with very low surface and strain energies. In the case of Price's experiments on zinc platelets, the strain energy may be reduced because of the small specimen thickness, and a rather small reduction in α and β will then give feasible values of ΔG_c. Nevertheless it is difficult to avoid the conclusion that the values assumed for α and β may already be too small. The alternative possibility seems to be that the true stress concentration factor at the re-entrant nucleation sites was higher than that calculated, or that the platelets contained some undetected defects which aided nucleation.

[1] A factor of 25 arises because Orowan's assumptions are equivalent to $\beta = 500$ erg cm^{-2}, and there is a further factor of 16/9 from the different geometry.

8.3. Preexisting Nuclei, and the Embryo Theory of Martensite

If twins are nucleated homogeneously, the process must clearly be athermal and the local stress required is probably of the order $\mu/50$. Bulk specimens are commonly observed to twin at much lower stresses, and this is due, at least partially, to stress concentrations. There is also the possibility that tiny twins are present in most crystals as a result of growth accidents, and that these are able to grow under an applied stress (Oliver 1952). If this is to be feasible, the preexisting twin must be larger than the critical nucleus size of the classical theory at the stress level at which it is to grow. Although there need be no strain energy associated with the original nucleus, it would not be appropriate to take $A = 0$ in calculating this minimum size, since the strain energy continues to provide a resistance to mechanical growth.

In classical nucleation theory, an embryo of subcritical size should tend to disappear. In principle this applies to an included twin under zero applied stress, but the process may be indefinitely slow if recrystallisation is involved. Mechanical detwinning will not normally be expected in such a situation since the decrease in surface free energy is opposed by the strain energy created. There is, however, a related problem of why a large twin should not spontaneously detwin when the stress is removed. Venables (1964) has suggested that a possible explanation of the observed twin stability is a growth resistance which becomes smaller under dynamic conditions. A simpler and perhaps more probable explanation is that as a twin grows some of the elastic accommodation energy is converted irreversibly into plastic work, thus destroying the coherency of the interface and reducing the back stress on the twinned region. The known phenomena of elastic twinning (Cahn 1953) and thermoelastic martensite (Christian 1965) support this view.

Reductions in the above values of α and A of about one order of magnitude would allow homogeneous nucleation theory to be applied to martensite, but many experimental results, e.g. the very low M_s temperatures in some iron-nickel alloys, could still not be explained. The embryo theory of martensite (Knapp and Dehlinger 1956, Kaufman and Cohen 1958, Cohen 1958), is based on the postulate that regions of martensite form in the parent phase at high temperatures, but are able to grow only when they become supercritical at M_s. A semicoherent interface of the dislocation type is formed from existing lattice dislocations (the mechanism is not specified in detail), and the strain energy of these dislocations is assumed to be released to aid nucleation. In effect, the usual nucleation calculation is made with the signs of the terms reversed. The chemical change in free energy is positive, since the process is assumed to occur at temperatures where bulk martensite is not stable, but the surface energy is effectively negative. This gives a minimum of free energy at a critical size which changes with temperature, so that the unspecified mechanism of embryo formation must also allow growth or coalescence as the temperature is reduced.

If this growth were continued, the "equilibrium" size would eventually become infinite but it is further assumed that the process is frozen in at some particular temperature. On subsequent cooling, there is no further change in embryo size, so that the M_s temperature is determined by the maximum size of embryo, r_m, at the freezing-in temperature. The condition for athermal growth is assumed to be that the embryo size exceeds some critical value, r_c', which is larger than the critical size of the classical theory r_c. In the temperature range where $r_c < r_m < r_c'$, thermally activated nucleation will be observed if the embryo can increase in size from r_m to r_c' by fluctuations. At slightly lower temperatures, r_c' will have decreased below r_m and athermal nucleation will take place, followed by isothermal activation of those embryos in the distribution that are initially smaller than r_c'.

The special assumptions of this theory appear to be very speculative and unsatisfactory, and there is little experimental evidence to support it. Attempts to detect embryos by electron microscopy have not been successful, and experiments on the effects of high hydrostatic pressures on M_s, with one exception (Kaufman, Leyenaar, and Harvey 1960), are not in agreement with the predictions of the theory (Radcliffe and Schatz 1962, Christian 1969).

8.4. Nucleation on Defects

In the first dislocation model of twinning, Cottrell and Bilby (1951) introduced the pole mechanism of growth and also proposed that a twin in a b.c.c. structure forms from the dissociation of a prismatic dislocation in (211) followed by the cross-glide of the $a/6$ [111] twinning dislocation into ($\bar{1}$21) or (2$\bar{1}$1). Later Venables (1961) showed how the pole mechanism might operate from a single fault in the f.c.c. structure by a process of repeated recombination of partials followed by prismatic glide. The twinning stress in all models of this type is usually written in the form

$$\tau = (\alpha_f/b_T) + (\Gamma/b_T r) + \tau_R \qquad (8.5)$$

where the symbols have already been defined except that r is now the radius of a semicircular loop of twinning dislocation and τ_R is an unspecified growth resistance, included for generality. The local stress τ is presumably some multiple of the externally applied stress, but Venables (1964b) has shown there is reasonably good correlation between measured twinning stresses and stacking fault energies in f.c.c. alloys.

In various alternative models of nucleation from dislocations, attention is confined to the formation of twinned regions only one or three atom

layers thick, and it is assumed that suitable generating nodes for operation
of the pole mechanism will be obtained from random encounters with
lattice dislocations. These models usually imply dissociation of lattice
dislocations in their glide planes, rather than of prismatic dislocations; in
f.c.c. metals for example, the back partials have been assumed to be held
in place by Lomer-Cottrell locks (Suzuki and Barrett 1958, Haasen and
King 1960). In b.c.c. crystals, Sleeswyk (1963) has proposed a model based
on the assumed dissociation of a $\langle 111 \rangle$ type screw dislocation on intersecting $\{\bar{2}11\}$ planes, and the transformation of the stable configuration
under stress into a 3-layer twin. An alternative proposal (Ogawa and
Maddin 1964) is that an edge dislocation can dissociate directly into three
twinning dislocations on successive planes, and Ogawa (1965) has justified
this by arguments designed to show that the interfacial energy of a three-
layer twin is appreciably less than that of a monolayer stacking fault.
Recent computer calculations by Vitek (1968) indicate that wide stacking
faults in b.c.c. metals are unstable rather than metastable configurations,
so that models requiring single-layer faults may be unrealistic.

Glide sources have the advantage that the required dislocation configuration is likely to occur more frequently, but there are other difficulties in
the mechanisms proposed for their operation (see Venables 1964a for
f.c.c. metals). The critical stage in the operation of such a source might be
the passing of opposite twinning partials on neighboring planes, which
would require a very high stress. If this is somehow avoided, however,
the necessary stress will again be similar to that obtained from Equation 8.5.
Sleeswyk estimates, for example, that the minimum stress needed to produce his twinning configuration from a lattice screw dislocation is ~ 0.33
α_f/b_T.

Dislocation models of martensite formation have not been presented in
such detail as the twinning models, except for the fully coherent f.c.c. $\rightleftharpoons$
h.c.p. transition. In this case there is reasonably good evidence that the
transformation is initiated by dissociation of dislocations in their glide
planes, although it is not clear whether growth occurs by a pole mechanism
(Seeger 1953, 1956, Basinski and Christian 1953) or by repeated two-
dimensional nucleation at grain boundaries, etc. (Christian 1951, Bollman
1961). No single dissociation of a lattice dislocation can produce a b.c.c. or
a b.c. tetragonal phase from a f.c.c. structure, but lattice shears that approximate to these structural changes have been expressed in dislocation terms
and suggested as possible nucleation processes for the formation of martensite in steels. An early theory of this type follows the Kurdjumov-Sachs
two-shear description of martensite formation, and involves a half-twinning
shear on a f.c.c. $\{111\}$ plane. This shear could be produced by "quarter-
dislocations" of type $a/12 \langle 11\bar{2} \rangle$ (Jaswon 1956). A more recent suggestion,
based on hard sphere models of the f.c.c. and b.c.c. structures, begins with
a one-third twinning shear produced by $a/18 \langle 11\bar{2} \rangle$ dislocations (Bogers

and Burgers 1964). Both theories require further shears and dimensional changes, and it is not obvious how these are produced, nor how the nucleus is evolved from the initial stacking fault.

In the simpler dislocation models of twinning, the condition of Equation 8.5 is assumed to be critical for all stages of subsequent growth. This could not be carried over to martensite formation because of the probable much greater complexity of any dislocation reactions which contribute to nucleation. When Equation 8.5 is compared with the theory of Section 8.2, it is apparent that the nucleus thickness y and radius r are regarded as selected by the mechanism, rather than chosen to minimize the energy. Provided $\alpha_f > \alpha$ (estimates of α_f/α for twins vary from 2 to 4), the assumption that Equation 8.5 represents a critical stage in twinning may well be justified. By analogy with the production of slip dislocations, the appropriate configuration is often described as a twin source.

8.5. Summary of Present Position

The importance in twinning of stress concentrations and/or defects first became widely recognised with the work of Bell and Cahn (1957). Frank and Stroh (1952) suggested that very high stress concentrations are provided by incomplete kink bands, or by existing twins or cracks, and Bilby and Entwisle (1954) considered in more detail the case of kink bands or bounded slip planes. Their theory rested on the assumption that nucleation of a twin is governed by the local value of the resolved shear stress on the K_1 plane, and was further developed by Bilby and Bullough (1954) who considered the formation of twins around a moving crack. Experimental results by Deruyttere and Greenough (1954) were consistent with the theoretical predictions.

Many experiments show similarly that existing martensite plates are probably the most effective agents yet identified in promoting the nucleation of new plates. This leads for example to the well-known burst phenomenon, and it appears that in iron-nickel alloys the most effective coupling takes place among groups of four plates with nearly parallel habit plane normals. Recent investigations of burst phenomenon in previously deformed single crystals (see Christian 1969 for summary) show that the deformation substructure also influences the preferred variants, although it is not clear at present whether this arises from particular nucleation sites or from anisotropy of growth resistance provided by lattice dislocations. The main significance of the burst effect, however, is the implication that high mechanical stresses are effective in assisting martensitic nucleation. Presumably such stresses can be important only when the mechanical work, Δg_{mech}, of the internal stress field in creating the nucleus is at least of the same order of magnitude as Δg_{chem} for the transformation.

Martensitic transformations have been studied in whiskers of cobalt

(Gedwill, Wayman, and Alstetter 1964) and iron (Zerwekh and Wayman 1965); in both cases the whiskers contained defects which were believed to assist nucleation. Perhaps the best evidence to support the hypothesis that defects are needed for nucleation comes from the observed suppression of transformation in very small, presumably defect-free, crystals. Small precipitates of iron in copper apparently remain f.c.c. on cooling to $1.4°K$ (Abrahams, Guttman, and Kasper 1962), even though the M_s temperature of the bulk iron-copper solid solution is estimated at $600°C$. The precipitates transform when the solid solution is deformed, and the transformation seems to be nucleated by matrix dislocations which have cut through the precipitates (Easterling and Miekk-oja 1967).

Recent work on twin initiation, especially in b.c.c. crystals, has been concerned largely with the question of whether localized slip precedes twinning, and this may be compared with the similar problem of crack nucleation. Local slip could be important simply in providing static stress concentrations, leading to twin nucleation when the slip is blocked, but there is also the possibility of dynamic effects arising, for example, because of the stress dependence of dislocation velocity in b.c.c. metals. Recent experiments which imply that slip precedes twinning have been described by Hamer and Hull (1964), Altshuler and Christian (1966), and Bolling and Richman (1965, 1967). However, there are also many experiments in which slip has not been detected, so that this question may be regarded as still open.

In summary, it appears that there are many difficulties in accepting the theory of homogeneous nucleation for either twinning or martensite, even if the parameters assumed in Section 8.2 are varied. Although the local stress may attain a very high value, it is also necessary to ensure that the stress gradient is not too large; otherwise a small nucleus will form but be unable to grow. The rather widely accepted embryo theory of martensite rests on special assumptions and in some respects appears to be as unsatisfactory as the theory of homogeneous nucleation. It appears probable that nucleation is defect-aided in most circumstances, and that the critical nucleus size is then reduced appreciably below its value in defect-free regions.

REFERENCES

Abrahams, S. C., L. Guttman, and J. S. Kasper, 1962, *Phys. Rev.*, **127**, 2052.
Altshuler, T. L., and J. W. Christian, 1966, *Acta Met.*, **14**, 903.
Basinski, Z. S., and J. W. Christian, 1953, *Phil. Mag.*, **44**, 791.
Bell, R. L., and R. W. Cahn, 1957, *Proc. Roy. Soc.*, **A234**, 221.
Bilby, B. A., and R. Bullough, 1954, *Phil. Mag.*, **45**, 631.
Bilby, B. A., and A. R. Entwisle, 1954, *Acta Met.*, **2**, 15.
Bogers, A. J., and W. G. Burgers, 1964, *Acta Met.*, **12**, 255.
Bolling, G. F., and R. H. Richman, 1965, *Acta Met.*, **13**, 723.

Bolling, G. F., and R. H. Richman, 1967, *Acta Met.*, **15**, 678.

Bollman, W., 1961, *Acta Met.*, **9**, 972.

Cahn, R. W., 1953, *Il Nuovo Cimento (Suppl.)*, **10**, 350.

Christian, J. W., 1951, *Proc. Roy. Soc.*, **A206**, 51.

Christian, J. W., 1958, *Acta Met.*, **6**, 377.

Christian, J. W., 1959, *Acta Met.*, **7**, 218.

Christian, J. W., 1965, *The Theory of Transformations in Metals and Alloys*, Oxford: Pergamon.

Christian, J. W., 1969, in *Symposium on Phase Transformations*, London: The Institute of Metals, p. 129.

Cohen, M., 1958, *Trans. Met. Soc. AIME*, **212**, 171.

Cottrell, A. H., and B. A. Bilby, 1951, *Phil. Mag.*, **42**, 573.

Deruyttere, A. E., and G. B. Greenough, 1954, *Phil. Mag.*, **45**, 624.

Easterling, K. E., and H. M. Miekk-oja, 1967, *Acta Met.*, **15**, 1133.

Eshelby, J. D., 1957, *Proc. Roy. Soc.*, **A241**, 376.

Fisher, J. C., J. H. Hollomon, and D. Turnbull, 1949, *Trans. Met. Soc. AIME*, **197**, 918.

Frank, F. C., 1953, *Acta Met.*, **1**, 15.

Frank, F. C., and A. N. Stroh, 1952, *Proc. Phys. Soc.*, **B65**, 295.

Gedwill, M., C. M. Wayman, and C. Alstetter, 1964, *Trans. Met. Soc. AIME*, **230**, 453.

Haasen, P., and A. H. King, 1960, *Z. Metallk.*, **51**, 700.

Hamer, F. M., and D. Hull, 1964, *Acta Met.*, **12**, 682.

Inman, M. C., and H. R. Tipler, 1963, *Met. Reviews*, **8**, 105.

Jaswon, M. A., 1956, in *The Mechanism of Phase Transformations in Metals*, The Institute of Metals: London, p. 173.

Kaufman, L., and M. Cohen, 1958, in B. Chalmers and R. King (eds.), *Progress in Metal Physics*, New York: Pergamon, vol. 7, p. 165.

Kaufman, L., A. Leyenaar, and J. S. Harvey, *Acta Met.*, 1960, **8**, 270.

Knapp, H., and U. Dehlinger, 1956, *Acta Met.*, **4**, 289.

Oliver, D. S., 1952, *Research*, **5**, 45.

Ogawa, K., 1965, *Phil. Mag.*, **10**, 217.

Ogawa, K., and R. Maddin, 1964, *Acta Met.*, **12**, 713.

Orowan, E., 1954, in M. Cohen (ed.), *Dislocations in Metals*, New York: A.I.M.E., p. 116.

Price, P. B., 1961, *Proc. Roy. Soc.*, **A260**, 251.

Radcliffe, S. V., and M. Schatz, 1962, *Acta Met.*, **10**, 201.

Satyanarayan, K. R., W. Eliasz, and A. P. Miodownik, 1968, *Acta Met.*, **16**, 877.

Seeger, A., 1953, *Z. Metallk.*, **44**, 247; 1956, *ibid*, **47**, 653.

Sleeswyk, A. W., 1963, *Phil. Mag.*, **8**, 1467.

Suzuki, H., and C. S. Barrett, 1958, *Acta Met.*, **6**, 156.

Venables, J. A., 1961, *Phil. Mag.*, **6**, 379.

Venables, J. A., 1964a, in R. E. Reed-Hill et al. (eds.), *Deformation Twinning*, New York: Gordon and Breach, p. 111.

Venables, J. A., 1964b, *J. Phys. Chem. Sol.*, **25**, 685, 693.

Vitek, V., 1968, *Phil. Mag.*, **18**, 773.

Zerwekh, R. P., and C. M. Wayman, 1965, *Acta Met.*, **13**, 99.

ABSTRACT. A close analogy is drawn between the development of the dislocation theory of crystal plasticity and the development of models of muscular contraction. The structure of insect muscle in rigor determined by Reedy may be derived from a perfectly crystalline structure by the introduction of a regular array of edge disclinations. A simple theoretical model predicts four structures as the energy of a cross-bridge is increased. The first is that of complete relaxation. The second may be identified with the state of active contraction. The third is that of rigor, while the fourth is an unobserved state of perfectly crystalline rigor.

9. The Disclination Structure of Insect Muscle

F. R. N. NABARRO

9.1. Plastic Deformation of Crystals

In this section we outline some aspects of the dislocation theory of crystal plasticity.

A perfect crystal, in which each plane of atoms is in perfect register with its neighbors, remains elastic up to a shear strain of the order of 10 percent (Frenkel 1926, see also Tamm 1962).

Masing and Polanyi (1923) showed in their "Biegegleitung" model (Figure 9.1) that one plane of atoms could slide over another relatively easily if, as a result of a macroscopic curvature of the crystal, the inter-

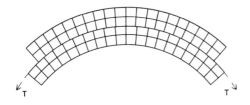

FIGURE 9.1. "Biegegleitung" makes crystal plasticity possible by introducing a difference in the spacing of atoms in adjacent layers (after Masing and Polanyi 1923).

atomic spacings in the two planes measured in the direction of relative motion were unequal.

Prandtl (1928) developed a mechanical model (Figure 9.2) which showed that plastic hysteresis could be explained if an atom in one plane could

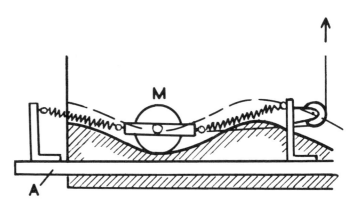

FIGURE 9.2. The irreversible motion of an atom of one layer in the sinusoidal potential field of the adjacent layer (detail from Figure 9 of Prandtl 1928).

move against the elastic constraints of its neighbors in that plane under the influence of the periodic potential of the atoms in a neighboring plane. The motion of an atom M from one trough in this potential to the next is then irreversible. Soon afterwards, Dehlinger (1929) introduced the idea of a "Verhakung" (a pair of dislocations of opposite sign), which drew attention to the existence of localized imperfections involving broken interatomic bands.

In 1934 appeared the celebrated papers of Orowan, Polanyi, and Taylor, which introduced the crystal dislocation, and in particular the edge dislocation (Figure 9.3) in which n atoms on one side of a glide plane face $n + 1$ atoms on the other side.

In the homogeneous crystal, dislocations are imperfections which cause

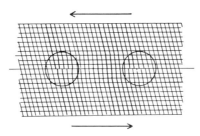

FIGURE 9.3. The dislocation as the mechanism of crystal plasticity (from Orowan 1934).

the dislocated crystal to have a higher energy than the perfect crystal. However (Frank and van der Merwe 1949a, b, van der Merwe 1963a, b, 1964), "misfit dislocations" may be present in the state of lowest energy of the interface between two crystals having different lattice parameters. Whether they are present or not depends on the misfit between the two lattices, and on the ratio between the forces between the two lattices and some suitable mean value of the elastic constants of the lattices.

9.2. Plastic Deformation of Muscle

In this section we outline some aspects of the theory of the contractile process in muscle, indicating a parallelism with crystal dislocation theory.

Striated muscle is pseudocrystalline (e.g. Hanson and H. E. Huxley 1955). The cellular repeat distance in the longitudinal direction is the sarcomere (Figure 9.4), some 3.1μ long, terminated by the Z bands, plates

FIGURE 9.4. The contractile element in muscle (from Hanson and Huxley 1955): $a =$ actin filament, $m =$ myosin filament, $Z = Z$ line.

which stretch perpendicular to the longitudinal axis to form quasirigid anchoring sheets. From each side of these sheets, filaments of the protein actin stretch into the sarcomere. Midway between the actin filaments, and centered in the cell, lie the thicker filaments of myosin, apparently not directly connected to the Z bands, but showing a high degree of lateral register. The process of contraction is effected by the sliding of myosin filaments between pairs of actin filaments. The actin and myosin filaments are linked at intervals by cross-bridges.

If muscle were truly crystalline, with frequent bonds between actin and myosin, one would expect it to sustain strains of a few percent elastically. In fact, muscle in a state of rigor is moderately elastic, but relaxed muscle extends and contracts plastically under very small stresses.

According to H. E. Huxley and Brown (1967): "One of the most basic features of the actin and myosin structures is so obvious that its significance tends to be overlooked. It is simply that the periodicities are different. . . . That such a difference might be a feature of muscles was suggested a considerable time ago . . . and it was noted that a difference in repeat would ensure that all the cross-bridges did not at any given moment occupy the same relative positions with respect to sites on the actin filament;

the arrangement would therefore allow a relatively steady tension to be developed in a given thin filament by the unsynchronised action of a number of cross-bridges attaching to it."

The first detailed model of the contractile process was developed by A. F. Huxley (1957). In this model (Figure 9.5) the cross-bridge M on a

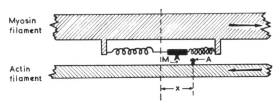

FIGURE 9.5. The irreversible motion of a myosin cross-bridge in the potential field of an adjacent actin filament (from A. F. Huxley 1957).

myosin filament is attached elastically to the myosin substrate. The active site A on the actin filament is the source of a long-range potential field. In conjunction with the potentials arising from other, equally spaced, active sites on the same actin filament, this produces a periodic potential in which M can move. The resemblance between Figures 9.2 and 9.5 is striking.

A. F. Huxley's model concentrated on the interaction between a single cross-bridge on a myosin filament and a single active site on an actin filament. The model of Spencer and Worthington (1960) (Figure 9.6)

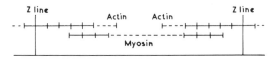

FIGURE 9.6. Muscular contraction is possible because there is a difference between the spacings of active sites in adjacent actin and myosin filaments (from Spencer and Worthington 1960).

showed how the continuous motion of a myosin filament past an actin filament could occur as the result of the vernier of alternate good and bad register arising from the difference between the spacings of myosin cross-bridges and of actin-active sites.

By analogy with the results of Frank and van der Merwe, one expects (Nabarro 1967a) that the forces between myosin cross-bridges and actin-active sites will distort the uniform vernier of Spencer and Worthington into a pattern in which there is register over most of the length of each filament, while misfit dislocations with "broken bonds," or a related kind of defects known as disclinations, appear. The separation S between dislocations is given by $S^{-1} = d_1^{-1} - d_2^{-1}$, where d_1 and d_2 are the repeat distances in actin and myosin.

9.3. Outline of a Model of the Disclination Structure of Insect Muscle

After mentioning the three states of muscle, we outline a model of insect muscle structure based on a deliberately oversimplified presentation of the experimental evidence on the structure of the contractile elements. Muscle exists in three states, relaxation, active contraction, and rigor. Relaxed muscle is easily extended irreversibly by external forces. Actively contracting muscle shortens, doing work against an external tension (isotonic contraction), or exerts a force on an immovable external object (isometric tetanus). Muscle in rigor does not spontaneously change its length, and can be stretched plastically only by a finite external force.

Both actin and myosin form helical filaments. The unstrained actin helix is a right-handed twist of two strands, the subunits in each strand being spaced 54.6Å apart on a helix with a pitch in the range 710Å–740Å. Since the two strands seem to be identical, active sites on actin filaments appear in opposed pairs, the sites pointing in any given direction being separated by 355Å–370Å (H. E. Huxley and Brown 1967). Myosin, at least in insect flight muscle, has a subunit spacing of 146Å. Each subunit has two symmetrically opposed pairs of cross-bridges. " The four bridges typically form the arms of a chiasmatic or ' flared X' configuration centered on the myosin filament and are shared out to four of the six surrounding actin filaments " (Reedy 1968). In insect muscle in rigor, each X is derived from the preceding by a left-handed screw of about 60 deg. In relaxed vertebrate muscle, the myosin helix is integral, successive opposed pairs of cross-bridges at a spacing of 143Å being rotated by 60 deg on a left-handed screw. (Crystallographically, this would give a helical period of 6 × 143Å. The least repeat of 3 × 143Å is obtained by taking the rotation to be 120 deg on a right-handed screw.)

The myosin filaments form a simple hexagonal array, with the actin filaments forming groups of six around each myosin. The compositions of vertebrate and insect muscles are different, as indicated in Figure 9.7. Vertebrate muscle has two actin filaments to each one of myosin. Since actin has two active sites at each level, the assumption that vertebrate myosin has a flared " X" of four cross-bridges at each level suggests that in rigor every cross-bridge could be linked to an active site. It seems that, in the actual configuration, a periodic structure must leave a quarter of the cross-bridges and active sites unbonded at each level, while it is not possible to connect the structures of successive levels by systematic screwing of the actin and myosin helices. It may be for this reason that vertebrate muscle in rigor is of low crystallinity (H. E. Huxley and Brown 1967). Although more is probably known about vertebrate muscle than about insect muscle, we shall concentrate on insect muscle because of its approximation to a simple structure when in rigor.

In order to produce a perfect simple crystalline structure, the actin helix

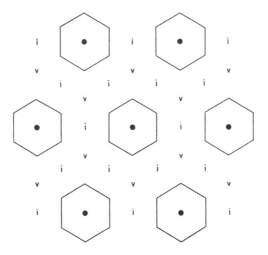

FIGURE 9.7. Arrangements of myosin and actin filaments in vertebrate and in insect muscles. The myosin filaments (heavy dots) form a simple hexagonal array in each type of muscle. The positions of the actin filaments in vertebrate muscle are indicated by v, and the positions in insect muscle by i. The closed hexagons represent the loops of edge disclinations in insect muscle in rigor.

must twist or untwist until it becomes integral, so that active sites are presented to a neighboring myosin filament at regular intervals. The vertebrate myosin filament already has the internal sixfold symmetry which is required in order that it may bond to its six symmetrically disposed neighbors, and we do not expect the myosin fibers to be twisted away from equilibrium in a perfectly crystalline structure.[1] It is also necessary for the actin and myosin filaments to stretch so that their axial periods are equal.

One deviation from perfect periodicity is likely to be permitted with only a slight increase in energy. In resting vertebrate muscle, the axes of neighboring actin and myosin filaments are separated by 238Å, and this distance is not likely to be very different in insect muscle in rigor. It is known that the cross-bridges are flexibly attached to the myosin filament. Suppose that the actin filament is strained so that one of the two helical lines of actin sites comes opposite to a myosin filament at the correct level for a cross-bridge to form a bond, but that no individual actin site is at exactly the right level to form a bond. The nearest actin site cannot be more than 27.3Å from this position, and the cross-bridge needs to bend by only 27.3/238 rad, or less than 7 degrees, in order to bond with an actin site. We therefore regard the actin filament as containing two continuous helices

[1] According to experimental evidence which was not available at the time this work was begun (Reedy 1967), insect myosin in the relaxed state has a twist of 67.5 deg. per layer. This change from our assumed 60 deg per layer would alter the details of our calculation but is not likely to alter the general conclusions.

of active site. When we speak of dislocation or disclination motion occurring by the breaking of bonds, we do not think of the jump of a cross-bridge from one actin subunit to the next, but of the jump of a cross-bridge from one actin filament to another, or from an actin filament to a nonbonded position.

On this basis we can devise a crystal structure, which we may call the state of crystalline rigor, for insect muscle (Figure 9.8). This differs from

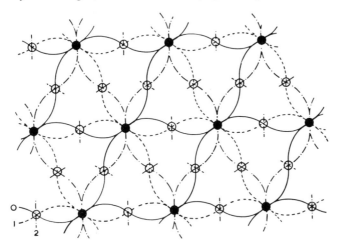

FIGURE 9.8. The cross-bridge structure in layers 0, 1, and 2 in insect muscle in crystalline rigor. Full circles represent myosin filaments, hollow circles, actin filaments. The bonds in layer 0 are represented by full lines, those in layer 1 by dashed lines, and those in layer 2 by chain-dotted lines.

the structure of insect muscle in glycerine rigor in that the filaments of the crystalline model rotate by ± 60 deg in passing from one level to the next, while the filaments in glycerine rigor rotate by ± 67.5 deg.

If we assume that the pitch of unstrained insect actin is close to that of actin in relaxed vertebrate muscle, which we take to be 710Å, we find that the twist of the unstrained actin helix in a longitudinal period of 146Å is $360 \deg \times 146/710 = 74$ deg. The actual twist in glycerine rigor of 67.5 deg lies between this value and the value of 60 deg which achieves complete bonding. This equilibrium twist represents a state of minimum energy in a situation in which the number of broken bonds increases with the deviation from 60 deg, while the strain energy increases with the deviation from 74 deg in actin and with the deviation from our assumed resting value of 60 deg in myosin.

It is now possible to specify more precisely the nature of the defects which convert insect muscle in crystalline rigor into insect muscle in glycerine rigor. If a crystal contains a line defect such that, associated with any circuit around the line, there is a displacement, this line defect is called a dislocation. If a crystal contains a line defect such that, associated with any circuit around the line, there is a rotation or misorientation, this line defect is called a disclination (Nabarro 1967b). Take a circuit which passes (Figure 9.9) from an actin to a myosin in layer 0, along the myosin from

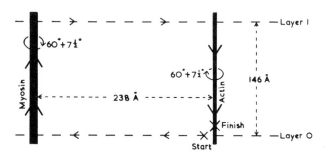

FIGURE 9.9. The Burgers circuit (arrowheads) which defines the strength of the edge disclination in insect muscle in glycerine rigor. The additional twists of $7\frac{1}{2}$ deg left-handed in myosin and $7\frac{1}{2}$ deg right-handed in actin are both anticlockwise when viewed by an observer looking from layer 0 to layer 1, and define the strength of the disclination as 15 deg.

layer 0 to layer 1, from the myosin back to the actin in layer 1, and then back to layer 0 in the actin. A twist of 60 deg per layer forms part of the ideal crystal structure, and must not be taken into account in assessing the defect structure. The defect corresponds to an additional twist of $7\frac{1}{2}$ deg left-handed in the myosin, which is anticlockwise as seen by an observer looking from layer 0 to layer 1. There is also an additional twist of $7\frac{1}{2}$ deg right-handed in the actin, which is also anticlockwise as seen by an observer looking from layer 0 to layer 1. Thus the circuit, of area $146 \times 238 = 34800\text{Å}^2$, is associated with a total rotation of 15 deg = 0.262 rad. The circuit which gives rise to this rotation is about the normal to the plane defined by the actin filament and the myosin filament, and this normal is the disclination line. The disclination is of edge type, because the resultant rotation is about an axis perpendicular to the disclination line. It is convenient to treat the disclination lines as forming closed loops around the myosin filaments, as indicated in Fig. 9.7, the loops being spaced $8 \times 146 = 1168\text{Å}$ apart on each myosin filament, and each representing a twist of 60 deg per filament, which is a symmetry operation of the crystal.

The effect of the disclination array is that, after 8 levels, the actin and myosin filaments have rotated through 9×60 deg instead of the 8×60 deg of the perfect crystal. As a result, the actin and myosin filaments are in rotational register at levels 0 and 8, but in the worst possible rotational register in layer 4, which corresponds to $4\frac{1}{2} \equiv 1\frac{1}{2}$ steps of 60 deg rotation. We may say that the core of the disclination lies in layer 4. In a disclination, as in a dislocation, the energy varies with the period of the lattice as the core moves through the crystal. There are usually two symmetrical configurations, which in our case are that in which the core lies in a layer of cross-bridges and that in which it lies midway between two layers of cross-bridges. The difference of energy between these positions is known as the Peierls potential, and leads to a hysteresis in the motion of a dislocation under an alternating applied stress. The magnitude and even the sign of the Peierls potential for a dislocation in a crystal depend very sensitively on the assumed interatomic forces, and the same will be true for the potential which governs the motion of a disclination through insect muscle. The bonding patterns when the disclination core lies just below, in, and just above, layer 4 are shown on the left, middle and right of Figure 9.10.

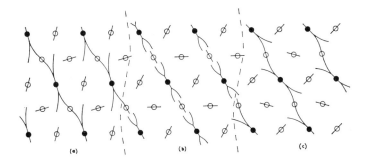

FIGURE 9.10. Layer 4 in insect muscle in glycerine rigor. The angular positions of the actin and myosin filaments correspond to layer $4\frac{1}{2}$ (equivalent to layer $1\frac{1}{2}$) in the state of crystalline rigor. (a) Configuration of layer 4 when the core of the disclination lies just below this layer. (b) Configuration of layer 4 when the core of the disclination lies in this layer. (c) Configuration of layer 4 when the core of the disclination lies just above this layer.

The configuration in layer 4 when the disclination lies well below layer 4 is the same as that shown for layer 1 in Figure 9.8; the configuration in layer 4 when the disclination lies well above layer 4 is the same as that shown for layer 2 in Figure 9.8. The symmetrical configuration in the middle of Figure 9.10 corresponds to the "speculative configuration at level 0" in Figure 6 of Reedy (1968). If double bonding to one actin site is permitted, this is probably a minimum of the Peierls potential; if double bonding is forbidden, this is probably the configuration of maximum energy.

9.4. Energetics

We now estimate the energy of muscle in a state of rigor as a function of its deviation from the state of perfect crystalline rigor, and of muscle in a state close to that of perfect relaxation.

We define the state of rigor as one in which actin and myosin have the same, not necessarily uniform, pitch. We measure distance z along the helices in units of the myosin spacing of 146Å, and take the helices to be in perfect register at $z = 0$. Suppose that after n layers the helices have turned by $60n + \theta(n)$ degrees, where $\theta(n)$ increases monotonically with n. Then the cross-bridges are not fully bonded, or are bonded only with strain, unless $\theta(n)$ is a multiple of 60 deg. Following the Peierls model of the core of a dislocation, we shall take the core of the disclination to be "wide," in the sense that $\theta(n)$ may be regarded as a differentiable function of a continuous variable n. We imagine that in the case of the disclination $\theta(n)$ increases only slowly when n is small. Then, when $\theta(n) \simeq 30$ deg, θ increases rapidly with n, the rate of increase again becoming small as $\theta(n)$ tends to 60 deg, where the helices are again in register.

The myosin filament is assumed to be unstrained when $d\theta/dn = 0$, and we may write its energy per unit length as $(1/2)k_m(d\theta/dn)^2$. The actin filament is unstrained when $d\theta/dn = 14$ deg/layer, and we may write its energy per unit length as $(1/2)k_a(d\theta/dn - 14)^2$. We assume the energy of the bonds to a given myosin filament in a given layer to be given by $-B\cos(2\pi\theta \deg/60)$, $B > 0$. We expect to find that when B is very large the state of lowest energy is close to that of crystalline rigor, with very few disclinations. For moderately large B, we expect to find one disclination every N layers, while for small B we expect to find a structure which has higher energy than the relaxed state, in which the actin and myosin helices do not twist equally. The pattern repeats itself after N turns. The average energy per layer is

$$U = \frac{1}{N} \int_0^N \left[\tfrac{1}{2}k_m(d\theta/dn)^2 + \tfrac{1}{2}k_a(d\theta/dn - 14)^2 - B\cos(2\pi\theta/60)\right] dn,$$

(9.1)

and this is to be minimized subject to the conditions $\theta(0) = 0$; $\theta(N) = 60$ deg.

The formal solution in terms of elliptic functions is given by Frank and van der Merwe, but we shall find it instructive to take a trial solution in which

$$\left.\begin{array}{ll} \theta = \alpha n \text{ degrees} \qquad \alpha > 0 \qquad 0 \leq n \leq N_1 < N/2 \\ \theta = \alpha N_1 + \beta(n - N_1) \text{ degrees,} \qquad \beta > \alpha \qquad N_1 \leq n \leq N - N_1 \\ \theta = \alpha(n - N + 2N_1) + \beta(N - 2N_1) \text{ degrees} \qquad N - N_1 < n \leq N \end{array}\right\}$$

(9.2)

Since we have taken $\theta(N) = 60$ deg, we must have

$$\beta = (60 - 2N_1\alpha)/(N - 2N_1).$$

(9.3)

Inserting Equation 9.2 into 9.1 and eliminating β by means of 9.3, we obtain

$$NU = (k_m + k_a)(NN_1\alpha^2 - 120N_1\alpha + 1800)/(N - 2N_1)$$

$$+ k_a(98N - 840) - \frac{60B\sin(2\pi N_1\alpha/60)}{\pi\alpha(60 - 2N_1\alpha)}(60 - N\alpha). \quad (9.4)$$

Minimizing U with respect to N, we find

$$0 = -(k_m + k_a)N^2N_1\alpha^2 + (k_m + k_a)(240N_1\alpha - 3600)(N - N_1)$$

$$+ 840k_a(N - 2N_1)^2 + \frac{60B\sin(2\pi N_1\alpha/60)}{\pi\alpha(60 - 2N_1\alpha)}60(N - 2N_1)^2.$$

$$(9.5)$$

This is a quadratic equation for N. We see that when B is large the last term dominates unless $N - 2N_1$ is small. Thus, when B is large, the distance N between the cores of successive disclinations is much larger than the width $N - 2N_1$ of the core of a disclination. In finding an approximate solution to Equation 9.5, we assume, and shall later verify, that the denominator $60 - 2N_1\alpha$, the number of degrees of twist in the core of a disclination, does not become small when B becomes large. We may then put

$$N = 2N_1 + \gamma B^{-1/2} \quad (9.6)$$

when the dominant terms in Equation 9.5 become of order unity.

We find that

$$\gamma = [(30 - N_1\alpha)^3(k_m + k_a)(2\pi N_1\alpha/60)/15\sin(2\pi N_1\alpha/60)]^{1/2}$$

$$+ 0[B^{-1/2}]. \quad (9.7)$$

Substituting Equation 9.6, with the leading term in 9.7, into 9.4, we obtain

$$U = (k_m + k_a)(30 - N_1\alpha)^2 B^{1/2}/N_1\gamma + \tfrac{1}{2}(k_m + k_a)N_1^2\alpha^2/N_1^2$$

$$+ k_a(98 - 420/N_1) - \frac{B\sin(2\pi N_1\alpha/60)}{2\pi N_1\alpha/60}$$

$$\times \frac{60 - 2N_1\alpha - \gamma B^{-1/2}N_1\alpha/N_1}{60 - 2N_1\alpha} \frac{N_1}{N_1 + \tfrac{1}{2}\gamma B^{-1/2}}. \quad (9.8)$$

Although the individual contributions to the total energy are not independent, it is easy to recognize the origin of the separate terms in Equation 9.8. Since γ is proportional to $(k_m + k_a)^{1/2}$, the first term is proportional to $[(k_m + k_a)B]^{1/2}/N_1$, and is the energy of the core of the disclination, averaged over all the layers. The second term, $\tfrac{1}{2}(k_m + k_a)\alpha^2$, is the energy required to twist the actin and myosin filaments, regarded as a single structure, while the third is the energy required to twist the actin filament into register with the myosin. The fourth term is the energy of the cross-bridges. We have expressed Equation 9.8 as a function of $N_1\alpha$, the total angle of twist between layer 0 and the beginning of the core, and N_1, the number of layers in which this twist occurs, as independent variables. We expect the core structure, and hence $N_1\alpha$, to change only slightly as $2N_1$, the distance between neighboring cores, changes from one large value to another.

When $N_1\alpha$ is constant, the contributions to U of the first, second and fourth terms in Equation 9.8 are all decreasing functions of N_1, while the contribution of the third term is an increasing function of N_1. In minimizing U with respect to N_1, it will be adequate to approximate the product of the last two fractions in Equation 9.8 by $1 - 30\gamma B^{-1/2}/(60 - 2N_1\alpha)N_1$.

Minimizing Equation 9.8 with respect to N_1, and using 9.7, we find

$$N_1 = \frac{(k_m + k_a)N_1^2\alpha^2}{420k_a - 2[\{15(k_m + k_a)B(30 - N_1\alpha)\sin(2\pi N_1\alpha/60)\}/(2\pi N_1\alpha/60)]^{1/2}}.$$
(9.9)

This value of N_1 may now be substituted in Equation 9.8 to give U as a function of $N_1\alpha$, and the resulting expression minimized with respect to $N_1\alpha$. However, the calculation is complicated, and we consider only the case in which $N_1\alpha$ does not lie close to either of its extreme values, 0 and 30 deg. If $N_1\alpha$ has a value in this range (we shall take $N_1\alpha = 15$ deg), Equation 9.9 leads to values of N_1 of the order of

$$N_1 \approx (k_m + k_a)/2k_a$$
(9.10)

when B is small, increasing to infinity when

$$B \approx 300k_a^2/(k_a + k_m).$$
(9.11)

When the bonding energy is greater than this, no disclinations are formed, and the stable state is the state of crystalline rigor.

With $N_1\alpha = 15$, we obtain from Equation 9.7

$$\gamma \approx 19(k_m + k_a)^{1/2},$$
(9.12)

so that the width of the core is

$$\gamma B^{-1/2} \approx 19[(k_m + k_a)/B]^{1/2}.$$
(9.13)

Finally, Equation 9.8 gives, for the energy per layer of the equilibrium state when disclinations are present,

$$U = -\frac{2B}{\pi} + 48k_a\left(\frac{B}{k_a + k_m}\right)^{1/2} + \frac{(98k_m - 292k_a)k_a}{k_a + k_m}$$
(9.14)

This is negative for large B, so that the state of rigor with disclinations has a lower energy than the state in which both filaments have their natural twists, and the mean bonding energy is zero. These two states have the same energy when $U = 0$, or

$$B^{1/2} = [38k_a + (981k_a^2 + 154k_a k_m)^{1/2}]/(k_m + k_a)^{1/2}. \tag{9.15}$$

Rigor sets in when $B^{1/2}$ exceeds this value.

We now consider the contractile process. We assume that in the relaxed muscle $B = 0$, and the myosin and actin filaments have their natural twists of 60 deg per layer left-handed and 74 deg per layer right-handed, respectively. In the presence of a bonding energy B, this configuration has no energy of twist and no mean bonding energy, because all regions of the bonding potential $-B \cos 2\pi\theta/60$ are sampled with equal frequency. A negative bonding energy is achieved, at the expense of a positive energy of twist, if the actin filaments twist more slowly where they are in good register with the myosin, and more rapidly where they are in bad register. Since the myosin cross-bridges already point toward the actin filaments when the myosin filaments are relaxed, we do not expect the myosin filaments to twist. If the angular coordinate of the relaxed actin filament at the nth level is $n60$ deg $+ n14$ deg, we take the angular coordinate in the presence of a bonding energy B to be

$$\theta = n60 \text{ deg} + n14 \text{ deg} - \lambda \sin 28n\pi/60, \tag{9.16}$$

where λ is small.

Then the average energy per layer is

$$U = \langle \tfrac{1}{2}k_a(28\pi\lambda/60)^2 \cos^2(28n\pi/60) - B \cos[2\pi(14n - \lambda \sin 28n\pi/60)/60] \rangle. \tag{9.17}$$

Thus
$$
\begin{aligned}
U &= \tfrac{1}{4}k_a(28\pi\lambda/60)^2 + \langle -B \cos[28n\pi/60]\cos[(2\pi\lambda \sin 28n\pi/60)/60] \\
&\qquad - B \sin[28n\pi/60]\sin[(2\pi\lambda \sin 28n\pi/60)/60] \rangle \\
&\approx \tfrac{1}{4}k_a(28\pi\lambda/60)^2 + \langle -B \cos[28n\pi/60] \\
&\qquad + B \cos[28n\pi/60]\tfrac{1}{2}[2\pi\lambda \sin(28n\pi/60)/60]^2 \\
&\qquad - B \sin[2n\pi\lambda/60][2\pi\lambda \sin(28n\pi/60)/60] \rangle \\
&= \tfrac{1}{4}k_a(28\pi\lambda/60)^2 - 2\pi\lambda B/120. \tag{9.18}
\end{aligned}
$$

Minimizing with respect to λ, we see that U is least when

$$\lambda = 15B/98\pi k_a, \tag{9.19}$$

corresponding to

$$U = -B^2/784k_a. \tag{9.20}$$

It would not be meaningful to say that rigor sets in for the value of B which makes Equation 9.14 more negative than 9.20, because 9.14 is an approximation for B large and 9.20 for B small.

9.5. Conclusion

A simple model of insect muscle, based on the electron microscope observations of Reedy (1968), passes from a state of relaxation to a state which may be identified with active contraction, to rigor, and finally to perfect crystalline rigor, as the assumed bonding energy increases.

I am grateful to Dr. H. E. Huxley, F.R.S., for a discussion on this subject, to Dr. M. K. Reedy for correspondence, to Mr. A. T. Quintanilha for checking the calculations, and to Professor B. I. Balinsky for reading parts of the manuscript.

REFERENCES

Dehlinger, U., 1929, *Ann. Phys.*, **2**, 749.

Frank, F. C., and J. H. van der Merwe, 1949a, *Proc. Roy. Soc.*, **A198**, 205.

Frank, F. C., and J H. van der Merwe, 1949b, *Proc. Roy. Soc.*, **A198**, 216.

Frenkel, J., 1926, *Z. Physik*, **37**, 572.

Hanson, J., and H. E. Huxley, 1955, in *Symp. Soc. Exp. Biol.*, **9**, 228.

Huxley, A. F., 1957, in *Progr. Biophys. Biophys. Chem.*, **7**, 255.

Huxley, H. E., and W. Brown, 1967, *J. Mol. Biol.*, **30**, 383.

Masing, G., and M. Polanyi, 1923, *Ergeb. exakt. Naturw.*, **2**, 177.

Nabarro, F. R. N., 1967a, *Theory of Crystal Dislocations*, Oxford: Clarendon Press, appendix. 1967b, *ibid.*, Chapter 3.

Orowan, E., 1934, *Z. Physik*, **89**, 634.

Polanyi, M., 1934, *Z. Physik*, **89**, 660.

Prandtl, L., 1928, *Z. angew. Math. Mech.*, **8**, 85.

Reedy, M. K., 1967, *Amer. Zoologist*, **7**, 465.

Reedy, M. K., 1968, *J. Mol. Biol.*, **31**, 155.

Spencer, M., and C. R. Worthington, 1960, *Nature*, Lond. **187**, 388.

Tamm, I. E., 1962, *Usp. Fiz. Nauk*, **74**, 397. (*Sov. Phys. Uspekhi*, **5**, 173).

Taylor, G. I., 1934, *Proc. Roy. Soc.*, **A145**, 362.

van der Merwe, J. H., 1963a, *J. Appl. Phys.*, **34**, 117.

van der Merwe, J. H., 1963b, *J. Appl. Phys.*, **34**, 123.

van der Merwe, J. H., 1964, in Francombe and Sato (eds.), *Single Crystal Films*, Oxford: Pergamon Press, p. 139.

2. Hardening Mechanisms and Dislocation Dynamics

ABSTRACT. The applied stress which will bend dislocations between strong precipitate particles, the Orowan stress, depends on the flexibility of a single dislocation, and is unlikely to be affected by cross-slip. The stress is calculated by attributing a line tension, like that of a taut string, to the dislocation line. This line-tension approximation is refined by allowing its magnitude to vary with the screw or edge character of the dislocation and with the scale of the precipitate particles. The spatial distribution of particles also influences the stress at which many dislocations move freely through the field of particles.

The yield stress of single crystals of copper is increased by particles of SiO_2 and BeO by an amount which agrees fairly well with the calculated Orowan stress.

10. On the Orowan Stress

M. F. ASHBY

10.1. Introduction

Materials can be made more resistant to deformation by dispersing particles of a second phase — even a gas — in them. Metals containing precipitates, oxides, or nitride particles are all stronger in several ways than the pure metal matrix. Liquids containing colloidal particles, such as paints, are more viscous than pure liquids.

One of the reasons for this increased resistance to deformation is very simple. Without the particles lamellar flow is possible; even in a metal crystal, glide occurs predominantly on one set of slip planes, and can be thought of as lamellar. Particles perturb the flow pattern; elements of the material must now move further in order to flow round the particles, and, in a liquid, for example, more energy is dissipated as viscous friction. Flow in a crystal is also perturbed by particles, provided they are sufficiently strong: this perturbed flow causes rapid work hardening, but the details are much more complicated than the mere dissipation of energy against viscous friction (Ashby 1969).

Unlike a liquid, a well-annealed crystal has a characteristic yield stress. This too is increased by the particles, by an amount which depends, among other things, on the strength of the particles. Weak particles tend to deform

113

when the crystal containing them deforms; sufficiently strong particles do not. As Orowan (1948) pointed out, strengthening due to strong particles is easy to understand physically, and it is amenable to clear-cut experimental investigation. The paper is concerned with the increase in yield stress of crystals containing strong particles such as oxides, or intermetallic compounds, an increase which has become known as the Orowan stress.

10.2. Experimental Facts

Alloys exist which contain particles that do not deform plastically, or fracture, at the yield stress of the alloy (Kelly and Nicholson 1963, Ebeling and Ashby 1966). Single crystals of these alloys obey the critical shear stress law (see, for example, Dew-Hughes and Robertson 1960, Al-CuAl$_2$, Ebeling and Ashby 1966, Cu-SiO$_2$, Jones 1968, Cu-BeO) are ductile, and show a critical shear stress which is much larger than that of the pure matrix. The critical shear stress increases with increasing volume fraction and decreasing size of particles, and is insensitive to the temperature and strain rate of testing (Dew-Hughes and Robertson 1960, Kelly and Nicholson 1963, Ebeling and Ashby 1966). Suitably oriented crystals containing a small volume fraction ($\gtrsim 1$) of particles deform, like pure single crystals, by a single slip on the primary glide system, but as the volume fraction of particles is increased, secondary slip seems to become more prevalent (Kelly and Nicholson 1963, Byrne et al. 1961, Price and Kelly 1962, Russell and Ashby 1968) although the crystals still obey the critical shear stress law. A stress-strain curve typical of a pure copper crystal, and of an identical crystal containing spherical SiO$_2$ particles, is shown in Figure 10.1.

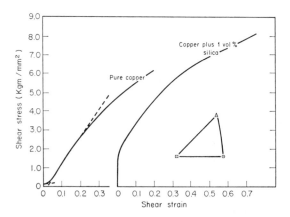

FIGURE 10.1. The stress-strain curves of identically oriented crystals of pure copper and of copper containing 1 volume percent of SiO$_2$ (from Ebeling and Ashby 1966).

As the degree of dispersion hardening, as measured by the yield stress, increases, slip lines become finer and more closely spaced, and finally become indetectable, at least at small strains. When visible, they bend round particles often leaving traces parallel to the cross-slip plane (Thomas and Nutting 1957, Ebeling 1968). Transmission electron microscopy shows that dislocations bow between and bypass particles, often in such a way as to leave prismatic loops round them (Ashby 1964, Humphreys and Martin 1967, Hirsch and Humphreys, this volume). These observations of slip lines and prismatic loops leave no doubt that the process of bypassing of a particle by a dislocation frequently involves local cross-slip; the details will be discussed later.

In one alloy it has been demonstrated that solution hardening and dispersion hardening are approximately additive (Ebeling and Ashby 1966). Lacking any further information, we therefore assume that the critical shear stress of the alloy is just the sum of critical shear stress of the matrix and a critical shear stress due to the particles.

We first calculate the contribution of hard particles to the critical shear stress, and then compare the result with recent experimental results.

10.3. Below the Critical Shear Stress

10.3.1. Self-Stresses

Consider an undeformed single crystal of a matrix metal containing a dispersion of hard spherical particles. No external stress acts on the crystal. In general the crystal will contain *self-stresses* associated with the particles. These stresses can be caused by coherency between particle and matrix, by difference in coefficient of thermal expansion between particle and matrix, or by a phase change in particle or matrix during some earlier heat treatment, or even by the condensation of point defects onto the particle. In many cases the magnitude of these self-stresses, and the way they vary with distance from a particle can be calculated (Mott and Nabarro 1940, Eshelby 1961). For example, suppose that the single crystal had been annealed at a high temperature and then cooled rapidly to room temperature, and that, due to a difference in coefficient of thermal expansion, the matrix shrinks more than the particles. Suppose the difference in thermal strain between particle and matrix is δ, then the matrix surrounding each particle contains shear stresses which reach a maximum value, at the surface of the particle, of roughly $2G_M \delta$, where G_M is the shear modulus of the matrix. Even for small values of δ, these are large stresses; for example, for metal containing oxide particles which had been quenched through a temperature interval of 500°C, δ is of the order 5×10^{-3}, giving shear stresses at the particle surface of $G_M/100$. Frequently the stresses are large enough to deform plastically the matrix metal surrounding each particle. Even when no plastic flow is nucleated, the stresses locally may be large, and can be "seen," and even measured on transmission electron micrographs where they appear as characteristic lobes of darker contrast. These stresses fall off quickly with distance from a particle, so that at a distance of just over two particle diameters the stresses have fallen to 1 percent of their value at the particle surface, and thus their *average* value in the matrix is small. For example, a volume fraction f of particles, in a crystal which had been cooled as before, results (Mott and Nabarro 1940) in an average stress opposing macroscopic slip of only about $2G_M f \delta$. Taking $f = 10^{-2}$ and $\delta = 5 \times 10^{-3}$, we obtain $10^{-4} G_M$, which is much less than the flow stress of most dispersion-hardened alloys.

10.3.2. *Perturbation of the Applied Stress Field by Particles*

Let us now apply a tensile stress to the dispersion-hardened crystal, and increase it to some point below, but near, the yield stress. Freeze the crystal in this state and consider the stresses inside it. Because the elastic constants of the matrix always differ from those of the particles, a second kind of local stress pattern is set up, which depends on this difference in elastic properties (Southwell and Gough 1926, Goodier 1933, Eshelby 1961). Once again, the maximum tensile and shear stresses occur at the surface of the particle, where, for a "hard" particle, they reach maximum values in an isotropic matrix of up to 2 and 2.18 times the value, respectively, that they would have if no particle were present. They fall off rapidly with distance from the particle, so that at a distance of 2 diameters, the stress field is perturbed by less than 1 percent. The particles perturb the stress field in such a way that the *average* stress over distances of the order of the particle spacing, acting on the slip plane, is altered very little; macroscopic slip is therefore not affected. Slip over large distances will thus start on the slip system on which the *resolved component of the applied stress is greatest*, just in pure single crystals.

All studies of slip systems in dispersion-hardened alloys have shown this to be true (examples are: Dew-Hughes and Robertson 1960; Ebeling and Ashby 1966; Humphreys and Martin 1967; Hirsch and Humphreys, this volume; Jones 1968, Russell and Ashby 1968).

A note on particle *shape*. The calculations deferred to above were for spherical particles. The stresses at the surface of an angular, but equiaxed particle (e.g., a cube, or a tetrahedron) differ from those of a sphere — particularly near the corners and edges. But, by St. Venant's principle, these differences must die out at a distance of about x (the particle "diameter") from the surface of the particle. Beyond this distance the stress field from a sphere and from an equiaxed shape of the same volume and misfit δ must be almost identical. Hence all the conclusions about the *average* stress acting in the region between spherical particles hold also for all equiaxed shapes.

10.4. The Microscopic Orowan Stress

In this section we calculate the *local* stress τ required to cause plastic flow on a microscopic scale. The problem reduces to that of calculating the local stress to bend one dislocation between a pair of particles, and thus to bypass them. This depends on the edge or screw character of the dislocation. In practice, also, a segment of dislocation bowing between one pair of particles is influenced by the neighboring segment bowing between the next pair of particles along the line; this we can allow for only in an approximate way. In subsequent sections we relate this microscopic yield stress, τ, to the applied shear stress σ required to cause measurable plastic flow.

10.4.1. *Dislocation Sources and the Bowing of Dislocations between Particles*
Let us now continue to increase the stress on the crystal described in the last section. Naturally occurring dislocations move forward under the applied stress, and press against particles which obstruct their slip planes. Dislocations do not enter strong particles like those which concern us here. Were the dislocation core to enter an oxide particle, for example, the increase in core energy would outweigh the small increase associated with the dislocation merely bending round the particle; this increase in core energy is closely related to the shape of the atomic force-displacement curves for particle and matrix. Equivalently, we may say that the particle exerts a strong, short-range, repulsive force on the dislocation, which prevents it entering and thus shearing it.

No source can operate until a stress is reached at which dislocations can bow between particles and thus bypass them, since the area of slip plane swept out during the operation of a source is itself studded with particles. According to Orowan (1948, 1954), the unit step in this micro-yielding is the bending of a segment of dislocation to a radius of curvature of half the particle separation, between a pair of particles. The segment can then expand beyond the particles, thus bypassing them and leaving a dislocation loop round each one.

Following Orowan, the local stress, τ, required to do this is roughly equal to the stress required to bend a dislocation to a semicircle of radius $D/2$, where D, strictly speaking, is the space between particles in the slip plane through which the dislocation bulges, but is usually replaced by the center-to-center particle spacing in the slip plane. This stress is

$$\tau \approx \frac{2T}{bD},$$

where T is the line tension of the dislocation and b the magnitude of the Burgers vector. T is often assumed to be equal to the line energy of the dislocation, roughly $G_M b^2/2$ per unit length, giving for the Orowan stress

$$\tau = \frac{G_M b}{D}.$$

10.4.2. Other Contributions to the Critical Shear Stress

Before proceeding with a more detailed calculation of this Orowan stress, we consider other possible contributions to the critical shear stress. There are theoretical reasons for supposing that strong particles contribute to the flow stress of the alloy containing them in other ways than that considered by Orowan. We shall consider two of them: one caused by elastic strain in the matrix surrounding the particle (Mott and Nabarro 1940), the other caused by difference in elastic constants between particle and matrix (Head 1953 a,b, Fleischer 1961, 1963). Both effect the flow stress in two distinct ways. First, there is a short-range interaction between a particle and a dislocation which is pressed against it. If this interaction is repulsive, the dislocation may "standoff" from the surface of the particle, increasing the effective particle size (Geisler 1951) and thus decreasing the separation, D, in the slip plane. The effect on the flow stress could be allowed for by reducing D in the Orowan equation by the appropriate amount. Second there is a long-range interaction, causing the potential energy of a segment of dislocation about D long to fluctuate with position, implying a contribution to the flow stress which must be added to the Orowan stress.

These contributions to the flow stress have been reviewed and discussed elsewhere (Ashby 1964). The conclusions are

1. A misfit δ, as defined in Section 10.3, will cause a dislocation to "stand-off" from a suitably positioned particle only if the maximum stress due to the misfit, roughly $2G_M \delta$ (Section 10.3), exceeds the effective stress forcing the dislocation against a particle, namely, D/x times the applied stress, where x is the particle diameter and D is the particle spacing. In the alloys with which we are concerned here, δ is usually small ($\simeq 10^{-3}$) and stand-off due to misfit, if it occurs at all, will be unimportant.

2. The long-range back stress due to misfit is of order $2G_M f\delta$ (Section 10.3) where f is the volume fraction of particles. When δ is small ($< 10^{-3}$), this contribution to the flow stress can be neglected in comparison with the Orowan stress. When δ is large, it should be added to the Orowan stress.

3. A positive difference in shear modulus $\Delta G = G_P - G_M$ between particle and matrix will cause a dislocation to "stand off" at a distance ζ given approximately by

$$\zeta = \frac{1}{6}\left(\frac{\Delta G}{G_P + G_M}\right)x,$$

where x is the particle diameter, G_p the shear modulus of the particle and G_M that of the matrix. Typical values of $\Delta G/(G_P + G_M)$, tabulated by Ashby (1964), indicate that $\zeta = 0.05x$ is a normal stand-off distance. This is too small to affect the flow stress in any important way. When ΔG is negative, as it sometimes is, the dislocation is attracted to the particle.

4. The long-range back stress due to difference in elastic constants is of the order

$$\frac{\Delta G}{4\pi} \frac{bf}{D}$$

and is almost always small compared with the Orowan stress.

In summary, if dislocations bypass particles by the Orowan mechanism, there exist other contributions to the microscopic yield stress besides that considered by Orowan. These will usually, but not always, be small compared to the Orowan stress; for an exact analysis, they should be considered.

10.4.3. Refinements of the Orowan-Stress Calculation

The concept of a "line tension" is an approximation. Each segment of a bent dislocation exerts a force on each other segment: it is the sum of these forces which we describe as a "line tension." Although approximate, it is a most useful concept, particularly in the calculation of dislocation shapes. Serious errors arise only when the dislocation is bent to extremely small radii of curvature, less than 100 Å for example. A good discussion of line tension is given by Brown (1964) and by Hirth and Lothe (1968).

Within the framework of the line-tension approximation, a more detailed calculation than that given above is possible. The energy of a dislocation $E(\theta)$ depends on its character, that is, on the angle θ between the Burgers vector and the tangent to the dislocation line (Foreman 1955) ($\theta = 0$, screw; $\theta = \pi/2$, edge). It then becomes important to distinguish between the energy and the line tension $T(\theta)$, which also depends on θ and is in fact given by (de Wit and Koehler 1959):

$$T(\theta) = E(\theta) + \frac{d^2 E(\theta)}{d\theta^2}.$$

Then a segment of dislocation, δl (Figure 10.2) is in equilibrium under the action of two forces, one due to the applied stress σ of magnitude $\sigma b\, \delta l$, the other due to its curvature K and equal to $T(\theta)K\, \delta l$. Both act normal to the segment, but in opposite directions. By expressing K in terms of Cartesian coordinates, a differential equation expressing the equilibrium of the segment is obtained which, when integrated, gives the dislocation shape (de Wit and Koehler 1959, Ashby 1968). Three such shapes are shown in Figure 10.3.

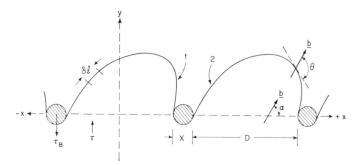

FIGURE 10.2. A dislocation bending between particles.

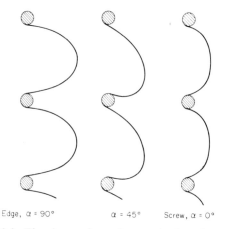

Edge, α = 90°   α = 45°   Screw, α = 0°

FIGURE 10.3. The shape of an edge, a mixed, and a screw dislocation which bulge between equally spaced particles such that the arms on each side of a single particle are just parallel. The interaction between arms has not been allowed for.

To obtain the critical stress required to just push a dislocation between particles, a *criterion for bypassing* must be adopted. The choice of an incorrect criterion has led to errors in previous work, but although the error in this old criterion is now clear (see below), the exact formulation of a correct one is not yet possible.

The incorrect criterion adopted by Ashby (1968) assumed that the bulging dislocation was in a critical configuration, requiring the maximum applied stress, when the two arms of the dislocation, one on each side of the particle, first became parallel. The dislocations shown in Figure 10.3 are in fact in this configuration. It can then be argued that a small forward displacement of the dislocation pulls out a dipole at each particle of width equal to the particle diameter x; and that a dislocation initially of screw character (when straight) therefore pulls out an edge dipole; and a dislocation initially of edge character pulls out a screw dipole. The critical, or Orowan, stress is then that stress which is just capable of elongating appropriate dipoles of width x spaced D apart. This leads to two extreme values for the local stress required for bypassing, one for edges, τ_e, and the other for screws, τ_s; assuming elastic isotropy, these are

$$\left.\begin{aligned} \tau_e &= \frac{1}{2\pi}\frac{Gb}{D}\ln\left(\frac{x}{r_0}\right) \\[2ex] \tau_s &= \frac{1}{2\pi(1-v)}\frac{Gb}{D}\ln\left(\frac{x}{r_0}\right) \end{aligned}\right\}, \tag{10.1}$$

where x is the particle diameter, r_0 the inner cut-off radius, b the Burgers vector, and G and v the matrix shear modulus and Poissons ratio. The ratio of passing stress for screws to that for edges is 1.5; anisotropy of the type found in copper increases this proportion.

The dipole argument is a way of allowing for interaction between adjacent bowed-out loops; in Figure 10.2, the sections marked 1 and 2 attract, tending to make bypassing easier. It is certainly a reasonable way of calculating the stress that will extend the assumed configuration, that is, one in which segments on either side of a particle or parallel, as they are shown in Figure 10.3. The error lies in assuming this to be the *critical* configuration. The following approximate argument shows that the critical configuration occurs earlier in the bypassing process, and that the critical stress is larger than the dipole argument assumes.

Consider a dislocation bulging between particles, as shown in Figure 10.4. The angle ϕ describes the progress of the bypassing (in the preceding

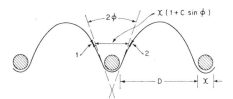

FIGURE 10.4. A dislocation bending between particles; the angle ϕ measures the progress of the bypassing

discussion we assumed the critical position to be that for which $\phi = 0$). For simplicity, assume that the energy per unit length of the dislocation, E, is independent of orientation and calculate the force exerted by the dislocation on the central particle. Either by regarding the dislocation as having a line tension $T = E$, or by a virtual work argument based on a small displacement of the particle, we obtain for this force, F,

$$F = 2E\cos\phi.$$

If the interaction between bowed-out segments is ignored, E is independent of ϕ; for the sake of example, set it equal to

$$E = \frac{Gb^2}{4\pi} \ln \frac{D}{r_0}.$$

The force as a function of angle ϕ appears as the broken uppermost line, in Figure 10.5 a and b. Now consider interaction between adjacent bowed-

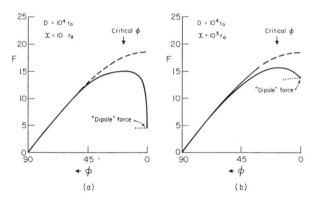

FIGURE 10.5. a, b. The force F (in units of $Gb^2/4\pi$) exerted on particle as bypassing, measured by the angle ϕ, proceeds. Note that the maximum in the force, and thus in the applied stress, occurs at $\phi \simeq 20°$, and that the maximum force is greater than the dipole force. The two sets of curves are for two values of particle diameter x at constant spacing D.

out loops; this interaction has the effect of lowering the energy by an amount which depends on the separation of the arms of the dislocation and thus on ϕ. The separation of two segments, 1 and 2 (Figure 10.4) is equal to $x(1 + C \sin \phi)$ where C is a constant which depends on where the separation is evaluated. The energy as a function of ϕ is roughly

$$E(\phi) = \frac{Gb^2}{4\pi} \ln\left\{ \frac{r_0}{x} \{1 + C \sin \phi\} \right\}.$$

But when $\phi = \pi/2$, we require the outer cut-off radius to reduce to D, as before; this fixes C so that

$$E(\phi) = \frac{Gb^2}{4\pi} \ln\left\{ \frac{x}{r_0} \{1 + \left(\frac{D}{x} - 1\right) \sin \phi\} \right\}.$$

Values of F calculated from this expression are shown as full line on Figure 10.5 a and b. For comparison, values obtained by setting the outer cut-off radius equal to x (the "dipole" argument) are also shown as dotted lines.

The local stress, τ', required for by-passing is obtained by setting the force, $\tau'bD$, on a segment of length D equal to the force F exerted on it by the obstacle. The result for an edge dislocation is thus

$$\tau'_e = \frac{1}{2\pi} \frac{Gb}{D} \cos \phi \ln \left\{ \frac{x}{r_0} \left(1 + \left(\frac{D}{x} - 1 \right) \sin \phi \right) \right\}$$

and varies with ϕ in the same way that F does (Figure 10.5). The approximate result for a screw dislocation is $1/(1 - v)$ times this. The critical angle ϕ^* is then given by the condition that $d\tau'/d\phi = 0$, and corresponds to the maximum of the F curve. The by-passing stress is obtained by resubstituting ϕ^* into the expression for τ'_e.

Recently Foreman et al. (1969) have used a computer method to simulate the shape and stability of a pair of loops which bow between obstacles. This technique correctly allows for the interaction between neighboring bowed-out segments. Comparison of these computer results with the approximate analysis given above shows substantial agreement. Both methods show (a) that the critical angle is always greater than zero (it lies between 0 and 30 deg) but tends toward zero as the particle spacing, D, increases, and as the ratio of size to spacing (x/D) increases; (b) that the by-passing stress lies between the value given by Equation 10.1 and the larger value obtained by replacing x by D in the log term; the approximate method differs from the computer method by less than 4 percent.

While a critical ϕ^* greater than zero tends to increase the Orowan stress above that given by Equation 10.1, it also reduces the "statistical factor," discussed in Section 10.5.2, because the dislocation touches fewer particles. As a result, Equation 10.1 is the best simple approximation for the local by-passing stress. Comparison with the refinement just described shows that, at worst, Equation 10.1 underestimates the Orowan stress by 20 percent.

Mitchell (private communication) has used a computer model of a dislocation bowing between pinning points to find the stable equilibrium shape and the Orowan stress. Bacon (1967) has used an improved analytical method to do the same thing. Both avoid the specific use of the line-tension approximation and consider instead the force on a segment of the bowed-out loop due to a finite number of other segments, plus the force due to the applied stress. The equilibrium shape is that for which the force on every segment is zero. Neither, however, adopts a correct criterion for bypassing.

Stacking fault energy and order, in an alloy, can also change the Orowan stress. Physically, the change comes about because the line tension of a dislocation, like its line energy, varies with the square of the magnitude of its Burgers vector, b. A dislocation of small b is therefore more flexible than one of larger b. To calculate the maximum change, we note simply that the Orowan stress is proportional to the magnitude of b. Dissociation of a dislocation in an f.c.c. crystal into two widely separated partials thus lowers the Orowan stress by a factor $1/\sqrt{3}$. Association of pairs of complete dislocations into one superdislocation, due to order, raises it by a factor 2.

10.4.4. *The Influence of Cross-Slip*

Local cross-slip occurs at particles in certain alloys. A careful study of slip lines on copper crystal containing SiO_2 particles and oriented for single slip (Ebeling, private communication), has proved that cross-slip occurs close to particles at shear strains of greater than 0.25. Transmission electron microscopy of copper alloys containing oxides, and containing cobalt particles (Ashby and Smith 1960; Humphreys and Martin 1967; Hirsch and Humphreys, this volume) has shown numerous prismatic loops with the primary Burgers vector, which can readily be explained by cross-slip.

The general features of the path followed by a dislocation as it cross-slips round a particle have been described by Hirsch (1957) and, with minor modifications, by others (Ashby and Smith 1960; Ashby 1964; Humphreys and Martin 1967; Gleiter 1967; Hirsch and Humphreys, this volume). The process is shown schematically in Figure 10.6. A section of a dislocation approaching a particle on the primary slip plane is stopped by the long-range and short-range repulsive forces exerted on it by the particle; these repulsive forces were discussed earlier. The dislocation stops in a configuration such that the net force on each segment, in the primary slip plane, is zero; this configuration is shown in the figure. A section of dislocation lies parallel to its Burgers vector, and thus to the cross-slip plane. In general, the net force on this segment in the cross-slip plane is *not* zero, and the local stresses do work if the section moves on the cross-slip plane. If the nucleation problem involved in transferring part of the dislocation from the primary to the cross-slip plane requires a sufficiently large local stress, then bypassing will be completed by the Orowan mechanism (Figure 10.6a). If, however, the local stress is sufficient to effect this transfer, then cross-slip occurs (Figure 10.6b, c). Further cross-slip lowers the energy of the resulting configuration, generating two prismatic loops, labeled I and V because one is of interstitial character, the other of vacancy character. One loop may be carried away with the dislocation, as shown at Figure 10.6b. In practice (Gleiter 1967; Hirsch and Humphreys, this volume) an Orowan loop may first form, and then cross-slip under the influence of a second, bypassing dislocation, as shown at Figure 10.6d.

A simple rule determines whether the loops on one side of a particle are of interstitial or vacancy type. Define x and y directions parallel to the Burgers vector and slip-plane normal, as shown in Figure 10.7. Then, if, in a positive shear (as shown), cross-slip of the screw dislocation occurs *upward* (i.e. with a component in the $+y$ direction), interstitial loops form on the $-x$ side of the particle and vacancy loops form on the $+x$ side. Cross-slip *downwards* $(-y)$ reverses the arrangement. The two directions of cross-slip are in general unequally stressed, so that a bias in the arrangement of loops is to be expected.

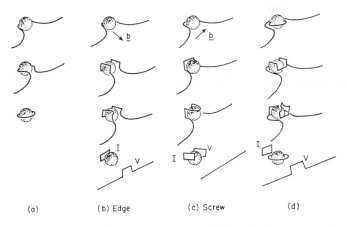

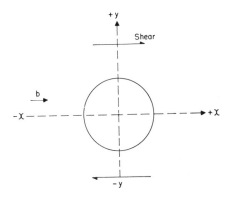

FIGURE 10.6. (a) A dislocation bypassing a particle by the Orowan mechanism; (b), (c) by cross-slip; and (d) by the Orowan mechanism followed by cross-slip.

FIGURE 10.7.

Both experiment and theory indicate that although cross-slip occurs, it does *not* permit bypassing at a stress significantly lower than the Orowan stress. Experimentally, the critical shear stress of dispersion-strengthened alloys depends only slightly on temperature (Ashby 1964, Ebeling and Ashby 1966, Jones and Kelly 1968), suggesting that the normally temperature-dependent cross-slipping process is not controlling it. Transmission microscopy (Gleiter 1967; Hirsch and Humphreys, this volume) indicates that one or more Orowan loops may form at a particle before cross-slip starts, in which case the critical shear stress is obviously controlled by the Orowan mechanism. Even when cross-slip occurs immediately, as sketched in Figure 10.6b and c, it is difficult to see how it lowers the bypassing stress. We shall neglect the influence of cross-slip in the rest of this paper.

10.5. The Macroscopic or General Yield Stress

We have calculated the local stress required to bow one dislocation between two particles—the microscopic yield stress, τ. We now ask, at what special value of applied stress, σ, will *general* bypassing occur, and so what is the applied stress for yielding?

10.5.1. *Stress Concentrations*

Stress concentrations cause the local stress τ to differ from the applied stress, σ. We assume cracks, holes, notches, and other large inhomogeneities are absent. But the particles themselves have elastic constants different from those of the matrix, and so concentrate stress. The discussion of Section 10.3 showed, however, that the resulting change in the *average* stress on a slip plane, between particles, was almost always negligible.

A second sort of stress concentration must be considered: that formed by glide. For certain special arrangements of particles, a pileup of dislocations can form before general yielding occurs, and bypassing by one dislocation is aided by the stress-concentrating effect of the pileup (Ashby 1968). However, cooperative effects of this kind will *not* aid in bypassing uniformly or randomly distributed, equiaxed particles. Accordingly, we shall restrict the further discussion to roughly equiaxed particles.

10.5.2. *Effect of Random Particle Spacing*

If all particles had the same spacing, D, the crystal would show a sharp yield stress at the value we have calculated. In all real crystals, however, a *distribution* of particle spacings will exist. Assume that the particles are randomly placed in the crystal, giving a distribution of spacings between them. The yield stress will be associated with some mean particle spacing: a few exceptionally wide spacings do not soften the crystal, nor do a few exceptionally narrow ones harden it.

Every spacing between a pair of particles has its own passing stress. At a given applied stress, σ, a dislocation segment which has successfully bowed through one "transparent" spacing then finds itself pressed against two others which may or may not be transparent. If *one* of these spacings is transparent, the dislocation is able to move on; if not, the segment stops. Kocks (1966) and Foreman and Making (1966) have considered this problem and have shown that extensive slip can first occur at the critical value of the applied stress, such that about one third of the spacings between particles are "transparent." Then the average segment can keep moving indefinitely. Kocks, and Foreman and Makin, conclude that *if* the critical configuration is that at which the angle ϕ (Figure 10.4) is zero, then a "statistical factor," about 0.85, relates the macroscopic flow stress σ of the random array to the average local Orowan stress τ (and thus to the average spacing $\bar{D}$), thus

$$\sigma = 0.85\,\tau$$

where $\bar{D}$ is defined by $\bar{D} = N_s^{-1/2}$ and N_s is the number of particles per unit area of the slip plane. When the critical value of ϕ is greater than zero the statistical factor is smaller (see Kocks, this volume). This change in statistical factor compensates to some extent for the increase in local Orowan stress from the value given by Equations 10.2 (below) as the critical value of ϕ increases, as discussed in Section 10.4.

Deviations from randomness can be important. Deviations in the direction of a *more uniform* distribution strengthen the alloy slightly. Deviations in the direction of a more *clustered* distribution may drastically lower the Orowan stress.

10.5.3. *The Best Estimate for the Orowan Stress*

The final expressions relating the increase in macroscopic shear stress due to nondeforming particles, to the number and size of the particles is given by

$$
\left.
\begin{aligned}
\sigma_{\text{edge}} - \sigma_0 &= 0.85 \frac{Gb}{2\pi\bar{D}} \ln\left(\frac{x}{r_0}\right) \\[2mm]
\sigma_{\text{screw}} - \sigma_0 &= 0.85 \frac{Gb}{2\pi(1-v)\bar{D}} \ln\left(\frac{x}{r_0}\right) \\[2mm]
\bar{D} &= N_s^{-1/2},
\end{aligned}
\right\} \tag{10.2}
$$

where b is the matrix Burgers vector, G is the matrix shear modulus, v Poissons ratio, x the particle diameter, r_0 the inner-cut-off radius, and N_s the number of particles intersecting a unit area of the slip plane.

The particle size x is usually much smaller than their spacing $\bar{D}$. When this is not so, $\bar{D}$ should be replaced by $(\bar{D} - x)$. Orowan (1954) has shown that the activation energy of the bypassing is so much larger than energies available from thermal vibrations that σ should change with temperature only as G. The term σ_0 includes any other frictional contributions to the flow stress, such as the critical shear stress of the matrix material, the hardening due to self-stresses that was described in Section 10.3, and work hardening (though in this last case the assumption of linear additivity is certainly incorrect).

Hirsch and Humphreys (this volume) point out that the best estimate of the actual flow stress should be the geometric mean of Equations 10.2. Their argument leading to this conclusion is valid only if the critical position of bulging dislocation, screw or edge, is that at which the angle $\phi = 0$ (Figure 10.4). The discussion of Section 10.4.3 indicated that this is probably incorrect. This, the other uncertainties of the theory (Sections 10.4.3 and 10.5.2) and the experimental errors involved in even the best measurements of D and x, the deviations of real particle distributions from randomness, and elastic anisotropy of most metals, mean that the calculated and measured values of σ are unlikely to agree to better than ± 30 percent.

10.6. Comparison with Recent Experimental Results

Recently Ebeling and Ashby (1966) have tested pure copper single crystals, all of one orientation, containing various sizes and volume fractions of silica particles. The specimens were prepared by internally oxidizing copper-silicon single crystals. Precautions were taken to ensure that the average particle diameter did not vary with position in any given crystal. The entire particle size distribution was measured in each crystal, and from this their average size and spacing was calculated.

The crystals deformed by single slip on the expected primary slip system to over 60 percent shear strain. Particles extracted from these deformed specimens were not measurably deformed. The critical shear stresses varied with temperature only as G in the range $-196°C$ to $20°C$, and were independent of strain rate, in agreement with Orowan's (1954) prediction.

Jones and Kelly (1968) have conducted similar, careful tests on single crystals of copper containing BeO particles. Their crystals contained a greater spread of particle spacings and so covered a wider span of stress than those just described, although their experimental uncertainties were a little larger.

Both sets of results are plotted in Figure 10.8 according to the following

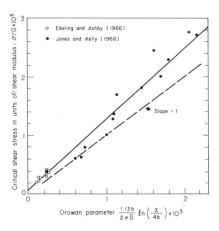

FIGURE 10.8. The measurements of critical shear stress of copper single crystals containing SiO_2 particles (Ebeling and Ashby 1966) and BeO particles (Jones and Kelly 1968), plotted according to the Orowan expression.

scheme. The critical shear stress, in units of the shear modulus G, is plotted against the geometric mean Orowan equation, according to

$$\frac{\sigma}{G} = 1.13 \frac{b}{2\pi\bar{D}} \ln \frac{x}{r_0}$$

(10.3)

When the anisotropy of copper is taken into account, the factor $1/1 - v$ in the expression for σ_{screw} (Equation 10.2) is replaced by 1.77 (Ashby 1968). The geometric mean of Equations 10.2 then leads to the expression above, containing the numerical factor 1.13. In plotting the data, the following values of constants were used: $r_0 = 4b$; shear modulus at $293°K = 4.3 \times 10^3$ Kg/mm^2; shear modulus at $77°K = 4.6 \times 10^3$ Kg/mm^2. The theoretical slope of unity is shown as a broken line on the figure. Since it is clear that further refinments of the theory will modify Equation 10.3, the raw data from which Figure 10.8 was plotted are presented in Tables 10.1 and 10.2.

TABLE 10.1. Data of Ebeling and Ashby 1966.

Identification number	σ at 293°K (kg/mm²)	$\bar{D}$ (cm × 10⁵)	ln x (Å)	Volume fraction $f \times 10^2$
E 1050	0.835	18.74	7.114	0.33
E 950	0.90	14.10	6.947	0.33
D 1050	1.203	13.88	7.090	0.66
C 1050	1.26	16.46	7.518	1.0
E 850	1.263	9.94	6.544	0.33
D 950	1.405	8.02	6.620	0.66
D 850	1.60	7.10	6.450	0.66
C 950	1.625	8.48	6.980	1.0
D 750	1.705	6.66	6.368	0.66
C 850	1.77	8.22	6.811	1.0

TABLE 10.2. Data of Jones and Kelly 1968.

Identification number	σ at 77°K (kg/mm²)	$\bar{D}$ (cm × 10⁵)	ln x (Å)	Volume fraction $f \times 10^2$
1	2.75	2.21	5.42	0.82
2	2.88	1.86	5.24	0.82
3	3.60	1.5	5.03	0.82
4	4.83	1.34	5.46	2.46
5	5.90	1.2	5.35	2.46
6	6.25	1.12	5.19	2.05
7	8.25	1.13	5.3	2.46
8	8.35	0.713	4.74	2.05
9	9.5	0.697	5.07	4.10
10	10.55	0.492	4.37	2.05
11	11.25	0.645	4.74	2.46
12	12.5	0.465	4.66	4.10
13	12.7	0.407	4.28	2.46

The agreement between theory and experiment is satisfactory and can be improved further by treating r_0 as an adjustable parameter. With the constants used here, the theory appears to underestimate the Orowan stress slightly.

I am grateful to U. F. Kocks and A. S. Argon for many helpful discussions and to Professor P. Hirsch, Dr. F. J. Humphreys, and R. Ebeling for sending me unpublished information and manuscripts. This work was supported in part by the Advanced Projects Research Agency, in part by the Office of Naval Research under Contract N00014–67–A–0298–0010, and by the Division of Engineering and Applied Physics, Harvard University.

REFERENCES

Ashby, M. F., 1964, *Z. Metalk.*, **55**, 5.

Ashby, M. F., 1968, in *Proc. Second Bolton Landing Conference on Oxide Dispersion Strengthening*, New York: Gordon and Breach.

Ashby, M. F., 1969, in A. Kelly and R. B. Nicholson (eds.), *Strengthening Mechanisms in Crystals*, Amsterdam: Elsevier, in press.

Ashby, M. F., and G. C. Smith, 1960, *Phil. Mag.*, **5**, 298.

Bacon, D. J., 1967, *Phys. Stat. Sol.*, **23**, 527.

Brown, L. M., 1964, *Phil. Mag.*, **10**, 441.

Byrne, J. G., M. E. Fine, and A. Kelly, 1961, *Phil. Mag.*, **6**, 1119.

Dew-Hughes, D., and W. D. Robertson, 1960, *Acta Met.*, **8**, 147.

deWit, G., and J. S. Koehler, 1959, *Phys. Rev.*, **116**, 1113.

Ebeling, R., 1968, private communication.

Ebeling, R., and M. F. Ashby, 1966, *Phil. Mag.*, **13**, 805.

Eshelby, J. D., 1961, in I. N. Sneddon and R. Hill (eds.), *Progress in Solid Mechanics*, Chapter 3. New York: Wiley, Interscience, vol. 2.

Fleischer, R. L., 1961, *Acta Met.*, **9**, 996.

Fleischer, R. L., 1963, in G. Thomas and J. Washburn (eds.), *Electron Miscroscopy and Strength of Crystals*, New York: Wiley, Interscience, p. 973.

Foreman, A. J. E., 1955, *Acta Met.*, **3**, 322.

Foreman, A. J. E., and M. J. Makin, (1966), *Phil. Mag.*, **14**, 911.

Foreman, A. J. E., 1969, private communication.

Geisler, A. H., 1951, in R. Smoluchowski et al. (eds.), *Phase Transformations in Solids*, Wiley, New York, p. 387.

Gleiter, H., 1967, *Acta Met.*, **15**, 1213.

Goodier, J. N., 1933, *J. Appl. Mechanics, Trans. A.S.M.E.*, **55**, A39.

Head, A. K., 1953a, *Phil. Mag.*, **44**, 92.

Head, A. K., 1953b, *Proc. Phys. Soc.*, **B66**, 793.

Hirsch, P. B., 1957, *J. Inst. Met.*, **86**, 7, appendix.

Hirsch, P. B., and F. J. Humphreys, 1969, private communication.

Hirth, J. P., and J. Lothe, 1968, *Theory of Dislocations*, New York: McGraw-Hill.

Humphreys, F. J., and J. W. Martin, 1967, *Phil. Mag.*, **16**, 927.

Jones, R. L., 1968, private communication (to appear in *Acta Met.*).

Jones, R. L., and A. Kelly, 1958, in *Proc. Second Bolton Landing Conference on Oxide Dispersion Strengthening*, New York: Gordon & Breach.

Kelly, A., and R. B. Nicholson, 1963, in B. Chalmers (ed.), *Progress in Materials Science*, New York: Pergamon, vol. 10, p. 149.

Kocks, U. F., 1966, *Phil. Mag.*, **13**, 541.

Kocks, U. F., 1967, *Canad. J. Phys.*, **45**, 739.

Mott, N. F., and F. R. N. Nabarro, 1940, *Proc. Phys. Soc.*, **52**, 86.

Nabarro, F. R. N., 1952, *Advances in Physics*, **1**, 269 ff.

Orowan, E., 1948, in *Symposium on Internal Stress in Metals and Alloys*, London: The Institute of Metals, p. 451.

Orowan, E., 1954, in Cohen (ed.), *Dislocations in Metals*, New York: AIME, p. 131.

Price, R. J., and A. Kelly, 1962, *Acta Met.*, **10**, 980.

Russell, K., and M. F. Ashby, 1968, Harvard University Technical Report No. 581.

Southwell, R. V., and H. J. Gough, 1926, *Phil. Mag.*, **1**, 71.

Thomas, G., and J. Nutting, 1957, *J. Inst. Met.*, **86**, 7.

ABSTRACT. A theoretical approach to the statistics of cutting randomly distributed point obstacles is presented. The motion of a dislocation can be analyzed in terms of nonuniformities in the breaking away from obstacles over the whole range of obstalce strengths. The elastic limit at 0°K is deduced as a function of the obstalce strength.

11. Nonuniformities in the Motion of a Dislocation Through a Random Distribution of Point Obstacles

J. E. DORN, P. GUYOT, AND T. STEFANSKY

11.1. Introduction

Most models for dislocation motion past localized obstacles have been erected by assuming that the obstacles can be represented by an ordered arrangement on the slip plane. In contrast, however, many types of obstacles are actually more or less randomly distributed over the slip plane. The statistical aspects of the problem have been clearly illustrated by the computer experiments of Foreman and Makin (1966) in which they studied, in detail, the movement of a dislocation through a random distribution of point obstacles. Their analysis revealed that the empirically deduced flow stress for cutting randomly dispersed obstacles is significantly different from that expected from a regular array of the same density. In addition, they noted that dislocation motion is not uniform but often takes place by an "unzipping" mechanism involving successive breakaway from obstacles on the dislocation. Nonuniform dislocation motion has been observed experimentally by Suzuki (1967).

We will present in this paper a simple statistical model of the unzipping process and from it deduce the elastic limit at 0°K. Some of the basic elements of the statistical theory were introduced by Kocks (1966, 1967). Although the elastic limit at 0°K as calculated by Kocks is in apparent agreement with the computer experiments of Foreman and Makin (1966), the theory is not formulated explicitly in terms of an unzipping mechanism. It is based instead on topological approximations. We believe that the present theory, based on an unzipping model, is more realistic.

11.2. Statistics of the Model

The following assumptions are made in an attempt to emphasize statistical features without encumbering the model with ancillary details:

1. Only one type of simple obstacle of width D is considered and these are assumed to be distributed at random on the slip plane. Let l_s^2 be the average area of the slip plane per obstacle, where $l_s^2 \gg D^2$.

2. The force to cut an obstacle at $0°K$ is taken as $F = \alpha \Gamma$, where α is the strength factor of the obstacle and Γ the average line tension. This force can also be expressed in terms of the breaking angle, ψ_c, between the arcs of the dislocation arrested at the obstacle as $F = 2\Gamma \cos \psi_c/2$.

3. Elastic anisotropy, the elastic interaction between the arcs of the dislocation line, and differences in line tension between edge and screw components of the dislocation are neglected.

4. Under an applied stress, τ, a dislocation moves in its slip plane until it is arrested at neighboring pairs of obstacles. Each arrested dislocation segment is assumed to bow out to a uniform radius of curvature

$$R = \frac{\Gamma}{\tau b}$$

regardless of the proximity of other dislocations (Figure 11.1).

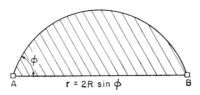

$$r = 2R \sin \phi$$

FIGURE 11.1. A neighboring pair of obstacles.

The obstacle B is a near neighbor of the randomly chosen obstacle A, when there are no obstacles in the cross-hatched area of Figure 11.1.

In terms of probability this leads to a distribution function

$$P(\phi)\, d\phi \propto \exp\left[-\frac{R^2}{l_s^2}(\phi - \sin \phi \cos \phi) \right] \cdot \frac{2R \sin \phi\, d(2R \sin \phi)}{l_s^2},$$

$$(11.1)$$

where the first term expresses the condition that there are no obstacles in the bowed-out area, and the second term is proportional to the probability of finding obstacle B between $r = 2R \sin \phi$ and $r + dr$ from obstacle A.

The statistics of the motion of a dislocation through a two-dimensional random distribution of point obstacles are now introduced: a dislocation will cut a contacted obstacle if the force on the latter, due to the line tension of the dislocation, exceeds $\alpha\Gamma$. A possible cutting configuration is shown in Figure 11.2a. The probability of occurrence of this cutting configuration

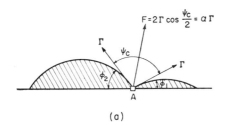

(a)

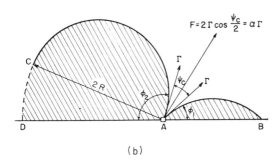

(b)

FIGURE 11.2. Possible cutting configurations (a) $\psi_c > \pi/2$; (b) $\psi_c < \pi/2$. For ϕ_1 less than $\pi/2 - \psi_c$.

is simply to have no obstacle in the cross-hatched area and therefore is proportional to

$$C(\phi_1) = \exp\left[-\frac{R^2}{l_s^2}(\phi_1 - \sin \phi_1 \cos \phi_1) \right]$$

$$\times \exp\left[-\frac{R^2}{l_s^2}(\phi_2 - \sin \phi_2 \cos \phi_2) \right]. \qquad (11.2a)$$

with

$$\phi_2 = \pi - (\phi_1 + \psi_c).$$

The average cutting probability, P_c, that is, the fraction of obstacles transparent to a dislocation at any given stress, is deduced by averaging $C(\phi_1)$ linearly over all permissible values of ϕ_1 as outlined in Appendix

11.1 (values of ϕ_1 (or ϕ_2) less than $\pi/2 - \psi_c$ lead to cutting configurations represented on Figure 11.2b). Therefore P_c is equal to

i.

$$P_c = \frac{\exp\left[-\frac{R^2}{l_s^2}(\pi - \psi_c)\right] \int_0^{\pi - \psi_c} \exp\left[\frac{R^2}{2l_s^2}\left[\sin 2\phi_1 - \sin 2(\phi_1 + \psi_c)\right]\right] d\phi_1}{\pi - \psi_c},$$

(11.2b)

for $\psi_c > \pi/2$;
and ii.

$$P_c = \frac{2\int_0^{\pi/2 - \psi_c} \exp\left[-\frac{R^2}{l_s^2}\left(\phi_1 - \frac{1}{2}\sin 2\phi_1\right)\right] \times \exp\left[-\frac{R^2}{l_s^2}\left[\frac{3\pi}{2} - 2\psi_c - 2\phi_1\right]\right] d\phi_1}{\pi - \psi_c}$$

$$+ \frac{\exp\left[-\frac{R^2}{l_s^2}(\pi - \psi_c)\right] \int_{\pi/2 - \psi_c}^{\pi/2} \times \exp\left[\frac{R^2}{2l_s^2}\left[\sin 2\phi_1 - \sin 2(\psi_c + \phi_1)\right]\right] d\phi_1}{\pi - \psi_c}$$

(11.2c)

for $\psi_c < \pi/2$.

The variation of P_c with the strength of the obstacle is a physical requirement and constitutes one of the principal points of departure from the formulation presented by Kocks (1967).

A dislocation is arrested at an obstacle when the force due to its line tension is less than $\alpha\Gamma$. Parameters defining these stable configurations can be evaluated using the distribution function given by Equation 11.1. As an example, the geometrical average link length, defined as the mean separation between neighbors averaged over all stable configurations (from Equation 11.1 and Figure 11.3), can be estimated by

$$\bar{r} = \sqrt{r_1 r_2} = \frac{\iint_{\substack{\text{stable} \\ \text{configurations}}} 2R\sqrt{\sin \phi_1 \sin \phi_2}\, P(\phi_1)P(\phi_2)\, d\phi_1\, d\phi_2}{\iint_{\substack{\text{stable} \\ \text{configurations}}} P(\phi_1)P(\phi_2)\, d\phi_1\, d\phi_2}.$$

(11.3)

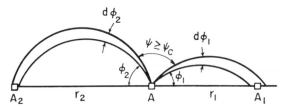

FIGURE 11.3. A stable configuration.

The integration limits are given in Appendix 11.1. As with P_c, the variation of $\bar{r}/l_s$ with α is physically reasonable and is another departure from the formulation presented by Kocks (1967).

An average cutting probability or an average link length cannot be uniquely defined in a random distribution of obstacles. They will depend on the models chosen for cutting configurations (Figures 11.2a and b) or stable configurations (Figure 11.3) and on the type of averaging. The elastic limit at 0°K, which we shall deduce in the next section, is quite sensitive to changes in P_c and $\bar{r}$.

11.3. The Unzipping Mechanism

The unzipping mechanism is approximated by the model given in Figure 11.4, where a dislocation is shown arrested along a line of obstacles. Under an applied stress, τ, each dislocation segment bows out with the

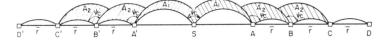

FIGURE 11.4. Unzipping model.

uniform radius of curvature $R = \Gamma/\tau b$. We assume that the cutting condition $2\Gamma \cos \psi_c/2 = \alpha\Gamma$ is satisfied at some obstacle S and that other obstacles are distributed symmetrically about S. To approximate this situation, we suggest that they be spaced symmetrically at A, B, C, ... A', B', C' ... and separated by the average link length. Once the dislocation cuts through the source obstacle S, it will automatically proceed further if the cutting conditions at A, B, C, ... A', B', C' ... are consecutively satisfied, i.e., if the dislocation can move by "unzipping."

In the formulation of the unzipping mechanism, two distinct cases arise:

1. Single Source (or Noninterfering Sources)

The dislocation has started to release at the source S and we consider, for the time being, its motion to the right of S. The upper boundary of the area A_1 * is an arc with radius of curvature $R = \Gamma/\tau b$ such that the cutting

* The areas A_1 and A_2 are evaluated in Appendix 11.2.

condition $2\Gamma \cos \psi_c/2 = \alpha\Gamma$ is satisfied at obstacle A. Thus A will be cut by unzipping if there are no obstacles in the area A_1. The probability of this event is

$$P_{u1} = \exp(-A_1/l_s^2) \tag{11.4a}$$

Similarly, the area A_2 * is such that the cutting condition is satisfied at obstacle B. The probability that there are no obstacles in A_2 is

$$P_{u2} = \exp(-A_2/l_s^2). \tag{11.4b}$$

Therefore, the probability that B is cut by unzipping is $P_{u1}P_{u2}$. Since the same area A_2 is associated with the remaining obstacles, the probabilities of succeeding events follow directly. Once the dislocation cuts the source obstacle, the total number of obstacles cut by unzipping is

$$i = \{1 + 2P_{u1} + 2P_{u1}P_{u2} + 2P_{u1}P_{u2}^2 + \cdots\}$$

$$i = \left(1 + \frac{2P_{u1}}{1 - P_{u2}}\right). \tag{11.5}$$

The factor of 2 appears since, once the source acts, the unzipping can proceed in both directions with equal probability.

2. *Interfering Sources*

When sources are a finite distance apart, the unzipping of each can overlap and the number of cut obstacles per source is no longer given by Equation 11.5. The fraction (P_c/l_s) of the obstacles are sources and therefore the distance between sources along a line is (l_s/P_c). The unzipping mechanism is now based on the model shown in Figure 11.5. The probability that A

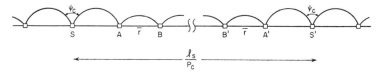

FIGURE 11.5. Unzipping model with interfering sources.

will be cut by unzipping is P_{u1} when the unzipping originates at S, and $P_{u1}P_{u2}(1/P_c - 2)$† when it originates at S'. Therefore the probability that A is cut by unzipping from the source S only, is $P_{u1}[1 - P_{u1}P_{u2}^{(1/P_c-2)}]$. Similarly, the probability that B is cut by unzipping from S only is $P_{u1}P_{u2}[1 - P_{u1}P_{u2}^{(1/P_c-3)}]$. The total number of obstacles cut by unzipping, per source, now is

$$i = 1 + 2\sum_K P_{u1}P_{u2}^{K-2}(1 - P_{u1}P_{u2}^{1/P_c-K},$$

with $K = 2, 3, \ldots, 1/P_c$. \tag{11.6}

* The areas A_1 and A_2 are evaluated in Appendix 11.2.
† Unzipping originating at S' reaches A by cutting A' with probability P_{u1} and each of the remaining $(1/P_c - 2)$ obstacles B', C' $\cdots$ C, B with probability P_{u2}.

Fractional values of $1/P_c$ are treated by averaging over integral values as outlined in Appendix 11.3. For large values of $1/P$, corresponding to sources far apart, Equation 11.6 reduces to Equation 11.5, which applies for noninterfering sources.

In order to calculate the total area swept out by a dislocation per source, in addition to unzipping along the line joining the sources, the forward motion of the dislocation has to be considered. As a first approximation, we assume that the number of new obstacles met by each unzipped dislocation segment resulting from its forward motion is proportional to the ratio of the unzipped arc length divided by the average arc length. Referring to Figure 11.5, the average arc length is $2R\phi$, where $\phi = \sin^{-1}(\bar{r}/2R)$. The unzipped arc length associated with a source S is $R(\pi - \psi_c) = 2R\phi_c$, with a neighboring obstacle of type A, $R(\phi_c + \phi)$, and with an obstacle of type B, $2R\phi$. Therefore, from Equation 11.7, the total number of obstacles contacted, per source S, becomes

$$j = \frac{\phi_c}{\phi} + 2P_{u1}(1 - P_{u1}P_{u2}^{1/P_c - 2})\left(\frac{\phi_c + \phi}{2\phi}\right)$$

$$+ 2\sum_{K=3}^{K=1/P_c} P_{u1}P_{u2}(1 - P_{u1}P_{u2}^{1/P_c - K}). \tag{11.7}$$

Equations 11.6 and 11.7 describe the dependence of the unzipping process on the stress and the obstacle strength. The total area swept out by a dislocation per successful cutting can now be written as

$$A = l_s^2\{i + (jP_c)i + (jP_c)^2 i + \cdots = l_s^2\left(\frac{i}{i - jP_c}\right), \tag{11.8}$$

the term il_s^2 being the area swept out by unzipping along the line connecting the sources. The term jP_c is the number of new sources among the j new obstacles contacted and $(jP_c)il_s^2$ is the area swept out by unzipping that originates at the P_c new sources. The meaning of the remaining terms follows directly.

As previously shown by Kocks (1966), yielding occurs when a dislocation sweeps across the entire slip plane. In terms of this model, the flow stress at $0°K$ is then given by $jP_c = 1$, and is readily determined from Equations 11.2a and 11.7, where the integrals must be evaluated numerically. Our values for the flow stress $(l_s/R) = \tau b l_s/\Gamma$ at $0°K$ are shown in Figure 11.6 together with the results of Kocks (1967) and of Foreman and Makin (1966), over the whole range of obstacle strengths. Beyond $\alpha = 1.5$ there is a small but increasing disagreement between our results and those given by the computerized experiments of Foreman and Makin. Our formulation of the unzipping mechanism is based essentially on a straight line model which might be somewhat in error, at high obstacle

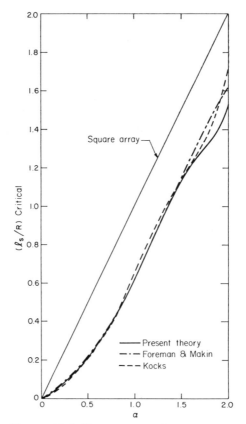

FIGURE 11.6. Flow stress at 0°K.

strengths, where dislocations are expected to be zigzag rather than straight. We have also neglected the formation of loops around groups of closely spaced obstacles which was observed by Foreman and Makin.

11.4. Conclusions

A new formulation of the statistical theory for the glide of a dislocation through a random array of point obstacles has been presented. The macroscopic flow stress at 0°K is deduced on the basis of nonuniformities in the motion of the dislocation, namely, an unzipping process. The overall agreement with the computer experiments of Foreman and Makin reinforce the following features:

1. The dependence of the statistical parameters such as the cutting probability, as well as the link length or the number of obstacles cut or contacted, on both stress and strength of the obstacles is physically reasonable.

2. A simple and essentially straight line unzipping model adequately describes the average motion of a dislocation through a random array of obstacles over the entire range of obstacle strength.

This research was conducted as part of the activities of the Inorganic Materials Research Division of the Lawrence Radiaton Laboratory of the University of California, Berkeley, and was done under the auspices of the U.S. Atomic Energy Commission.

REFERENCES

Foreman, A. J. E., and M. J. Makin, 1966, *Phil. Mag.*, **14**, 911.

Kocks, U. F., 1966, *Phil. Mag.*, **13**, 541.

Kocks, U. F., 1967, *Canad. J. Phys.*, **45**, 737.

Suzuki, T., 1967, in A. R. Rosenfield et al. (eds.), *Dislocation Dynamics*, New York: McGraw-Hill, p. 551.

APPENDIX 11.1

The limiting values of ϕ_1 and ϕ_2 are readily obtained from Figure 11A.1. Equation $\phi_1 = \pi - \psi_c - \phi_2$ represents the critical cutting condition. The shaded areas in Figure 11A.1 correspond to the bowing-out process (Figure 11.2b) and are forbidden ranges for stable configurations.

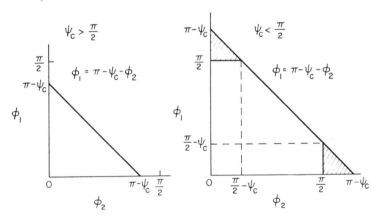

FIGURE 11A.1. Limiting values of ϕ_1 and ϕ_2.

APPENDIX 11.2

The areas A_1 and A_2 are obtained from Figure 11A.2. For $\psi_c < \pi/2$,

$$A_1 = R^2 \left[\pi - \frac{3}{2} \psi_c - 2\phi + \frac{1}{2} \sin \psi_c - 2\left(1 - \cos \frac{\psi_c}{2}\right)^2 \cot \frac{\psi_c}{2} \right]$$

$$(11A2.1)$$

$$A_2 = R^2 [\sin 2\phi + 2 \sin \phi \cos (\phi + \psi_c)],$$ $$(11A2.2)$$

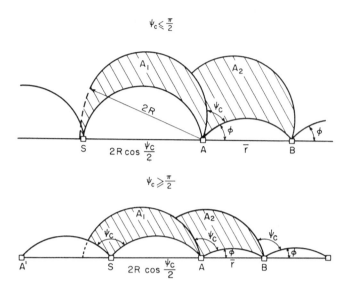

FIGURE 11A.2. Areas A_1 and A_2.

where $\phi = \sin^{-1}\bar{r}/2R$. For $\psi_c > \pi/2$,

$$A_1 = R^2\left[\frac{\pi}{2} - \frac{\psi_c}{2} - \phi\right] + \frac{R^2}{2}\left[\sin\psi_c + \sin 2(\psi_c + \phi)\right]$$

$$- 4R\frac{\left[\sin(\psi_c + \phi) - \cos\frac{\psi_c}{2}\right]^2 \cos\frac{\psi_c}{2}\sin(\psi_c + \phi)}{2\cos\left(\frac{3\psi_c}{2} + \phi\right)}.$$

$$(11A2.3)$$

A_2 is still given by Equation 11A.2.2.

APPENDIX 11.3

(a) For $1/P_c = 1$, all sources lie adjacent to each other and therefore Equation 11.6 reduces to $i = 1$.

(b) For $1/P_c = 2$, all sources again lie adjacent to each other. Equation 11.6 once more reduces to $i = 1 + 2P_{u1}(1 - P_{u1}) = 1$, because $P_{u1} = 1$. In order to distinguish between cases (a) and (b), we will consider all configurations characterized by $1 \leq 1/P_c \leq 3$ as the arithmetic mean of configurations with $1/P_c = 1$ and $1/P_c = 3$. Thus,

$$\{j\}_{1 < 1/P_c < 3} = f\{j\}_{1/P_c = 1} + (1 - f)\{j\}_{1/P_c = 3},$$

with $f.1 + (1 - f)3 = 1/P_c$.

In practice, this averaging procedure is used for $1 < 1/P_c < 3$. In all other cases, graphical interpolation yielded satisfactory results.

ABSTRACT. Applications of the Statistical Theory of Flow Stress to a material with randomly dispersed particles of a second phase, deformable or not, or with randomly distributed solute atoms, are reviewed and discussed. In general, the more quantitative aspects of this theory have served to sharpen various points of disagreement between theory and experiment for dispersion hardening, precipitation hardening, and solution hardening. In some instances, aperiodicity leads to qualitatively new effects. One of these is the existence of two flow stresses for penetrable obstacles, one for screws and a higher one for edges. The flow stress is thus controlled by the edges, while the screws can move very rapidly.

In a significant range of strengths, the edges have to propagate by an Orowan mechanism, although all the precipitates are sheared by the screws.

12. A Statistical Theory of Alloy Hardening

U. F. KOCKS

12.1. The Statistical Theory of Flow Stress

If a volume element at position $\mathbf{x}$ in in a specimen contains two obstacles r apart, the stress $\tau(\mathbf{x})$ necessary in this volume element to drive dislocations through or between these obstacles is described by

$$\tau(\mathbf{x}) = \frac{\alpha G b}{r(\mathbf{x})} \tag{12.1}$$

where b is the Burgers vector and G is the shear modulus of the matrix. Equation 12.1, with $\alpha \approx 1$, was derived by Orowan (1948) for obstacles so strong that the dislocation has to bow out between them, but it holds for weaker obstacles as well if α is allowed to depend on the strength of the obstacles.

If the whole slip plane contained obstacles located on a square array of lattice spacing r, Equation 12.1 would not only describe the local process but also the macroscopic flow stress τ; below this stress only elastic and anelastic strains could occur.

Real obstacles in real materials are not regularly spaced; they are found to be more copious in some regions than in others. This aperiodicity may be due to some fluctuations away from regularity, or to the absence

143

of any interactions (which would lead to a random arrangement); or it may be the consequence of a physical tendency toward clustering. The changes that any such deviation from periodicity produces in the relation between macroscopic properties and local interactions have been the subject of this study (Kocks 1966, 1967, 1968). In some instances, the results provide proportionality factors of order one, in others qualitative changes and new effects.

For a derivation of the basic relation, assume that N segments of dislocation per unit volume are hung up at obstacles at a given stress and that a fraction dP of these segments is released because, for example, the stress has been raised or thermal activation has taken place; then the resulting increase in strain will be

$$dy = bNa \cdot dP \tag{12.2}$$

where a is the average area swept out by one segment at that stress.[1] This area is larger than the average area per obstacle point, l^2, because the dislocation may be able to overcome one or more of the *next* obstacle pairs at the same stress. Thus, a depends sensitively on the applied stress σ. Figure 12.1 shows this dependence as derived from the Statistical Theory for a random arrangement of strong obstacles (Kocks 1967).

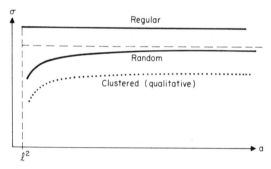

FIGURE 12.1. Applied stress σ versus mean free slip area a of one dislocation loop, for strong obstacles of area density $1/l^2$, in a square and a random array and, qualitatively, in a clustered array. The stress level at the asymptote is called the flow stress τ.

It is seen that there exists a stress at which the average segment that has been released can keep moving indefinitely. This asymptotic stress corresponds to the macroscopic flow stress τ at this obstacle distribution. (Both the material property τ and the applied stress σ are shear stresses.) In analogy to Equation 12.1, it may be described by

$$\tau = \frac{\alpha G b}{\lambda}, \tag{12.3}$$

[1] Equation 12.2 also follows if one lets the number of segments N vary during the increment, but keeps the area a independent of the stress (Kocks 1967).

where λ is the "effective spacing" of the obstacles at the flow stress. λ may be visualized as the width of the smallest gateway in the mean free slip area a when this a has just become a continuous area in the slip plane. Below the flow stress, a dislocation loop emitted by any source will sooner or later be stopped along its entire periphery and will, by its back stress, stop the source.

As the degree of clustering of the obstacles increases, so does the width of the smallest gateway, λ. Thus, the flow stress will, at the same average obstacle density, be smaller for a random arrangement than for a regular one, and smaller yet for a clustered arrangement. This is shown qualitatively in Figure 12.1.

Since screw and edge dislocations do not usually move with equal ease, there will be a range of stresses within which one kind of dislocation can move indefinitely but the other cannot. The dislocation loops then degenerate into an array of parallel dislocations that are stopped along their entire length. Their back stress will stop the sources just as well as closed loops do. The macroscopic flow stress is therefore determined by those dislocations which are harder to move.

The effective spacing λ is proportional to some spacing parameter of the obstacle structure in two[2] dimensions. We use as this "average spacing" l, the square root of the mean area per obstacle point in the slip plane, decreased by the mean obstacle diameter in the slip plane. For a random spatial arrangement of spheres of uniform radius R occupying a volume fraction c,

$$l = R \cdot \left\{ \sqrt{\frac{2\pi}{3c}} - \frac{\pi}{2} \right\} \simeq R \cdot \frac{1 - \sqrt{c}}{\sqrt{c/2}}, \tag{12.4}$$

setting $\sqrt{3\pi/8} \simeq \sqrt{3/\pi} \simeq 1$.

The dynamics of the process are described by writing Equation 12.2 as a time derivative, and by expressing explicitly that the penetration probability P may increase by virtue of thermal fluctuations, by virtue of a raise in the applied stress σ, or by virtue of a decrease in the flow stress τ (Kocks 1969):

$$\dot{\gamma} = \frac{b}{d} v^* \cdot \exp\left\{ -\frac{H_0 - V\sigma}{kT} \right\} + bN\alpha \left. \frac{\partial P}{\partial \sigma} \right|_\tau \cdot \left\{ \dot{\sigma} - \frac{\sigma}{\tau} h\dot{\gamma} - \frac{\sigma}{\tau} \dot{\tau} \right\} \tag{12.5}$$

The last two terms describe changes in flow stress due to work hardening h and (dynamic or static) recovery, $-\dot{\tau}$ ($\dot{\tau}$ is the *partial* derivative

---

[2] The one-dimensional mean free length (unfortunately often called "mean free path" in a completely different meaning from that used in Equation 12.16) of dislocations in a fixed direction is an irrelevant parameter because of the flexibility of dislocations, regardless of the *degree* of flexibility. This parameter only provides an upper limit for the effective spacing λ when $l \simeq R$; i.e. at unrealistic concentrations.

of the flow stress with respect to time at constant strain, and $\sigma/\tau = -(\partial P/\partial \tau)_\sigma/(\partial P/\partial \sigma)_\tau)$.

In a non-workhardening material, the stress-strain curve is described by the term on the left of Equation 12.5 balancing that proportional to $\dot\sigma$; this curve has a horizontal asymptote at the flow stress which follows from $a(\sigma)$. The number of mobile dislocation segments N enters into the transition region, but not into the flow stress where a single source is enough to activate an entire slip plane.

The first term on the right in Equation 12.5 describes the influence of thermal activation on lowering the flow stress from the value H_0/V at $0°$K, with the usual kinetic parameters v^* and kT and an activation volume V which may depend on stress and structure. Here again, an entire slip plane is activated by a single source, and only the number of active slip *planes* enters through their average spacing d (Kocks 1968). The use of a single Arrhenius term to describe the effect of thermal fluctuations has been shown (Kocks 1968) to be a good approximation even in this heterogeneous distribution of flow stresses.

Equation 12.5 describes creep by the balance of the left-hand side with the first term on the right (and possibly the last); it describes stress relaxation by the balance of $\dot\sigma$ with the term due to thermal activation and that due to recovery; and it describes work hardening, i.e. the plastic stress strain curve without the elasto-plastic transition region, by the balance of the last three terms with each other.

12.2. Dispersion Hardening

In a single crystal with very hard obstacles, the dislocations have to bypass the obstacles by the Orowan (1948) mechanism. Under the assumption of a constant line tension E, this means that they have to bow out to a semicircle whose diameter is equal to the width of the critical gateway, λ. In a random array, this value λ is about 20 percent larger than the average obstacle spacing l, resulting in a flow stress (Kocks 1966)

$$\tau = \frac{2E}{b\lambda} = 0.84 \cdot \frac{2E}{bl} \tag{12.6a}$$

In reality, the line tension is not constant for two reasons. First, it depends on the orientation of a dislocation element with respect to the Burgers vector and the crystallographic axes. As a result, the configuration of instable equilibrium is no longer a semicircle, but to a good approximation an elliptical half-loop (deWit and Koehler 1959). The average line tension of the bowed-out edge is the line *energy* of a straight screw E_s, the average line tension of the bowed-out screw is the line *energy* of the straight edge E_e. Thus it is harder to bow out a screw at a given spacing and we would have

$$\tau = \frac{2E_e}{bl} \qquad (12.7)$$

for a periodic array.

In a random array, however, the flow stress is defined (Kocks 1966) by a critical probability of finding "neighboring" obstacle pairs above a certain spacing. Since this probability depends on the *area* of the critical half-loop only, it is the same for edge and screw,[3] although the critical *spacings* are the minor and major axis, respectively: $\lambda_e = E_s/\sqrt{E_e E_s} \cdot \lambda$ and $\lambda_s = E_e/\sqrt{E_e E_s} \cdot \lambda$. The flow stress would then be the same for edge and screw and would be given by Equation 12.6a with[3]

$$E = \sqrt{E_e E_s}. \qquad (12.6b)$$

There is, however, a second correction to be applied to the constant-line-tension approximation, and that is due to the elastic interaction between different branches of the dislocation. In the average, this effect is well accounted for (Li and Liu 1968) by making the outer cut-off radius in the line energy proportional to the distance between roughly parallel branches.[4] Ashby (1966) proposed that the two branches to be considered should be those on either side of an obstacle, rather than those at the two ends of one segment.[5] With this and Equations 12.4, 12.6a, and 12.6b, using Foreman's (1955) terminology:

$$\tau = 0.58 \cdot \sqrt{K_e K_s} \cdot \frac{b}{R} \cdot \frac{\sqrt{c}}{1 - \sqrt{c}} \cdot \frac{\ln(2R/r_0)}{2\pi}, \qquad (12.8a)$$

where r_0 is the effective core radius — another rather little known quantity. If one defines $G = K_s$ which, for fcc crystals, is the geometric mean of the minimum and maximum shear moduli,[6] sets $K_e \gtrsim 1.5 \cdot K_s$ and $2R/r_0 \gtrsim 100$, one obtains for $c < 0.1$:

$$\tau \gtrsim \frac{Gb\sqrt{c}}{2R}, \qquad (12.8b)$$

which is only a little lower than Orowan's (1948) value Gb/l.

---

[3] This was pointed out by Hirsch in this volume. The author (Kocks 1967, 1968), as well as Ham (1968), have stated incorrectly that under the same assumptions the screw would be harder to move, at a stress described by Equation 12.6a, with $E = E_e$.
[4] The proportionality factor is larger for the screw than for the edge (Bacon 1967).
[5] See, however, Ashby's reconsiderations in this volume.
[6] The shear modulus resolved on the slip system is irrelevant for dislocation bowing.

Having, in some way, accounted for the *average* elastic interaction of dislocation branches, one has yet to consider the variation of this interaction with the bow-out angle. This is by no means a trivial effect, since it may lead to a critical configuration *before* the half-loop has been obtained. The Orowan stress would then be lowered. Unfortunately, no calculation of sufficient rigor appears to be available for this problem. In any case, the effect is bound to be different for edge and screw dislocations, and whenever this is the case, the harder one controls the flow stress. In connection with precipitation hardening, we have shown in Figure 12.2c a

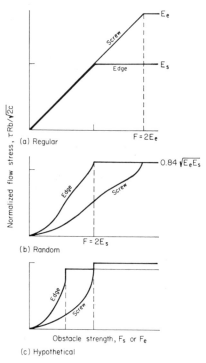

FIGURE 12.2. Normalized flow stress versus obstacle strength relations : (a) in a regular array, assuming an anisotropic but constant line tension; (b) in a random array under the same assumption; (c) in a random array, assuming a hypothetical dependence of the dislocation self-interaction on angle. The ratio E_e/E_s was taken to be 1.77 as in Cu.

case in which the Orowan level for the screw is arbitrarily assumed to be slightly higher. In this case, it is conceivable that both Orowan levels are observed, for example as a function of temperature, depending on whether screw dislocations can cross-slip or not.

The comparison of any of these more quantitative theories with experiments presents a number of problems. First, one has to establish that the distribution of obstacles is random; otherwise one has to modify the theory. In the experiments of Ebeling and Ashby (1966) on SiO_2 in Cu, it was shown that the area density of particles does obey the Poisson distribution. Second, however, there are usually a number of different processes contributing to the flow stress; the superposition rule has to be known before the Orowan contribution can be separated out. This problem becomes less acute when dispersion hardening is very severe, such as in the experiments of Jones (1969) on BeO in Cu. Both of the quoted sets of experiments agree with Equation 12.8a to within the experimental scatter of ± 15 percent. However, for these various reasons, it seems impossible at present to definitely establish any such equation over a number of other suggested relations.

12.3. Precipitation Hardening

If a quasistraight dislocation, such as one moving in steady state, can break through obstacles before it surrounds them, its local angular deviation from a straight line cannot exceed that given by the breakthrough angle. The average spacing λ of obstacles along such a dislocation line is a function of this maximum angular deviation, which in turn depends on the ratio of the obstacle strength F to the dislocation line energy E. The macroscopic flow stress is thus

$$\tau = \frac{F}{b \cdot \lambda(F/E)}. \tag{12.9}$$

In order to lump the F-dependence into one term we can write

$$\tau = s\left(\frac{F}{2E}\right) \cdot \frac{2E}{bl}. \tag{12.10}$$

This relation is analogous to Equation 12.6a if 0.84 is the limiting value of s for large F. Here, l is again the average two-dimensional spacing between obstacles, which may be expressed in terms of the volume fraction and the obstacle radius as in Equation 12.4.

The function $s(F/2E)$ has been derived from the Statistical Theory of Flow Stress (Kocks 1967), in good agreement with computer experiments by Foreman and Makin (1966). A first notable result is that the Friedel (1956) relation

$$s = \left(\frac{F}{2E}\right)^{3/2} \quad \text{for weak obstacles,} \tag{12.11}$$

holds over a rather large range of obstacle strengths, viz. up to stresses of about 1/4 of the Orowan stress, corresponding to an obstacle dislocation interaction of many electron volts.

Yet there is a considerable span of obstacle strengths that can be described neither by the Friedel relation (Equation 12.11) nor by the Orowan relation (Equation 12.6a). The Statistical Theory has essentially solved the problem of the effective neighbor spacing for these cases also. Figure 12.2b shows the result, in comparison with Figure 12.2a in which

$$s = F/2E \quad \text{for a periodic array.} \tag{12.12}$$

The most striking difference between these two figures is that the random array gives two separate curves for screws and edges while in the regular array there is just one (Ham 1968). This is exactly opposite to the observation made for impenetrable obstacles in the preceding section. The difference can be qualitatively explained by saying that the flanks of a bowed-out edge dislocation have screw character and thus exert less force on the obstacle at a given angle than the flanks of a bowed-out screw.

There may be special cases in which screw dislocations nevertheless control the flow stress, namely, when their interaction with the obstacle is much stronger than that of the edge (by more than the ratio E_e/E_s). No interaction mechanism has yet been proposed that would lead to such a possibility. We conclude, then, that edge dislocations always control the flow stress in precipitation-hardened alloys.

It is particularly interesting to notice that there is a significant range of strengths for which the flow stress is controlled by an Orowan mechanism for the propagation of edge dislocations, although all particles are sheared by the screws. Experimental evidence showing a change in particle shape can thus not be used to prove the absence of an Orowan mechanism of yielding.

Difficulties remain, in the case of penetrable obstacles as in the case of impenetrable ones, with respect to the dependence of the dislocation line energy on the bow-out angle and the angle included between neighboring branches on either side of an obstacle. To get a qualitative idea of the influence of such effects, Kocks (1968) introduced a heuristic funtion describing the interaction of the two neighboring dislocation branches at one obstacle. The result is shown in Figure 12.2c. The assumed increase of the attraction with increasing parallelism of the dislocation branches leads to a decrease in the force exerted on the obstacle by the dislocation. If the Orowan mechanism is not yet possible at the finite included angle corresponding to this maximum force, one gets vertical jumps in the stress versus obstacle strength curves, as indicated in Figure 12.2c.

The uncertainties concerning the influence of dislocation branch interactions would make it very desirable to have a precipitation-hardening system in which the the strength of the particles can be varied continuously over the entire range. It is sometimes claimed that this is in fact done in aging and overaging experiments. However, with the more quantitative theoretical tools we have at hand now, it appears unlikely that the Orowan stress has been reached in any precipitation-hardening system without a change in the structure of the precipitate. The widest continuous range of strengths has probably been reached in copper-cobalt alloys (Bonar 1962, Kelly 1963). It is possible that, at maximum strength, the Orowan stress might just have been reached for edge dislocations; however, the product $\tau R/\sqrt{c}$ after longer aging varied by more than a factor 2. This discrepancy may be connected with Bonar's (1962) metallographic observation that the particles tended to cluster (in strings) after these long aging treatments. This is a departure from randomness in the direction that *should* lower the flow stress.

The relation between obstacle strength and particle diameter is not known. Using Equation 12.11, one can derive such a relation experimentally. This has been attempted for the widely investigated alloy aluminum-zinc (Gerold and Haberkorn 1966, Harkness 1967), with the result that $F(R)$ is definitely not linear (as it had sometimes been assumed to be, on the basis of coherency stresses).

Voids might possibly be obstacles that span the entire range. While they can always be cut, the dislocations have to remain essentially perpendicular to the surface of large voids and may thus have to be bowed out to the position of maximum force, which corresponds to the stress at the lower end of the vertical jumps in the curves of Figure 12.2c.

12.4. Solution Hardening

For weak obstacles, the predictions of this theory coincide with those by Friedel (1956) and, as a consequence, with those by Fleischer (1963) for "symmetrical defects." For the stronger atomic defects with tetragonal distortions, Fleischer (1963) postulated that the dislocations are "fully flexible" and that this would lead to Equation 12.12, as in a periodic array. With the quantitative results of our theory, the strength of these obstacles can, however, be shown to be within the range of applicability of the Friedel relation (Equation 12.11).

The small size of solute atoms should give rise to a significant temperature dependence of the effective obstacle strength F. Figure 12.2 may then be looked upon as a flow-stress temperature diagram, the temperature rising rapidly to the left (Kocks 1968). While the absolute magnitude of the flow stress of strong solutes ("tetragonal distortions") should depend strongly on temperature, the *relative* change over $1000°K$ should not be more than about 30 percent. On the other hand, weak ("symmetrical") solutes should provide a strengthening that vanishes at reachable temperatures.

None of the above predictions appears to be compatible with any significant body of experimental observations (see Haasen 1968). For example, the temperature dependence is generally strong on a *relative* basis (suggesting weak obstacles), although at high temperatures there is frequently a temperature-independent " plateau stress " that still depends on concentration

(possible only for strong obstacles). Furthermore, the Friedel relation
(Equation 12.11) is not well obeyed (Haasen 1968) and even a dependence
on the square root of the concentration is the exception rather than the
rule. In the light of all these contradictions, it appears unwarranted to
continue (Haasen 1968) to apply the Fleischer analysis to solution
hardening.

All of these difficulties may be due to a basic hypothesis of all of these
theories, namely, that solutes and dislocations do not interact at zero
stress. At least for tetragonal distortions this assumption seems unrealistic,
but no model has yet been proposed that takes into account both this
effect and the aperiodic distribution of solutes. In principle, the Statistical
Theory may be extended to cover this interaction.

One further difficulty is that solution-hardened alloys often deform by
the spreading of a Lüders band. If Figure 12.2 is applicable, this could be
merely a consequence of the fact that screw dislocations are very easy to
move through the slip plane and have thus no reason to transfer slip to
neighboring planes. In any case, it is quite possible that the macroscopic
flow stress is controlled not by the propagation of slip through the slip
plane, but by the nucleation of slip in a sufficient number of slip planes —
in which case this entire theory is inapplicable.

12.5. Superposition of Mechanisms

It was stated in Orowan's classic paper (1954), and it is still
generally accepted, that the superposition of any of the above hardening
mechanisms with a "lattice friction" stress would be linearly additive.[7]
On the other hand, the superposition rule for two obstacle-controlled
mechanisms, even if they do not interact, is not generally agreed upon.
Since additivity is a trivial first approximation if the flow stress is dominated by one of the two mechanisms, there are only two interesting cases:
either many weak obstacles and few strong ones, or about equal densities
of roughly equally strong obstacles. We shall treat these two in turn.

Imagine a dislocation segment that is held up between two strong
obstacles. As the stress is increased, the dislocation will cut through many
weak obstacles in its path. The more it bows out, however, the closer to
180 degrees gets the angle between neighboring branches at each one of the
weak obstacles, while the angle included at the strong obstacle continuously decreases. Thus, the force on the strong obstacles increases while
that on weak ones decreases. The critical situation is reached when the
dislocation segment can just cut through weak and strong obstacles simultaneously. Then, the force applied to this dislocation segment of effective

[7] Argon (1968) has correctly pointed out that this is not true for a drag stress that
vanishes at small dislocation velocities.

length λ_1 by the stress τ is balanced by the sum of the strengths of all obstacles then touched by the segment:

$$\tau b \lambda_1 = F_1 + F_2 \cdot \lambda_1/\lambda_2, \tag{12.13}$$

where λ_1/λ_2 is the initial number of weak obstacles on a quasistraight line between two strong ones. Thus the flow stress is the sum of the flow stresses that each of the two mechanisms would demand separately:

$$\tau = \tau_1 + \tau_2 \quad \text{for } F_2 \ll F_1. \tag{12.14}$$

In addition to[8] $F_2 \ll F_1$, this derivation presupposed $\lambda_2 \ll \lambda_1$; but when this is not true, the flow stress is dominated by obstacle set 1, and Equation 12.14 holds as a first approximation. Note that the relative *ranges* of the two obstacles did not enter into the derivation at all.

When the densities of the two sets of obstacles are similar, the obstacles may be overcome one at a time. For the case that the obstacle strengths are the same, additivity of the two-dimensional obstacle densities and thus of the *squares* of the flow stresses follows (Koppenaal and Kuhlmann-Wilsdorf 1964). As a first approximation, one may thus write

$$\tau^2 = \tau_1^2 + \tau_2^2 \quad \text{for} \quad F_1 \simeq F_2. \tag{12.15}$$

Computer experiments by Foreman and Making (1967) have confirmed these two limits and covered the range in between. The experiments by Ebeling and Ashby (1966) on solute Au and dispersed SiO_2 in Cu are among those confirming Equation 12.14.

For cases in which the flow stresses are additive, it also follows from the necessity of breaking through all obstacles at the same time, that the reciprocal activation volumes are additive. These two relations together provide a convenient mechanism of separating two independent processes by means of strain-rate change experiments (Kocks 1968).

12.6. Work Hardening of Alloys

Work hardening must, in some way, be a consequence of the fact that some length $d\rho$ per unit volume of those dislocations that provided the strain $d\gamma$ gets stored in the crystal. The storage rate may formally be described (Nabarro et al. 1964) by a mean free path L (see footnote 2):

$$d\rho = \frac{1}{Lb} \cdot d\gamma. \tag{12.16}$$

With the flow stress contribution due to dislocation interactions,

$$\tau_d \propto \frac{Gb}{l_d} \propto Gb\sqrt{\rho}, \tag{12.17}$$

[8] This condition was not stated in Kocks (1968), but was suggested by Diehl in a discussion to that paper.

the work-hardening rate becomes (Kocks 1966)

$$\theta \propto l_d/L \cdot G, \tag{12.18}$$

where the proportionality constant is of the order $1/10$. Now it was a fundamental result of the Statistical Theory that, at the flow stress, dislocations keep moving indefinitely and are not stopped after a finite "mean free path." However, the definition of the "mean free path" by Equation 12.16 is also applicable to the case where a moving dislocation leaves debris behind (Kocks 1966).

In dispersion-hardened alloys, a loop is left around every single particle (Orowan 1954) and dipoles should be left around many closely spaced pairs of particles. Whether these primary dislocation residues themselves act as obstacles, or whether they cause secondary dislocation movement which causes the work hardening (Ashby 1968), does not matter to the argument. In either case, the total length of dislocation stored makes $1/L$ in Equation 12.16 proportional to $R\tau_0^2/(Gb)^2$, where τ_0 is the yield stress. Since this is independent of strain, Equation 12.16 can be integrated directly and inserted into Equation 12.17. Since dislocations and non-deforming particles are obstacles of similar strength, we use the superposition rule expressed in Equation 12.15 and find

$$\tau \simeq \tau_0 \sqrt{1 + \frac{R}{b}\gamma}, \tag{12.19}$$

where a proportionality factor of about 1 or less has been omitted in front of γ.

In the range where dislocations can cut through the particles, work hardening has often been expected to be absent. However, there is still the possibility that dislocation loops are left around groups of more than two closely spaced particles. Again, L would be independent of strain, leading to a parabolic stress-strain curve. However, L would be an order of magnitude larger and independent of R. Setting a factor $G/(1000 \cdot \tau_0) \simeq 1$, and again using Equation 12.15, we find

$$\tau \simeq \tau \sqrt{1 + \gamma}. \tag{12.20}$$

In solution-hardened materials, individual loops or dipoles cannot be supported by the strength of the solutes, and loops around groups of solutes cannot be expected either: the dislocations stay so straight during plastic flow that they cannot surround any groups of appreciable depth. Thus, there is no evident reason for any work hardening in connection with solution hardening in good agreement with experimental observations.

A noteworthy feature of all these primary loops of different kinds that may have been stored during straining is that they tend to *shrink* under the applied stress and are thus stable upon unloading. In reverse loading, they provide additional sources which should give rise to an extensive transition region, although the asymptotic flow stress would not be affected. The proposed mechanism of work hardening thus contains the elements necessary for an explanation of the Bauschinger effect and, by similar arguments, of polarized latent hardening (Kocks 1960).

12.7. Summary

The realization that obstacles to dislocation motion, of whatever kind they may be, are never arranged on a regular lattice, has led to an analysis of the importance of deviations from periodicity. Such an analysis had been undertaken previously only for the case of very weak obstacles such as solute atoms. Friedel's results for this case have been confirmed. The more quantitative treatment has in fact shown that they are applicable in the range of the moderately strong tetragonal distortions, in contradiction to Fleischer's assumptions. On the other hand, the sharpening of these predictions has served to emphasize the disagreement in important points between all of these theories and some experimental facts for the very weak obstacles.

Other results, in the nature of a change in proportionality constants in known equations, are that the flow stress in a crystal with a random dispersion of hard particles is lower by a factor of 0.84 than it would be in a material with a periodic dispersion of the same density. In a material with a tendency toward clustering it would be lower yet. On the other hand, all flow stresses should be higher than previously assumed because they should always relate to those dislocations that are harder to move. This results in an effective rise in the proper elastic constant to which the flow stress is proportional.

In some instances, the difference between aperiodic and periodic arrangements leads to major qualitative changes. For example, the difference between screw and edge dislocations, when their interaction strength with the obstacles is roughly the same and not too large, does not enter into the flow-stress relation for a periodic obstacle arrangement; but in an aperiodic arrangement it leads to the striking effects displayed in Figure 12.2. These predictions are as yet unconfirmed by experiment.

Work hardening in pure materials, or in materials with strong deformable precipitates, would be impossible in a periodic arrangement; only the existence of groups of relatively closely spaced obstacles provides the basic possibility for any dislocation storage mechanism. When particles of a second phase cause the dislocation storage, work hardening should be parabolic; when the dislocations themselves are the controlling obstacles, it should be linear. Dynamic recovery may lower the rate in either case.

While many of the details of the theories reviewed here, and of the contributions from the Statistical Theory of Flow Stress, may have to be revised or worked out in more quantitative form, the qualitative statements made in this summary should stand and form a link between physical considerations of dislocation obstacle interactions and the macrocopic strength and plasticity of metals and alloys.

At present, the author knows of no case of alloy hardening in which there is definitive and quantitative agreement between theory and experiment.

This work was performed under the auspices of the United States Atomic Energy Commission.

REFERENCES

Argon, A. S., 1968, *Materials Science and Engineering*, 3, 24.

Ashby, M. F., 1966, *Acta Met.*, **14**, 679.

Ashby, M. F., 1968, in *Proc. Second Bolton Landing Conference on Oxide Dispersion Strengthening*, New York: Gordon and Breach.

Ashby, M. F., 1969, this volume.

Bacon, D. J., 1967, *Phys. Stat. Sol.*, **23**, 527.

Bonar, G., 1962, Ph.D. thesis, University of Cambridge, England.

DeWit, G., and J. S. Koehler, 1959, *Phys. Rev.*, **116**, 1113.

Ebeling, R., and M. F. Ashby, 1966, *Phil. Mag.*, **13**, 805.

Fleischer, R. L., 1963, in *The Strength of Metals*, New York: Reinhold Press.

Foreman, A. J. E., 1955, *Acta Met.*, 3, 322.

Foreman, A. J. E., and M. J. Makin, 1966, *Phil. Mag.* **14**, 911.

Foreman, A. J. E., and M. J. Making, 1967, *Canad. J. Phys.* **45**, 511.

Friedel, J., 1956, *Les Dislocations* (English ed. Reading, Mass: Addison-Wesley, 1964) p. 224.

Gerold, V., and H. Haberkorn, 1966, *Phys. Stat. Sol.*, **16**, 675.

Haasen, P., 1968, *Proc. Int'l Conf. on Strength of Metals and Alloys, Trans. Jap. Inst. Met.*, **9** (Suppl.), xL.

Ham, R. K., 1968, *Proc. Int'l Conf. on Strength of Metals and Alloys, Trans. Japan Inst. Met.*, **9** (Suppl.), 52.

Harkness, S. D., 1967, Ph.D. thesis, University of Florida, Gainesville.

Hirsch, P. B., 1957, *J. Inst. Met.*, **86**, 13.

Jones, R. L., 1969, *Acta Met.*, **17**, 229.

Kelly, A., 1963, in G. Thomas and J. Washburn (eds.), *Electron Microscopy and Strength of Crystals*, New York: Wiley-Interscience, p. 1948.

Kocks, U. F., 1960, *Acta Met.*, **8**, 345.

Kocks, U. F., 1966, *Phil. Mag.*, **13**, 541.

Kocks, U. F., 1967, *Canad. J. Phys.*, **45**, 737.

Kocks, U. F., 1968, *Proc. Int'l Conf. on Strength of Metals and Alloys, Trans. Jap. Inst. Met.*, **9** (Suppl.), 1.

Kocks, U. F., 1969, in *Fundamental Aspects of Dislocation Theory*, Washington: National Bureau of Standards, in press.

Koppenaal, T. J., and D. Kuhlmann-Wilsdorf, 1964, *Appl. Phys. Let.*, **4**, 59.

Li, J. C. M., and G. C. T. Liu, 1968, *Proc. Int'l Conf. on Strength of Metals and Alloys, Trans. Jap. Inst. Met.*, **9** (Suppl.), 20.

Nabarro, F. R. N., Z. S. Basinski, and D. B. Holt, 1964, *Adv. Phys.*, **13**, 193.

Orowan, E., 1948 in *Symposium on Internal Stresses in Metals and Alloys*, London: Inst. Met., p. 451.

Orowan, E., 1954, in *Dislocations in Metals*, New York: AIME.

ABSTRACT. Of the direct effects on mechanical properties of high dislocation mobility in soft metals and dilute alloys, two distinct phenomena relating to the yield strength are described theoretically. One phenomenon is the effect of the randomness in the distribution of impurities or solute atoms, which yields a spacial variation of the dislocation speed, and which gives a yield stress smaller than the theoretical stress, obtained from a simple average distribution of solute atoms as in the previous theories of solid-solution hardening. The second phenomenon is the effect of the dynamic interaction of a moving dislocation with solute atoms, which becomes effective at lower temperatures where the quasistatic description of the yield stress no longer holds. These results are compared with experiments on copper and copper-nickel alloys, where satisfactory quantitative agreement is obtained.

13. Dynamic Yielding of Metals and Alloys

T. SUZUKI AND T. ISHII

13.1. Introduction

Although many measurements on dislocation mobility in various materials have been made so far, it was only very recently that similar measurements were made on soft metals of face-centered cubic or hexagonal close-packed structures. The measurements were made by one of the present authors and their co-workers: on copper (Marukawa 1967, dilute copper-nickel alloys (Suzuki 1967, Suzuki and Ishii 1968); and also by Vreeland and his group: on copper (Greenman et al. 1967), and zinc (Pope et al. 1967). It has become clear from these measurements that the mobilities of dislocations in these soft metals are larger by orders of magnitude in comparison to those in other materials studied before. For instance, the velocity of dislocations in copper at the yield point is as high as 10 m/sec.

The discovery of very large mobilities associated with dislocations has raised entirely new problems concerning the mechanical strengths of soft metals and alloys. One of these problems is that the yield strengths are possibly determined by the phonon and electron drags, because the dislocations are easily accelerated to the range of speed to be controlled by these drags. This has been predicted by Greenman et al. (1967) on copper

159

deformed at a usual strain rate, and shown by Ferguson et al. (1967) by studying high-speed deformation of aluminum. It can be expected that where impurities or solute atoms are randomly distributed, they determine the dislocation mobility. Owing to their high speed, dynamic interactions between moving dislocations and impurity atoms become important. These effects have usually been ignored in the discussion of yield strengths of metals and alloys. Two such problems are to be described in the present paper.

One of the effects which is ignored or, at least, not explicitly considered in conventional treatments of solution hardening is the randomness in the distribution of solute atoms. The dislocation mobility or yield strength has been described only in terms of an average distribution of solute atoms. In reality, however, the dislocation velocity determining the yield strength is the maximum local velocity and not the usually assumed average velocity over the whole area of the slip plane (Suzuki and Ishii 1968). Because of this, the theories predict higher yield stresses by about one order of magnitude than the observed values.

The second effect is also new. Some years ago, Ookawa and Yazu (1963) discussed theoretically a dynamic interaction between moving dislocations and solute atoms. According to their theory, the velocity of dislocations becomes proportional to the inverse of stress above a certain critical velocity. The transition from a quasistatic state of motion to this unstable state could give rise to a dynamic yielding of a new type to be discussed here.

The above-mentioned dynamic yielding will appear at low temperatures, where the dislocation velocity at the stress level of interest becomes larger than the critical speed mentioned above. Dislocation mobility and the absolute magnitude of the yield strength should depend on the phonon drag as well as the electron drag, in addition to the dynamic resistance of impurities or solute atoms.

In the present paper, first, the dislocation mobility in a dilute alloy is calculated for both the quasistatic and dynamic motions, and then the results are compared with observations on copper-nickel alloys. An anomalous yielding has been observed in these alloys below 100°K. The theoretical prediction of yield strength in this range of temperature is found to coincide satisfactorily with experiment both in magnitude and in dependence on temperature.

The above two phenomena associated with yielding are related to the motion of dislocations; in this sense both are dynamic phenomena. The term *dynamic yielding* is, however, used often in the present paper in a narrow sense for the discrimination between the above-mentioned yielding and the quasistatic one. Qausistatic yielding is described in Section 13.3, dynamic yielding in Section 13.4. Although the major part of Section 13.3 is the same as that in a preceding paper by one of the present authors

(Suzuki, 1968), it is briefly redescribed as it is relevant for the discussion in Section 13.4. Comparisons of our results with the Friedel-Kocks-Foreman-Makin relation and with computer experiments (Kocks 1966, Foreman and Makin 1966), as well as a comparison between theory and experiment on dislocation mobility, are treated for the first time in this paper.

13.2. Dislocation Mobility

The dislocation mobilities in various materials previously measured by many workers (Cu: Marukawa 1967, Suzuki 1968, Suzuki and Ishii 1968, Greenman et al. 1967, Pope et al. 1967; LiF: Johnston and Gilman 1959; W: Schadler 1964; Mo: Prekel and Conrad 1967; Si: Suzuki and Kojima 1966; Fe-Si: Stein and Low 1960; InSb: Mihara and Ninomiya 1967) are illustrated in Figure 13.1. Other measurements not shown in Figure 13.1 are Ni: by Rhode and Pitt (1967); NaCl: by Gutmanas et al. (1963) and

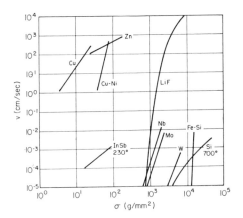

FIGURE 13.1. Dislocation velocities as a function of stress in various metals and compounds. All the data relate to the relations at room temperature except for InSb at 230°C and Si at 700°C.

some semiconductor materials: by Chaudhuri et al. (1962), and Kabler (1963). The curves in this figure all relate to the relations at room temperature except on silicon and indium antimonide. It is clearly seen that the mobilities in copper, copper-nickel (0.35% alloy), and zinc are very large as compared with the others. Vreeland's data on copper almost coincide with those on zinc. Their results on copper, therefore, are different from those of Marukawa's illustrated in Figure 13.1 with respect to the dependence on stress, although the absolute magnitudes of the velocities almost agree with each other. In the case of Marukawa, the average yield stress was 26 g/mm², and the velocity of dislocations, reached 8 m/sec at yield point. The velocities of dislocations at the yield point in various mterials are shown in Table 13.1.

TABLE 13.1. Dislocation velocities at yield point in various materials.

Material	Temperature (°K)	Yield Stress (kg/mm²)	Dislocation Velocity (cm/sec)	Author
Si	873	33.4	1×10^{-4}	Suzuki and Kojima (1966)
LiF	300	0.9	1×10^{-3}	Johnston and Gilman (1959)
W	300	19	0.07	Schadler (1964)
Fe-(3.35%) Si	298	14	4×10^{-5}	Stein and Low (1960)
Cu	298	0.026	8×10^2	Marukawa (1967)
	77	0.026	8×10^2	
Cu-(0.35%) Ni	298	0.080	6×10^2	Suzuki and Ishii
	77	0.134	3×10^3	(1968)

Figure 13.2 shows detailed results of measurements on copper and copper 0.35 % nickel alloy at room temperature and at liquid nitrogen temperature. The mobility in copper is found to be rather insensitive to

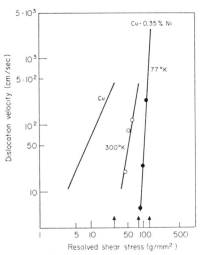

FIGURE 13.2. Dislocation velocities as a function of stress in copper and copper-0.35%- nickel alloy.

temperature in this range. If one writes the relation between the velocity and shear stress in our experiments as

$$(v/v_0) = (\sigma/\sigma_0)^m, \tag{13.1}$$

$m = 2$ for copper at $300°K$, and in copper-nickel alloy $m = 5.4$ at $300°K$, and 12 at $77°K$. On the other hand, Vreeland and his co-workers give $m = 1$ for copper. From this, they have concluded that the dislocation velocity is controlled by phonon drag. No firm evidence indicating that phonon drag controls the yield stress, however, has been obtained so far by Suzuki (1968), and Schüle et al. (1962).

13.3. Quasistatic Theory of Dislocation Velocity and Yield Stress in Dilute Alloys

13.3.1. Regular Lattice Distribution of Solute Atoms

It is generally accepted that the yield strength of face-centered cubic metals and also the yield strength related to basal slip of hexagonal close-packed metals are controlled by a small amount of impurities or by the interaction between dislocations. It has been generally accepted that the Peierls-Nabarro force in these soft metals is small when compared with the above resistive forces. This is clearly seen, for example, from the data on copper single crystals of 99.999 percent purity reported by many workers: 50 g/mm² by Schüle et al. (1962), 40–100 g/mm² by Basinski and Basinski (1964), 22 g/mm² by Rosenfield and Averbach (1960), 30–60 g/mm² by Young (1961a, 1961b), 25 g/mm² by Marukawa (1967), 13 g/mm² (zone-refined) by Suzuki (1968), 14–19 g/mm² by Young (1967), and 31–110 g/mm² by Argon and Brydges (1968). A round value for these comparatively recent data could be taken as 20 g/mm². However, there exist other data larger than 100 g/mm², for example, reported by Thornton et al. (1962), Mitchell and Thornton (1963), and H. Suzuki, et al. (1956). Comparatively low strengths seem to be obtained on good crystals of low dislocation density, and without any important substructure. It can be expected that the yield stress will be still lowered if both the purity and the dislocation density are decreased. At present, it is more difficult to reduce the total impurities (including gaseous impurities) below one ppm than to reduce the density of dislocations below 100 cm^{-2}. Impurity atoms are thus predominant obstacles for the motion of dislocations in such soft metals as copper.

Moving dislocations meeting with solute impurity atoms become curved at each obstacle. The external stress increases the component of the line tension, which is balanced by the resistive force due to the point obstacle. In soft metals, the external stress of interest is rather small in accordance with a low yield stress, and therefore the dislocation lines should be comparatively smooth, as described by Mott (1952), Friedel (1962), and

Fleischer (1962). Figure 13.3 shows a configuration of the dislocation line, where A, B_1, C, and B_2 are solute atoms and the solid line indicates the dislocation line. If the dislocation overcomes the point barrier B_1, it moves

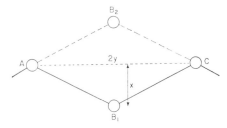

FIGURE 13.3. Unit of the triangular lattice distribution of solute atoms. A, B_1, B_2, and C are solute atoms; the solid line indicates a dislocation line which moves to the dotted line and, finally, to the broken line by the application of external stress. This is taken as a unit-activation process.

to the dotted line and, finally, to the broken line. This is taken as a unit-activation process. The softness of metals is reflected in the condition $2y \gg x$. In the following first-order approximation the result is independent of whether the interaction between the dislocation and the point barrier is positive or negative. Furthermore, only the impurity or solute atoms of the same species located on either side of the slip plane of the dislocation are taken into account. The maximum binding energy is designated by $|W|$, and the sign for the absolute magnitude will be hereafter omitted for the sake of brevity.

x and y defined in Figure 13.3 obey a relation

$$xy = b^2/(\alpha c), \tag{13.2}$$

where b is the magnitude of the Burgers vector, and c is the concentration of solute atoms. The symbol α is a geometrical factor, which for substitutional solutes in fcc metals, has a magnitude of

$$\alpha = 4.61. \tag{13.3}$$

It is necessary to determine x and y to calculate the yield stress or the dislocation mobility. For this one needs another condition for x and y. In a previous paper, one of the present authors (Suzuki, 1968) used the maximum barrier given by (x, y) for the calculation of the yield stress. This gave

$$x = \left(\frac{2}{3}\frac{W}{\alpha\mu b}\right)^{1/2} c^{-1/2} \tag{13.4}$$

$$y = \frac{b^2}{\alpha}\left(\frac{3}{2}\frac{\alpha\mu b}{W}\right)^{1/2} c^{-1/2}, \tag{13.5}$$

where μ is the shear modulus, which for simplicity is assumed to be independent of the concentration for dilute alloys.

Next, it is assumed that the triangular lattice distribution of (x, y) given by Equations 13.4 and 13.5 is uniformly spread in the whole crystal. This is equivalent to taking an average distribution as given by (x, y) above. The distribution thus defined is called the regular lattice distribution, following Kocks (1966).

The stress necessary to expand the dislocation AC to $AB_2 C$ is taken as τ_0. This stress is independent of temperature except for the dependence through μ and b. The plastic strain rate $\dot{\varepsilon}$ produced by the external stress τ at the temperature T is given as

$$\dot{\varepsilon} = \frac{n}{2y} 2xyb \frac{vb}{y} \exp\left[-\frac{V - (\tau - \tau_0)b^2 y}{kT} \right], \tag{13.6}$$

where n is the effective number of moving dislocations in unit volume, k is the Boltzmann constant, V the height of the barrier given by $V = W/3$, and vb/y is the effective frequency of vibration of the dislocation, where v is the Debye frequency. Equation 13.6 is a rather general formula introduced by Seeger (1954a,b) and Friedel (1956) for the case of dislocation forest cutting, and by Friedel (1962) for dislocation motion through impurity barriers.

On the other hand, the plastic strain rate is generally written in terms of the dislocation velocity (Cottrell 1953) as

$$\dot{\varepsilon} = bnv. \tag{13.7}$$

From Equations 13.6 and 13.7 it follows that

$$v = \frac{x}{y} vb \exp\left[-\frac{V - (\tau - \tau_0)b^2 y}{kT} \right],$$

or, using Equations 13.4 and 13.5, it can be written as

$$v = \frac{2}{3}\frac{W}{\mu b^2} v \exp\left(-\frac{W}{3kT} \right) \exp\frac{(\tau - \tau_0)b^2 y}{kT}. \tag{13.8}$$

The yield stress now is obtained as follows. When a hard testing machine is used, the strain rate $\dot{\varepsilon}_{ext}$ forced by the constant motion of the crosshead becomes equal to the plastic strain rate of the specimen at the yield point. That is,

$$\dot{\varepsilon} = \dot{\varepsilon}_{ext} = \text{const.} \tag{13.9}$$

Thus, the yield stress τ_c is given by taking the logarithm of Equation 13.6 and from Equation 13.9 as

$$\tau_c = \tau_0 + \frac{1}{3}\left(\frac{2}{3}\right)^{1/2} \alpha^{1/2} \frac{W^{3/2}}{\mu^{1/2}b^{9/2}} c^{1/2}\left(1 - \frac{T}{T_c}\right), \tag{13.10}$$

and

$$T_c = (W/3k)/\ln\left(\frac{2}{3}\frac{W}{\mu} nv/b\dot{\varepsilon}_{ext}\right).$$

On the other hand, from the definition, τ_0 satisfies the relation

$$\tau_0 \, b \, x \, y = 2E_3 y, \tag{13.11}$$

where

$$E_3 = \frac{\mu b^2}{4} \frac{x^2}{y^2}.$$

Therefore, it gives

$$\tau_0 = \frac{\mu b^3}{2} \frac{1}{\alpha c} \frac{1}{y^3}.$$

Using Equation 13.5, τ_0 can be written as

$$\tau_0 = \frac{1}{3} \left(\frac{2}{3}\right)^{1/2} \alpha^{1/2} \frac{W^{3/2}}{\mu^{1/2} b^{9/2}} c^{1/2}. \tag{13.12}$$

From Equations 13.10 and 13.12, it follows that

$$\tau_c(0^\circ K) = 2\tau_0 = \left(\frac{2}{3}\right)^{3/2} \alpha^{1/2} \frac{W^{3/2}}{\mu^{1/2} b^{9/2}} c^{1/2}. \tag{13.13}$$

13.3.2. *Comparison with the Friedel-Kocks-Foreman-Makin Relation*

The determination of (x, y) which maximizes the barrier height is based on the prediction that the yield stress is determined by the maximum barrier produced by possible regular lattice distributions. It is of interest to see that this hypothesis gives the same result as Friedel's result (1962), in which Friedel made a different assumption to determine the size of the effective average distribution of solute atoms. His assumption is that the yield stress at $0^\circ K$ is equal to the stress necessary to expand the dislocation AC to $AB_2 C$, in terms of Figure 13.3, and that this stress is equal to the stress necessary to pass a dislocation through the point barriers without thermal agitation. It is to be noted that τ_0 in Friedel's case is taken as the internal stress independent of solution hardening.

The yield stress at $0^\circ K$ due to Friedel, therefore, should be written in terms of $\tau_c(0^\circ K)$, as given by Equation 13.13 as

$$\tau_F(0^\circ K) = \tau_c(0^\circ K) - \tau_0 = \tfrac{1}{2} \tau_c(0^\circ K). \tag{13.14}$$

This will be verified as follows. According to the first assumption by Friedel, it follows that

$$\tau_F(0^\circ K) b \, x \, y = 2E_3 y. \tag{13.15}$$

Therefore,

$$\tau_F(0^\circ K) = \frac{\mu b^2}{2} \frac{x}{y^2} = \frac{\mu b^3}{2} \frac{1}{\alpha c} \frac{1}{y^3}. \tag{13.16}$$

Friedel's second assumption gives

$$\tau_F(0°K) = V/b^2y. \tag{13.17}$$

Thus, from Equations 13.16 and 13.17, and using the relation $V = W/3$, it follows that

$$\tau_F(0°K) = \frac{1}{2} \left(\frac{2}{3}\right)^{3/2} \alpha^{1/2} \frac{W^{3/2}}{\mu^{1/2}b^{9/2}} c^{1/2}, \tag{13.18}$$

which satisfies Equation 13.14. In other words, Friedel's assumption gives the same result as the theory of the present authors, except for τ_0.

Based on Friedel's assumption, Foreman and Makin (1966) and also Kocks (1966) obtained theoretically the so-called FKFM relation for the yield stress at $0°K$. It is found that the above relation agrees with the results of the computer experiments of Kocks, and Foreman and Makin. They expressed both theoretical and experimental relations in terms of the angle made by the dislocation line at a point barrier. The above-mentioned computer experiments attempt to see directly from the static viewpoint the effect of the random distribution of point barriers instead of taking the average regular distribution such as those given by Equations 13.4 and 13.5.

According to experiments made on crystals of copper-nickel alloys (Suzuki 1968, Suzuki and Ishii 1968), the FKFM relation does not agree with the observed yield stress. One of the reasons is thought to be due to the τ_0 assumed by Friedel. It is found that τ_0 in dilute alloys is proportional to the square root of the alloy concentration. Thus, in good alloy crystals the internal stress cannot be independent of the alloy concentration. The second and rather more essential reason is related to the static approach used both in the theory and in the computer experiments. The yield stress of real crystals is defined by the dynamic condition of Equation 13.9. The velocity v in Equation 13.9 may not be the average velocity over the whole slip plane, but rather a local average velocity, for which experimental data were given in our previous paper (Suzuki and Ishii 1968). This is because the latter velocity is as high as 10^3 cm/sec, so that a very small fraction of the total mobile dislocations, called the effective number, is sufficient to produce a macroscopic strain rate equal to the external strain rate given by Equation 13.9. This is ignored not only in the derivation of the FKFM relation and in the computer experiments, but also in the derivation of Equations 13.10–13.13 in the present theory. To account for the observed strengths it is necessary to consider the dynamic effects of the random distribution of solute atoms.

13.3.3. Dynamic Effects of Random Distribution of Solute Atoms on Dislocation Mobility and Yield Stress of Real Alloy Crystals

In general, the yield stress σ_c of real alloy crystals is smaller than the yield stress τ_c for the regular lattice distribution of solute atoms in the mathematical model. For instance in the triangular lattice distribution given by (x, y),

$$\tau_c = q\sigma_c,\tag{13.19}$$

where q is a numerical factor larger than unity. In the following, the physical meaning of q is discussed and also determined semiempirically.

The random distribution of solute atoms in real crystals is assumed to be expressed by the spatial variation of the size of the triangular lattice (x, y) such as (qx, qy), where q varies slowly as a function of position. In other words, the triangular lattice varies its size by uniform expansion and contraction from place to place. A dislocation can move in the crystal through the expanded lattice of solute barriers bypassing the contracted lattice. This is similar to Orowan's (1948, 1954) idea for the motion of dislocations through precipitate particles. The maximum velocity of dislocations in the expanded lattice regions as defined in the previous paper (Suzuki and Ishii 1968) is larger by orders of magnitude than the average velocity to be defined over the whole area of the slip plane. The effective value of q to be considered is, accordingly, larger than unity. The present working hypothesis must be justified experimentally.

The real yield stress is, therefore, given by replacing x and y by qx and qy, respectively, in the former equations, and it follows that

$$\sigma_c = \sigma_0 + \frac{1}{q}\frac{1}{3}\left(\frac{2}{3}\right)^{1/2}\alpha^{1/2}\frac{W^{3/2}}{\mu^{1/2}b^{9/2}}c^{1/2}\left(1 - \frac{T}{T_c}\right)\tag{13.20}$$

and

$$\sigma_0 = \frac{1}{2}\sigma_c(0°\,\mathrm{K}),\tag{13.21}$$

where T_c is the same as given by Equation 13.10. The dislocation velocity of Equation 13.8 is also rewritten as

$$v = \frac{2}{3}\frac{W}{\mu b^2}v\exp\left(-\frac{W}{3kT}\right)\exp\left[\frac{(\sigma - \sigma_0)b^2 yq}{kT}\right].\tag{13.22}$$

13.3.4. *Comparison with Experiments*

Since comparison of the present theory with experiments on copper-nickel alloys was explained in detail in the previous papers, they are only briefly summarized here. Further a comparison concerning the dislocation velocity is newly added.

1. q and W: From the comparison of theory and experiment on the yield stresses at 0°K of dilute alloys of five different concentrations it follows[1] that

$$1/q = 0.13 \text{ and } W = 0.11 \text{ eV}.$$

[1] In previous papers, we determined $1/q = 0.11$, and $W = 0.10$ eV. By renewed examination after the addition of new data, however, the values are refined as described above.

Taking as examples the 0.17 percent and 0.35 percent alloys, a numerical comparison of yield stress at 0°K between theory (Equations 13.20, 13.21), and experiment is shown in Table 13.2.

TABLE 13.2. Comparison of yield stress at 0°K between theory and experiment

Concentration (%)	Calculated* (g/mm²)	Observed (g/mm²)
0.17	90.7	106
0.35	130	160

* Calculation is made by the use of above-described q and W.

It is found that W thus determined is in good agreement quantitatively with the theory of elastic-size and modulus-misfit interactions (Fleischer 1963).

2. *Critical Concentration c_0*: The present theory predicts the yield stress proportional to the square root of the solute concentration. This holds for alloys below the critical concentration c_0. Theory predicts as $c_0 = 1.1 \times 10^{-2}$, while experiment shows as $c_0 = 3 \times 10^{-2}$.

3. *Critical Binding Energy W_{max}*: The present theory should hold for $W \leq W_{max}$, where W_{max} is found to be about 1 eV for copper alloys. Hence, the present theory is applicable for almost all substitutional alloys of copper. For instance, from comparison with the data on copper-aluminum alloys obtained by Koppenaal and Fine (1962), it gives $W = 0.23$ eV. This coincides with the theoretical value equal to 0.24 eV calculated by assuming the elastic-size effect.

4. *Dislocation Velocity as a Function of Stress*: Values of the velocity calculated from Equation 13.22, using $q = 7.69$, $W = 0.11$ eV, $c = 0.35$ percent, and $T = 77°K$ are shown by the solid line in Figure 13.4, which agrees amazingly well with experimental results (Suzuki and Ishii 1968) expressed by white circles. The calculated curve for 0.17 percent alloy and for the phonon-drag mechanism in the same alloy at 77°K are also illustrated in the same figure, which will be explained later.

The agreement between theory and experiment as seen in Figure 13.4 suggests that the dislocation velocity below the yield point is controlled by randomly distributed solute atoms, and that the interaction of a moving dislocation with these solute atoms can be described satisfactorily in terms of the quasistatic theory. This is also true for the yield stress at least above 100°K. It is of great interest, however, to note that the observed yield stress below 100°K becomes smaller than the quasistatic yield stress extrapolated from higher temperatures. This anomalous decrease of yield stress below

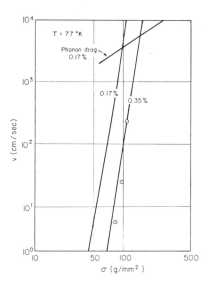

FIGURE 13.4. Comparison between theory and experiment on dislocation velocity as a function of stress in copper-nickel alloys. Open circles are measured points and solid curves are calculated as described in the text.

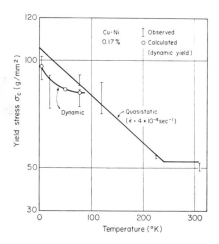

FIGURE 13.5. Yield stress as a function of temperature. The solid line is calculated in terms of the quasistatic theory, while the separate solid curve at lower temperatures is calculated for dynamic yielding. Open circles are calculated points as described in connection with Figure 13.9. The vertical lines indicate measured yield stresses, the lower bounds of which are proportional limits and the upper bounds of which are the upper yield stresses or macroscopic yield points determined coventionally, being the intersections of the extrapolations of stress-strain curves from both sides.

100°K produces a peak in the curve of yield stress plotted against temperature. The peak becomes more distinct as the strain rate increases.

Figure 13.5 is an illustration of the yield stress plotted as a function of temperature for the 0.17 percent alloy. The thin solid line is the theoretical curve corresponding to Equation 13.20, where the curve is displaced to fit the observed yield stress at 0°K as listed in Table 13.2. The observed value is designated by a vertical segment of a line, of which the lower bound is the proportional limit and the upper bound is the upper yield stress, or the macroscopic yield stress conventionally defined by the intersection between extrapolations of the stress-strain curve from above and below the yield point. As seen clearly in Figure 13.5, the observed yield stresses become systematically lower than the quasistatic values.[2]

In the following, it will be shown that the above anomaly is not due to the kind of the static interactions between dislocations and solute atoms, but due to the dynamic nature of the interaction.

13.4. Dynamic Resistance for the Motion of Dislocations

13.4.1. General

The dynamic resistances for the motion of dislocations which have been discussed theoretically in the past are the phonon drag (Leibfried 1950, Mason 1960, 1964, and others), the electron drag (Mason 1965, 1966, Huffman and Louat 1967, and others), and the impurity drag (Ookawa and Yazu 1963, and Takamura and Morimoto 1963). Apart from the etch-pit measurements of high-speed dislocations in LiF, Cu, Cu-Ni and Zn already mentioned above, experimental measurements of the phonon and electron drags have also been made via internal friction measurements (Stern and Granato 1962, Alers and Thompson 1961, Hutchison and Rogers 1962, Mason 1966, 1967, Suzuki, Ikushima, and Aoki 1964, and Hikata and Elbaum 1967). There seems to be only one flow stress experiment showing the effect of the phonon drag, which is a study of high speed deformation of aluminum by Ferguson, Kumar, and Dorn (1967). Very recently, a direct phenomenon reflecting the dynamic resistance due to the electron drag in the strength of metals has been observed on lead and niobium by Kojima and Suzuki (1968). They observed an increase of flow stress in both types of superconducting metals by the application of an external magnetic field above their critical magnetic field at 4.2°K. These superconductors show a smaller flow stress in the superconducting state than in the normal state. This is no doubt due to the drag of conduction electrons.

---

[2] Recently, Kuramoto and H. Suzuki (1968) (private communication) have also found a similar anomalous behavior of flow stress at low temperatures in copper-aluminum alloys.

Since the phonon drag rapidly decreases at lower temperatures, the dynamic resistances remaining at lower temperatures in good crystals of low dislocation density are due to conduction electrons and impurities. According to theories, the electron drag is inversely proportional to the electrical resistivity. In dilute alloys, therefore, the electron drag is generally small as compared to the dynamic resistance due to solute atoms.

13.4.2. Dynamic Resistance Due to Solute Atoms

A theory of the dynamical interaction between a moving dislocation and solute atoms was given by Ookawa and Yazu (1963). Although this is a classical treatment of the problem, the result is not so much different quantitatively from the wave mechanical treatment according to Ninomiya (private communication). For discussion of the transition of the dislocation motion from a quasistatic to a dynamic case, the Ookawa-Yazu theory will be extended as follows.

When an edge dislocation moving at a speed equal to v_0 meets a point barrier, the part of the dislocation in the local field of range equal to $\lambda(\sim 2b)$ is accelerated or decelerated, and it emits elastic waves. A part of the kinetic energy is thus lost. Take the dislocation line as the ζ axis, the normal of the slip plane as the η axis, and the normal to the line in the slip plane as the ζ axis. The energy lost during the pass through the local field is given by Ookawa and Yazu as

$$\Delta E = \frac{1}{10\pi} \frac{b^2 \rho}{v_c} \int_{\Delta t} \left[\int_\lambda \dot{v}(\xi, t) \, d\xi \right]^2 dt, \qquad (13.23)$$

where v_c is the velocity of the shear wave, ρ the density of material, and t is the time. Δt is given by λ/v_0.

The disturbance due to the point barrier propagates along the dislocation line at the shear-wave velocity. When the drift velocity of the dislocation v_0 is sufficiently fast, the dislocation passes through the effective range of the local field due to a solute atom before the above disturbance travels over the distance $2y$ between neighboring local fields, as shown in Figure 13.6. In this case, the quasistatic description of the interaction with solute

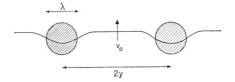

FIGURE 13.6. Schematic illustration of a dislocation line passing dynamically through the local fields due to solute atoms. λ is the effective range of the local field, and v_0 is the average drift velocity of the dislocation.

atoms can no longer be valid. The critical velocity of the dislocation is thus given by[3]

$$v^* = \frac{\lambda/2}{2y} v_c. \tag{13.24}$$

Now, taking the force due to a solute atom as F_ζ, it follows that

$$\int_\lambda \dot{v}\, d\xi = \frac{1}{m} \int_\lambda F_\zeta(\xi, \zeta)\, d\xi,$$

where m is the effective mass of the dislocation. If we assume that F_ζ is due to the elastic size effect, it is given by

$$F_\zeta = \frac{4}{3} Wb \frac{\eta\zeta}{(\xi^2 + \eta^2 + \zeta^2)^2}. \tag{13.25}$$

The solute atom of interest is at a postion $(0, \eta, 0)$ and η is taken as equal to $b/2$. The integration of Equation 13.23 gives

$$\Delta E \simeq \frac{1}{10\pi} \frac{W^2}{\beta^2 \mu b^3} \frac{v_c}{v_0}, \tag{13.26}$$

where β is a factor of the order of unity which is involved in the Frank relation (Frank 1949) such as for $m = \beta\mu b^2/v_c^2$. In conclusion, the stress σ_d necessary to keep the drift velocity v_0 constant is given from $-dE/dx = \Delta E/2xy = b\sigma_d$ and $xy = b^2/\alpha c$ by

$$\sigma_d = \frac{1}{20\pi} \frac{\alpha}{\beta^2} \left(\frac{W}{\mu b^3}\right)^2 \mu \frac{v_c}{v_0} c. \tag{13.27}$$

As Equation 13.27 shows, the dynamic resistance is inversely proportional to the drift velocity of the dislocation, independent of temperature and proportional to the solute concentration. Putting $\beta = 1$, $W = 0.11$ eV, and $c = 0.17$ percent, Equation 13.27 gives

$$\sigma_d = 4.1 \times 10^9/v. \tag{13.28}$$

On the other hand, taking $v_c = 1.62 \times 10^5$ cm/sec (Kittel 1956), the critical velocity is calculated from Equation 13.24 as, shown in Table 13.3. The critical velocity is proportional to the square root of the solute concentration. The ratio $\sigma_d(v^*)/\sigma_c(0°K)$ between the dynamic stress at the critical velocity and the quasistatic stress at $0°K$ is independent of the concentration and found to be 0.55. Further, it should be remarked that the phonon drag due to the uniform motion of the dislocation must be taken into account to estimate the total dynamic resistance except for very low temperatures.

[3] For the dynamic case, the zipper effect will be no longer important. Hence, the effective distance between the neighboring barriers along the dislocation line is given by $2y$ where y is given by Equation 13.5.

TABLE 13.3. The critical velocity of a dislocation for the transition from quasistatic to dynamic motion.

Concentration (%)	$y \times 10^6$ cm	v^* cm/sec	$\sigma_d(v^*)/\sigma_c(0°K)$
0.17	2.5	8.4×10^2	0.55
0.35	1.8	1.2×10^3	

13.4.3. Dynamic Yielding: Comparison between Theory and Experiment

From the fact that the critical resistance $\sigma_d(v^*)$ is only half the quasistatic yield stress at 0°K, it can be expected that the motion of a dislocation at lower temperatures begins to be controlled by the dynamic interaction with solute atoms at a certain stress below the quasistatic yield stress. Furthermore, following the nature of this dynamic interaction, the acceleration of dislocations can be achieved without an increase of stress as explained by Equations 13.27 or 13.28. As soon as the unstable motion of dislocations starts, the specimen crystal yields. This definition of yielding, which is called dynamic yielding, is used here in a narrow sense to discriminate from quasistatic yielding.

Figure 13.7 is a schematic drawing of the dynamic yielding at low temperatures, where the phonon drag and electron drag can be neglected.

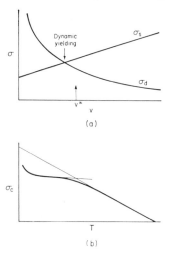

(a)

(b)

FIGURE 13.7. Schematic illustration of the relation between dislocation velocity and stress, σ_s vs. v is the quasistatic relation, σ_d vs. v is the dynamical relation, and v^* is the critical velocity. The intersection gives a dynamic yield stress, where the phonon drag can be neglected (a). Yield stress σ_c is plotted against temperature (b), in which it is shown that the yield stresses at lower temperatures become smaller than the quasistatic values on account of dynamic yielding.

Figure 13.7a shows the relation between stress and dislocation velocity, where σ_s and σ_d are the quasistatic and dynamic resistive stresses, respectively. The figure corresponds to a case where the velocity at the intersection between σ_s and σ_d curves is smaller than the critical velocity v^*. When v^* is smaller than the former, the stress at the intersection can be simply taken as a dynamic yield point. Figure 13.7b is the diagram of yield stress vs. temperature, in which the observed yield stresses are expected to vary along the thick line.

As explained before, the critical velocity v^* is proportional to $\sqrt{c}$, while the quasistatic velocity at a quasistatic yield point as a function of temperature is independent of the concentration. Therefore, the dynamic yielding will cease to occur in alloys of high concentration. The concentration at which the dynamic yielding disappears of course depends on the strain rate used.

Figure 13.8 illustrates the case where phonon drag cannot be neglected. It will be valid at intermediate low temperatures. The dynamic resistance is simply taken as a resultant of the dynamic resistance due to solute

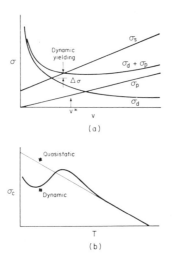

FIGURE 13.8. Schematic illustration of the relation between dislocation velocity and stress. In contrast to Fig. 7, here the phonon drag σ_p is taken into account. It gives rise to an increase of yield stress by $\Delta\sigma$ as illustrated (a). Yield stress σ_c is plotted against temperature (b).

atoms σ_d and the phonon drag σ_p. The intersection generally occurs at a stress higher by $\Delta\sigma$ than that in Figure 13.7. Correspondingly, a peak will appear in the curve of yield stress plotted against temperature. This peak will become distinct as the test is performed at higher strain rates.

Figure 13.9 shows theoretical curves, calculated for a copper-0.17-percent nickel alloy at 77°K, in which σ_s is the same as already given in

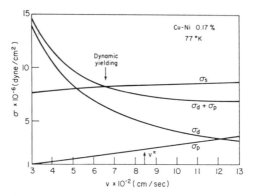

FIGURE 13.9. Calculation of the relations between dislocation velocity and stress in case of copper-0.17% nickel alloy at 77°K.

Figure 13.4, σ_d is calculated from Equation 13.28 and σ_p calculated by using $B = 0.8 \times 10^{-4}$ cgs. The drag coefficient used is an observed value for a copper-0.13% manganese alloy (Suzuki et al. 1964). Judging from the comparison of room-temperature values between copper-nickel and copper-manganese alloys, this value may be slightly larger for the copper-nickel alloy. The result shown in Figure 13.9 corresponds to Figure 13.8.

Comparison between theory and experiment on the relation between yield stress and temperature is illustrated in Figure 13.5. The straight solid line is the theoretical curve for quasistatic yielding as described in Section 13.4, and the curved solid line at lower temperatures corresponds to the

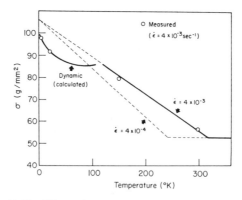

FIGURE 13.10. Effect of strain rate on the yield stress as a function of temperature. Open circles are measured at the strain rate equal to 4×10^{-3} sec^{-1}. Note that yield stresses at lower temperatures are independent of strain rate as expected from the theory of dynamic yielding. The yield stresses at higher temperatures increase with increasing strain rate as expected from the quasistatic theory (solid and dotted curves). The curve for $\dot{\varepsilon} = 4 \times 10^{-4}$ sec^{-1} is the same curve as shown in Figure 13.5.

theoretical curve for dynamic yielding. Phonon drag is neglected below 50°K. The theoretical stresses indicated by open circles for dynamic yielding are obtained from the intersections as shown in Figure 13.9. The effects of strain rate on yield stress is shown in Figure 13.10. Two dotted curves are related to quasistatic yield stresses. The lower is the same as illustrated in Figure 13.5, i.e., corresponding to $\dot{\varepsilon} = 4 \times 10^{-4}$ sec^{-1}, and the upper curve is for $\dot{\varepsilon} = 4 \times 10^{-3}$ sec^{-1}. Open circles are observed values measured at $\dot{\varepsilon} = 4 \times 10^{-3}$ sec^{-1}. It is as expected that measured points at lower temperatures do not depend on strain rate and remain on the theoretical curve for dynamic yielding indicated by a solid curve, which is the same as shown in Figure 13.5. While, measured points at higher temperatures move to higher stress levels as the strain rate increases, they lie on the curve as expected from the quasistatic theory.

13.5. Conclusions

It has been established by recent measurements that the mobilities of dislocations in soft metals of fcc and hcp structures are very high in comparison to those in other metals. As direct effects on mechanical properties of this high velocity, two distinct phenomena relating to yield stress are described theoretically and compared with experiments made on dilute copper-nickel alloys. The quantitative agreement between theory and experiment is quite satisfactory.

The yield stress of real crystals, σ_c, and the theoretical stress, τ_c, obtained from a proper average distribution (e.g., the triangular lattice distribution given by Equations 13.4 and 13.5) for random point obstacles are related to each other by $\tau_c = q\sigma_c$, where $q = 7.69$. The computer experiments of Kocks or Foreman and Makin do not give σ_c but τ_c because of their static nature of treatment. In the present paper it is discussed that the factor q that determines the yield stress of real crystals is due to one of the dynamic effects on the dislocation velocity.

The second effect is due to a dynamic interaction between moving dislocations and isolated point obstacles.

This is expected in soft impure metals and alloys at lower temperatures. Before the quasistatic yield points are attained, dislocations reach higher speeds than the critical speed, which leads them to an unstable motion and, accordingly, to dynamic yielding. The dynamic yield stress depends on phonon drag in addition to dynamic resistance due to solute atoms discussed above. Accordingly, it is expected that a peak appears at an intermediate temperature in the σ_c vs. temperature curve.

REFERENCES

Alers, G. A., and D. O. Thompson, 1961, *J. Appl. Phys.*, **32**, 283.
Argon, A. S., and W. T. Brydges, 1968, *Phil. Mag.*, **18**, 817.
Basinski, Z. S., and S. J. Basinski, 1964, *Phil. Mag.*, **9**, 51.

Chaudhuri, A. R., J. R. Patel, and L. G. Rubin, 1962, *J. Appl. Phys.*, **33**, 2736.

Cottrell, A. H., 1953, *Dislocations and Plastic Flow*, Oxford: Clarendon Press.

Ferguson, W. G., A. Kumar, and J. E. Dorn, 1967, *J. Appl. Phys.*, **38**, 1863.

Fleischer, R. L., 1962, in *Strengthening Mechanisms in Solids*, ASM, Metals Park, p. 93.

Fleischer, R. L., 1963, *Acta Met.*, **11**, 203.

Foreman, A. J. E., and M. J. Makin, 1966, *Phil. Mag.*, **13**, 911.

Frank, F. C., 1949, *Proc. Phys. Soc.*, **A62**, 131.

Friedel, J., 1956, *Les Dislocations*, Paris: Gauthier-Villars.

Friedel, J., 1962, in G. Thomas and J. Washburn (eds.), *Electron Microscopy and Strength of Crystals*, New York: Wiley-Interscience, p. 605.

Friedel, J., 1964, *Dislocations*, Reading, Mass.: Addison-Wesley.

Greenman, W. F., T. Vreeland, Jr., and D. S. Wood, 1967, *J Appl. Phys.*, **38**, 3595.

Gutmanas, E. U., E. M. Nadgornyi, and A. V. Stepanov, 1963, *Soviet Phys. Solid State*, **5**, 743.

Hikata, A., and C. Elbaum, 1967, *Phys. Rev. Letters*, **18**, 750.

Huffman, G. P., and N. P. Louat, 1967, *Phys. Rev. Letters*, **19**, 518.

Hutchison, T. S., and D. H. Rogers, 1962, *J. Appl. Phys.*, **33**, 792.

Johnston, W. G., and J. J. Gilman, 1959, *J. Appl. Phys.*, **30**, 129.

Kabler, M. N., 1963, *Phys. Rev.*, **131**, 54.

Kittel, C., 1956, *Introduction to Solid State Physics*, 2nd ed., New York: Wiley.

Kocks, U. F., 1966, *Phil. Mag.*, **13**, 541.

Kojima, H., and T. Suzuki, 1968, *Phys. Rev. Letters*, **21**, 13.

Koppenaal, T. J., and M. E. Fine, 1962, *Trans. Met. Soc.* AIME, **224**, 347.

Kuramoto, E., and H. Suzuki, 1968, private communication.

Leibfried, G., 1950, *Z. Physik*, **129**, 344.

Marukawa, K., 1967, *J. Phys. Soc. Japan*, **22**, 499.

Mason, W. P., 1960, *J. Acoust. Soc. Amer.*, **32**, 458.

Mason, W. P., 1964, *J. Appl. Phys.*, **35**, 2779.

Mason, W. P., 1965, *Appl. Phys. Letters*, **6**, 111.

Mason, W. P., 1966, *Phys. Rev.*, **143**, 229.

Mason, W. P., 1967, *J. Appl. Phys.*, **38**, 1929.

Mihara, M., and T. Ninomiya, 1967, private communication.

Mitchell, T. E., and P. R. Thornton, 1963, *Phil. Mag.*, **8**, 1127.

Mott, N. F., 1952, in W. Shockley, *et al.* (eds.) *Imperfections in Nearly Perfect Crystals*, New York: Wiley, p. 173.

Ninomiya, T., 1968, private communication.

Ookawa, A., and K. Yazu, 1963, *J. Phys. Soc. Japan*, **18**, Supplement 1, 36.

Orowan, E., 1948, in *Symposium on Internal Stress in Metals and Alloys*, London: Institute of Metals, p. 451.

Orowan, E., 1954, *Dislocations in Metals*, New York: AIME, p. 131.

Pope, D. P., T. Vreeland, Jr., and D. S. Wood, 1967, *J. Appl. Phys.*, **38**, 4011.

Prekel, H. L., and H. Conrad, 1967, *Acta Met.*, **15**, 955.

Rhode, R. W., and C. H. Pitt, 1967, *J. Appl. Phys.*, **38**, 876.

Rosenfield, A. R., and B. L. Averbach, 1960, *Acta Met.*, **8**, 625.

Schadler, H. W., 1964, *Acta Met.*, **12**, 861.

Schüle, W., O. Buck, and E. Köster, 1962, *Z. Metallk.* **53**, 172.

Seeger, A., 1954a, *Phil. Mag.*, **45**, 771.

Seeger, A., 1954b, *Z. Naturforsch.*, **9a**, 758, 856, 870.

Stein, D. F., and J. R. Low, Jr., 1960, *J. Appl. Phys.*, **31**, 362.

Stern, R. M., and A. V. Granato, 1962, *Acta Met.*, **10**, 358.

Suzuki, H., S. Ikeda, and S. Takeuchi, 1956, *J. Phys.*, *Soc. Japan*, **11**, 382.

Suzuki, T., 1968, in A. R. Rosenfield et al. (eds.), *Dislocation Dynamics*, New York, McGraw-Hill, p. 551.

Suzuki, T., A. Ikushima, and M. Aoki, 1964, *Acta Met.*, **12**, 1231.

Suzuki, T., and T. Ishii, 1968, *Proc. Int'l Conf. on Strength of Metals and Alloys*, *Trans. Jap. Inst. Met.*, **9** (Suppl.), p. 687.

Suzuki, T., and H. Kojima, *Acta Met.*, **14**, 913.

Takamura, J., and T. Morimoto, 1963, *J. Phys. Soc. Japan*, **18**, Supplement 1, 28.

Thornton, P. R., T. E. Mitchell, and P. B. Hirsch, 1962, *Phil. Mag.*, **7**, 337.

Young, F. W., Jr., 1961a, *J. Appl. Phys.*, **32**, 1815.

Young, F. W., 1961b, *J. Appl. Phys.*, **32**, 963.

Young, F. W., 1967, in A. R. Rosenfield et al. (eds.), *Dislocation Dynamics*, New York, McGraw-Hill, p. 313.

ABSTRACT. The pure "chemical" hardening, due to coherent spherical precipitates with the same crystal structure but different bonding energy than the matrix, is studied as a function of temperature. It is shown that Orowan's law should be followed at low temperatures for large interparticle distances.

14. On the Chemical Hardening by Coherent Precipitates

J. FRIEDEL

14.1. Introduction

It was first pointed out by Orowan that the presence in a plastic matrix of finely distributed and hard precipitates should produce a typical increase σ of the elastic limit. For precipitates of size D at average distance L along a moving dislocation,

$$\sigma \equiv \frac{2\tau}{b(L - D)}.$$

(14.1)

τ is here the line tension of the dislocation ((Figure 14.1).

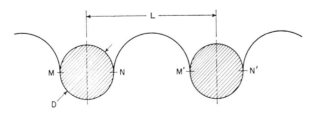

FIGURE 14.1. Definition of particle diameter D and distance L along a dislocation line.

The way L is related to the average distance Λ between precipitates has been extensively studied. If one neglects long-range interactions between dislocation and precipitates, $L^2 D \simeq \Lambda^3$ to a good approximation. Thus

$$\sigma \simeq \frac{2\tau f^{1/2}}{bD(1 - f^{1/2})}, \tag{14.2}$$

where $f = (D/\Lambda)^3$ gives the volume fraction of precipitates.

In order to reduce D and thus increase σ, it is tempting to use solid solutions able to produce coherent precipitates by spinodal decomposition: their easy nucleation produces precipitates with an average size D varying over wide limits down to atomic sizes.

Coherent precipitates have however, by definition, at least some atomic planes in common with the matrix. In simple cases such as **Al Ag**, the precipitates have indeed all their slip systems in common with the matrix. In most cases, some common slip systems do exist. Except for large Peierls frictional force in the precipitates, the dislocation loops MN dragging behind a precipitate can then slip through it, if the applied stress is large enough. This possibility of slip through precipitates will clearly limit the possible hardening by precipitates.

The difficulty for dislocations to cut through precipitates can be due to various factors: the "chemical" factor, which stabilizes the precipitate and thus opposes its shearing; an elastic factor and a size factor, due respectively to differences in elastic constants and in lattice parameters between the precipitates and the matrix; and eventually a higher Peierls stress or a stress due to ordering in the precipitate.

This note assumes the *chemical factor* to be predominant, and the precipitates to be *spherical*. This probably applies to a good approximation to **Al Ag** alloys. We want to point out that the analysis of hardening is not quite straightforward in such a case. Similar but more obvious complications have been discussed at some length when precipitates are thin platelets. More precisely, we want to show that the chemical factor makes it difficult for edge dislocations to penetrate into large precipitates, at least at low temperatures. The resulting increase in elastic limit is larger than is usually assumed. We shall first consider processes at $0°K$.

14.2. Hardening at 0°K

When a dislocation slips through a precipitate, the area of contact between the precipitate B and the matrix A is extended by edges E, E' (Figure 14.2). The surface tension γ that opposes the creation of these edges is related to the free energy of precipitation. It is typically of the order of 100 ergs/cm^2; then $\gamma b \simeq 1/50\tau$.

We want to show that this tension γ prevents the loops MN Figure 14.1

to penetrate into large precipitates. Under an increasing applied stress, these large precipitates will then be bypassed before they are sheared, and Orowan's law (Equation 14.1) will hold.

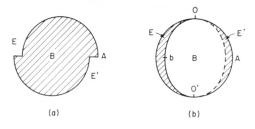

(a)   (b)

FIGURE 14.2. Edges $E\,E'$ of a particle sheared by a dislocation.

We must show that, under the critical condition of Figure 14.1, there is no tendency for the dislocation line to penetrate into the precipitate. We shall consider two types of penetration: *local* (Figure 14.3) and *general*

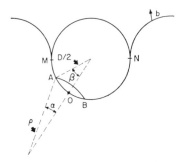

FIGURE 14.3. Local penetration.

(Figure 14.4). We shall show that a local penetration is never possible at 0°K, while a general one is possible only for small precipitates.

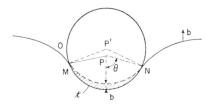

FIGURE 14.4. General penetration.

A *local* penetration would be most easy around point O, where the dislocation has a pure screw character, and thus only infinitesimal edges would be created. Under the applied stress σ, the small dislocation loop

AB would take a radius of curvature $\rho = 1/2(L - D)$ (Figures 14.1 and 14.3). The change of energy of the loop when it penetrates into the precipitates is easily obtained as

$$\Delta E(\beta) = \tfrac{1}{2}\gamma b D \beta^2 - \tfrac{1}{6}D\tau\beta^3 + O_4(\beta),\tag{14.3}$$

where 2β is the arc of local penetration. This energy is positive for small θ's; local penetration is impossible at 0°K.

In a *general* penetration, as pictured in Figure 14.4, the dislocation slips into the precipitate over a wide angle 2θ. The decrease in line tension energy can then help the penetration significantly for small precipitates. Penetration can be assumed effective when the central part of the loop MN has moved forward by a distance b. We shall obtain an upper limit of the change of energy $\Delta E'$ involved in this motion by assuming the loop MN to have edge character in its middle (Figure 14.4); this is actually an order of magnitude for the general case. Taking a circular form for the final position, one then finds, for large angles θ,

$$\Delta E' = b(\sin\theta - \theta\cos\theta)(\gamma D - \sigma b D - 2\tau).$$

The dislocation penetrates into the precipitates at 0°K if $\Delta E' < 0$. This occurs for stresses larger than

$$\sigma' = \frac{1}{b}\left(\gamma - \frac{2\tau}{D}\right).\tag{14.4}$$

Figure 14.5 compares this stress σ' to cut through the precipitates to

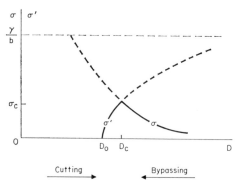

FIGURE 14.5. Orowan stress $\sigma(D)$ fand shear stress $\sigma'(D)$ for penetration.

the Orowan stress σ to bypass them. The abscissae give the size D of the precipitates, at constant volume fraction $f = (D/\Lambda)^3$.

Cutting through precipitates should therefore become easier than bypassing them when precipitation occurs on a fine enough scale: $D \le D_c$. Hardening should be maximum at the boundary between the two types of processes: $\sigma \leq \sigma_c = \sigma(D_c)$. For not too large volume fractions of precipitates, Equations 14.1 and 14.4 give

$$D_c \simeq D_0 = 2\tau/\gamma \tag{14.5}$$

and

$$\sigma_c = 2\tau f^{1/2}/bD_c = \gamma f^{1/2}/b. \tag{14.6}$$

As emphasized earlier, these formulae give only an order of magnitude. Equation (14.4) for $\sigma'(D)$ is valid only for large angles θ, thus near to D_c.

14.3. Hardening at Finite Temperature

The previous analysis, leading to Figure 14.5, is valid only at 0°K. At finite temperatures, dislocations can penetrate into the precipitates by thermal activation at stresses below σ, especially if they have nearly screw character.

14.3.1. *Thermally Activated Hardening*

If the applied stress σ is large enough for the loops MN to contain the point O of pure screw character (Figures 14.3 and 14.6), penetration into

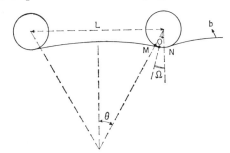

FIGURE 14.6. Limiting stress and angle Ω for easy thermal activation of penetration.

the precipitates is practically immediate except at very low temperatures. The condition for this to occur is $\theta \geqslant \Omega$ (Figure 14.6), thus

$$\sigma \geq \frac{2\tau \sin \Omega}{b(L - D \sin \Omega)}. \tag{14.7}$$

Penetration is then easily activated thermally by the local process of Figure 14.3. It is easy to show that Equation (14.4) can be used for any stress σ below Orowan's stress (Equation 14.1). At a critical angle

$$\theta_c = 2\gamma b/\tau, \tag{14.8}$$

this energy goes through a maximum value

$$\Delta E_c = \frac{2\gamma^3 b^3 D}{3\tau^2}. \tag{14.9}$$

These quantities are very small indeed for reasonable values of the chemical energy γ and the size D of precipitates. Thus $\gamma b/\tau \simeq 1/50$ and $D < 200\,b$ lead to $\theta_c \simeq 1/25$ and $\Delta E_c/k_B T \geqslant 1$ if $T < 10°K$.

If, on the other hand, the condition of Equation 14.7 is *not* fulfilled (Figure 14.7), a thermally activated penetration requires a definitely

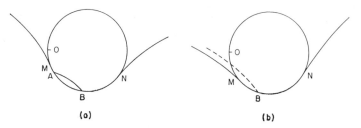

FIGURE 14.7. Two types of thermal activation of penetration at lower stress.

higher energy, because more edge energy is spent in the activation. Estimates for both kinds of processes pictured in Figures 14.7a and b lead to energies larger by a factor of the order of $(\tau/\gamma b)^{3/2} \simeq 500$ than the estimate of Equation 14.9, thus in the range of one eV.

In conclusion, nearly screw dislocations, for which angle Ω is small, will penetrate easily into precipitates under small stresses σ given by Equation 14.7. Except at very high temperatures, dislocations with more edge character penetrate appreciably into precipitates only under stresses σ which, according to Equation 14.7, are not much below Orowan's criterion (Equation 14.1). *Thus, at least for large precipitates $D > D_c$ where it is valid at 0°K, the Orowan criterion should remain approximately valid for most dislocations up till fairly high temperatures.*

14.3.2. *Temperature-Independent Hardening for Screw Dislocations*

Screw dislocations, which are expected to penetrate easily into precipitates by thermal activation, should experience a further temperature-independent hardening when crossing them. It is clear that an equilibrium configuration can be reached under a small applied stress when the dislocation has a position such as pictured in Figure 14.8: the small kinks produced at the

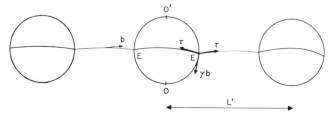

FIGURE 14.8. Athermal frictional force for screw dislocations cutting through precipitates.

edges of the precipitates allow the line tensions τ of the dislocation to equilibrate the chemical tension γb of the edges EE'.

As $\gamma b \ll \tau$, an order of magnitude of the friction stress can be obtained by considering a straight dislocation and balancing the energy gained under the applied stress by the energy spent to extend the edges. This gives

$$\sigma'' = 2\gamma/L'. \tag{14.10}$$

L' is here the average distance between precipitates along a *straight* dislocation. Thus $L'D^2 \simeq \Lambda^3$, and

$$\sigma'' = 2\gamma f/D \tag{14.11}$$

where $f = (D/\Lambda)^3$ is again the volume fraction of precipitates.

This stress is much smaller than Orowan's criterion, since $\gamma b \ll \tau$. Thus, nearly screw dislocations should pass precipitates at a much reduced stress. But the criterion of Equation 14.11 applies only when below the limit given by Equation 14.7, thus for very nearly screw dislocations:

$$\Omega < \gamma b f^{1/2}/\tau \simeq 1/100.$$

In conclusion, we see that the usual criterion of Equation 14.11 for chemical hardening should give only a lower limit for all but nearly screw dislocations. The real hardening should in general show a maximum, at low temperatures, for a critical size D_c; it should decrease somewhat with increasing temperature, but with quite appreciable activation energies.

ABSTRACT. The mechanism of plastic deformation of two-phase alloys containing small nondeformable particles in a ductile matrix is discussed. The effect of cross-slip on the Orowan bypass mechanism is considered; the experimentally observed microstructure and its variation with alloy composition is interpreted in terms of the cross-slip mechanism. The factors controlling the yield stress of the undeformed material, and the flow stress of the work-hardened crystals, are reviewed. Work hardening for small strains is considered to be due to the interaction between screw dislocations and prismatic loops with the primary Burgers vector to form helices. A theory is developed applicable to very small strains. Possible mechanisms to account for the strong recovery effect reported for copper-base alloys at and above room temperature are considered. Above about 100°C the density of loops and the work-hardening rates are very small. This effect is interpreted in terms of a pipe-diffusion mechanism.

15. Plastic Deformation of Two-Phase Alloys Containing Small Nondeformable Particles

P. B. HIRSCH and F. J. HUMPHREYS

15.1. Introduction

In 1948 Orowan made an important contribution to the theory of the yield stress of an alloy consisting of nondeformable particles of a second phase in a ductile matrix (Orowan 1948). He proposed that if the stress is large enough to bend the dislocations in the matrix into a semi-circular arc between the particles (Figure 15.1), the dislocations can bypass the particles, leaving behind glide loops encircling the particles. Orowan suggested that the yield stress τ of the alloy should be given by

$$\tau = \tau_m + \frac{2T}{bD_s} \tag{15.1}$$

where τ_m is the yield stress of the matrix, T is the line tension, b is the Burgers vector, D_s is the mean planar spacing of the particles. Taking $T = Gb^2/2$, where G is the shear modulus of the matrix, then

$$\tau = \tau_m + \frac{Gb}{D_s}. \tag{15.2}$$

Since this theory was first put forward, the following relevant developments have taken place:

189

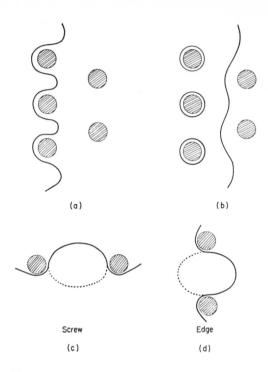

FIGURE 15.1. The Orowan mechanism.

1. The theory itself has been modified by taking into account the effect of the variation of line tension with dislocation character on the shape of the dislocation loop (Ashby 1968), the interaction of the two arms of the dislocation on opposite sides of the particle (Ashby 1968), and a statistical treatment of the effective spacing between obstacles along the dislocation (Kocks 1966, Foreman and Makin 1966).

For edge and screw dislocations the "bypass" stress (for a regular lattice of obstacles) can be obtained very simply by realizing that for these two symmetrical configurations, the stress must be sufficient for the dislocation to support a dipole consisting of the two arms of the dislocation on opposite sides of the particle (diameter $2R$), i.e., for edges, assuming isotropic elasticity,

$$\tau_e = \tau_m + \frac{Gb}{2\pi D_0} \ln \frac{2R}{r_0} \tag{15.3}$$

and for screws,

$$\tau_s = \tau_m + \frac{Gb}{2\pi(1-v)D_0} \ln \frac{2R}{r_0} \tag{15.4}$$

where D_0 is the appropriate interparticle spacing, and r_0 is the effective core radius, assumed to be the same for edges and screws, and v is Poisson's ratio. Thus, as pointed out by Ashby (1968), if $v = 1/3$, then

$$\tau_s/\tau_e \sim (1 - v)^{-1} = 1.5.$$

For copper, if the anisotropy is taken into account, this ratio is increased to 1.76.

Assuming a random array of point obstacles, the statistical theory of Kocks (1966) and the computer calculatoins of Foreman and Makin (1966) show that $D_0^{-1} = 0.842N_s^{1/2}$ or $0.81\ N_s^{1/2}$, respectively, where N_s is the number of particles intersecting a unit area of the slip plane. If $2R \ll D_0$, it is reasonable to use these expressions for D_0, and corrections for a finite particle size can be made (Kocks 1969). It should be pointed out, however, that the statistical theory has been developed on the assumption that the line tension of the dislocation is independent of dislocation character. It does not necessarily follow that the difference between the yield stress for edges and screws for a random array of obstacles is the same as for a regular lattice. This point will be considered again in Section 15.2.5.

Ebeling and Ashby (1968) and Jones and Kelly (1968) obtained good agreement between the yield stress of single crystals of copper containing silica particles and beryllia particles respectively, and the Orowan theory as modified by Ashby (1968)—the yield stress values lying between the values expected for edges and screws if isotropic elastic constants are used, and being closer to those for screws if anisotropy is taken into account.

2. It was suggested by Hirsch (1957) that dislocations can bypass particles by cross-slip. There is now considerable experimental evidence that this process takes place in a number of copper alloys containing different types of particles (Humphreys and Martin 1967, Humphreys, to be published).

It is the object of this paper to discuss the conditions under which bypassing by the Orowan mechanism or by cross-slip takes place, the effect of cross-slip on the yield stress, the microstructure to be expected if cross-slip occurs, and the hardening effect of this microstructure. The possibility of climb processes will also be considered.

15.2. The Cross-Slip Mechanism

15.2.1. Geometry of the Cross-Slip Mechanism

We shall consider first the basic mechanism for edge and screw dislocations. In Figure 15.2 an edge dislocation approaches the particle, but before cross-slip can take place, as pointed out by Humphreys and Martin (1967), the Orowan stress must be exceeded so that screw segments AB and CD are formed (Figure 15.2a). If these double cross-slip, jogs A, B, C, and D

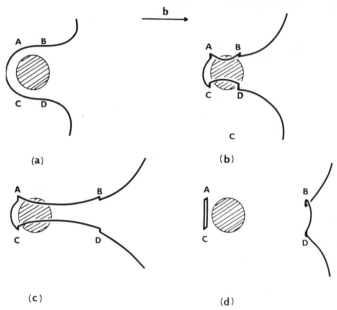

FIGURE 15.2. The cross-slip mechanism for edge dislocations.

are formed (Figure 15.2b); the loops AB, CD bow out, and the jogs B and D are pulled by glide along the screw direction (Figure 15.2c). Eventually, AB and CD annihilate, forming a glissile double jog in the edge dislocation and leaving a prismatic loop behind the particle.

Suppose now that cross-slip does not take place because the activation energy is too high. In that case an Orowan loop is formed around the particle (Figure 15.3a). A second dislocation may now be able to force the inner Orowan loop to cross-slip (Figure 15.3b). The screw arms near B and D will now bow inwards and touch at B and D (Figure 15.3c); the arms BF and DG will now link up with AB and CD respectively, the jogs B and D are free to travel along the screw arms (Figure 15.3d), the dislocation segments AB and CD bow out further and annihilate, leaving a prismatic loop AC and an Orowan loop at the particle, while the edge dislocation glides forward, carrying with it a double superjog (Figure 15.3e). Finally, the Orowan loop is likely to shrink; the screw arms touch the prismatic loop at A and C, the part of the prismatic loop on the same plane as the glide loop joins with it, collapsing onto the particle and leaving behind a similar prismatic loop (Figure 15.3f). The net effect is that the Orowan loop has passed through the prismatic loop.

It is clear that this mechanism can be extended easily to the case when several Orowan loops must be left around the particle before cross-slip can occur. It should also be noticed that we have assumed that if a screw segment cross-slips, it does so twice, thereby returning to a plane parallel

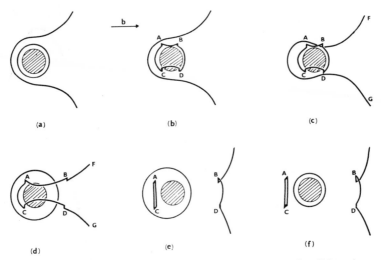

FIGURE 15.3. Combined Orowan and cross-slip mechanism for edge dislocations.

to the original primary plane. It is possible of course that the second cross-slip act takes place only under the action of a stress concentration due to a second or subsequent dislocation looping around the first. We also assume, when considering the cross-slip of a loop (e.g., Figure 15.3b) that both screw arms cross-slip together and annihilate. However, detailed calculations on the cross-slip process (Duesbery and Hirsch, to be published), suggests that one side of the loop cross-slips, and this goes around the particle and annihilates the other screw segment, which remains in the slip plane. This effect will not significantly alter the interactions proposed in this paper.

Consider now the mechanism for a screw dislocation. Suppose that the screw cross-slips over a length AB (Figure 15.4a), determined by the equilibrium condition that the arms of the dislocation at the jog (e.g., AD and AC at jog A) make equal angles with AB. The screw could now bypass either by pulling out two dipoles from the jogs A and B, or by forming an Orowan loop. Comparing the stresses required to form the two dipoles,

$$\tau = \frac{Gb}{\pi(1 - v)D_0} \ln \frac{\sqrt{2}\,h}{r_0}, \qquad (15.5)$$

where h is the height of the jog, with the Orowan stress for screws (Equation 15.4), it is clear than unless $h \ll 2R$, the Orowan stress will be smaller and the screw will bow round the particle rather than pull out the dipole. In making this estimate we have assumed the dipoles to be in their minimum energy configuration, i.e., each edge dislocation pair being in the 45 deg configuration. The interaction energy between the dipoles is relatively small if $h < 2R$, and has been neglected. However, if double cross-slip is easy, then the screw segments of the Orowan loop may cross-slip and bow

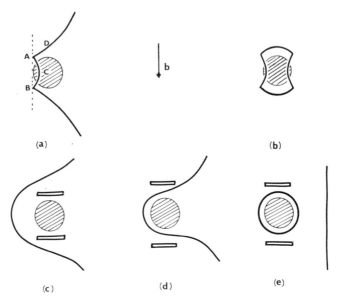

(a)   (b)

(c)   (d)   (e)

FIGURE 15.4. The cross-slip mechanism for screw dislocations.

out, as shown in Figure 15.4b. This process would be assisted and even completed by the bowing out of a second screw dislocation, in which case an interchange process similar to that described for edge dislocations will occur, leading to the sequence of events shown in Figures 15.4c–e. Again, this process can easily be extended to the case when several Orowan loops must be formed before double cross-slip can occur at the innermost loop.

Consider now the case when the spacing of the obstacles along the screw is not uniform (Figure 15.5a). If the Orowan stress is exceeded for EF, but not for FG, the dislocation loops around as shown in Figure 15.5b. If cross-slip occurs easily, cross-slipped arcs AB and CD are produced (Figure 15.5c), the jog D now runs along the line CD (Figure 15.5d), and the bowing arcs AB and CD meet and annihilate, forming the prismatic loop AC and the double jog BD which travels along with the edge part H of the bowing loop. Whether this mechanism or that of Figure 15.4 occurs will depend on the relative lengths EF and FG.

Comparison of Figures 15.2 and 15.3 for edges, and Figure 15.4 for screws when the obstacle spacing is uniform, shows that for edges, prismatic loops are formed on only one side of the particle, namely, that on which the edge is approaching the particle, while for screws (for uniform obstacle spacing) equal numbers of prismatic loops are formed on either side of the particle. On the other hand, if the obstacle spacing is nonuniform, the screw may advance by effectively forming an edge super double kink (Figure 15.5), in which case only one prismatic loop is left behind the particle. For dislocations with intermediate character, similar considerations apply; either

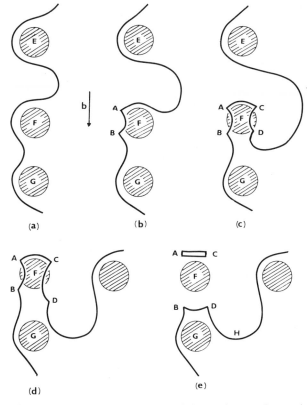

FIGURE 15.5. The effect of nonuniform particle spacing on the motion of a screw dislocation.

edge-type or screw-type behavior can occur, depending on the character of the dislocation and the relative magnitudes of adjacent spacings of the particles.

15.2.2. *Effect of a Particle on the Activation Energy for Cross-Slip*

In this section we shall discuss briefly three situations which may arise when a screw dislocation approaches an impenetrable particle. First, if the particle has a strain field due to misfit, if the screw is assumed to be undissociated, there is no net force, but only a couple exerted on it. For this reason it has been proposed (Gleiter 1967) that cross-slip would not take place when the screw is in this position. However, if the screw is dissociated into two partials (Figure 15.6a) there is a net force on the dislocation (either attractive or repulsive) due to the difference in the stress on the two edge components of the partials. Furthermore, the forces due to the particle on the partials are opposite in sign, consequently, depending on the sign of the dislocation, the screw ribbon is either constricted or expanded. The

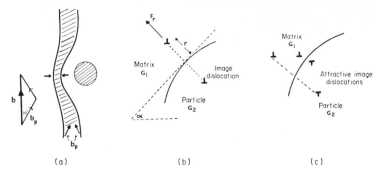

FIGURE 15.6. The interaction of an extended screw dislocation with a particle.

constriction effect depends of course on the size of ths misfit parameter, but it can be shɔwn that in practical cases the ribbons can be constricted considerably due to this effect, and the activation energy lowered correspondingly. Detailed calculations on this effect will be published elsewhere (Duesbery and Hirsch, to be published).

Second, if there is no misfit, the main interaction may be caused by the image force due to the different elastic constants of the matrix and the particle (Ashby 1964). The problem of the image stress on a dislocation loop around a particle has not yet been solved, but in practice the dislocations are so close to the particle that the image construction for a plane interface is a reasonable approximation. Figure 15.6b shows a typical situation. Suppose that the particle has a modulus of elasticity (G_2) larger than that of the matrix (G_1), the screw is repelled, and the radial force of interaction per unit length is

$$F_r \sim \left(\frac{G_2 - G_1}{G_1 + G_2}\right) \frac{G_1 b^2}{4\pi r}.$$

The force in the slip plane has to be equal to the force on the dislocation due to the dislocation bowing out between neighboring particles, spaced at D_0, i.e., $\tau b D_0$. The image force acts over a distance $\sim R$ so that in equilibrium

$$\tau b D_0 \sim \left(\frac{G_2 - G_1}{G_1 + G_2}\right) \frac{G_1 b^2 R \sin \alpha}{4\pi r}. \tag{15.6}$$

The component of the image force along a cross-slip plane (assumed, for simplicity, to be at 90 deg to the slip plane) is

$$F_r \cos \alpha \sim \frac{\tau b D_0 \cot \alpha}{R}. \tag{15.7}$$

The effective stress at the dislocation on the cross-slip plane is therefore $\sim (\tau D_0 / R) \cot \alpha$. For small volume fractions, and many experimental situations, $D_0 / R \sim 10$, so that the effective stress on the cross-slip plane is very large. This effect should also markedly reduce the activation energy for cross-slip. Detailed calculations still have to be carried out for this case.

It is interesting to note the equilibrium distance at which the dislocation rests outside the particle. For alumina particles in copper, $G_2 \sim 4G_1$ and with $D_0/R = 10$, $\alpha = 45$ deg, $G_1 = 4.3 \times 10^{11}$ dynes cm^{-2}, $\tau = 1$ Kg/mm^2, we find $r/b \sim 15$. The dislocations therefore sit quite close to the particle interface.

If the elastic modulus of the particle is less than that for the matrix, the dislocation will be attracted to the particle, until a short-range core repulsion prevents the front partial from entering the particle (Figure 15.6c). The second partial, however, will be attracted to the particle, and this effect gives rise again to a constriction effect. In addition there is a very large effective short-range repulsive force, acting on the front partial (larger than the attractive image force) and in this case it is likely that cross-slip takes place by the Fleischer (1959) mechanism. No detailed calculations have yet been carried out.

Although this discussion has been essentially qualitative, it is clear that in all three cases cross-slip is particularly favorable at the particles, partly because in some cases the dislocations are constricted, and partly because of the large effective stresses acting on the dislocations.

15.2.3. Cross-slip microstructure

With increasing deformation a number of dislocations pass on the same slip plane. Applying the principles discussed above, we expect the structures shown in Figure 15.7 to occur. Figures 15.7a and b show the structures expected for screws and edges if cross-slip occurs without leaving Orowan loops; Figures 15.7c and d show the structures expected if two Orowan loops are left around the particle before cross-slip occurs. If cross-slip occurs after deformation but before examination by microscopy, the structures of Figures 15.7e, f, or g, h will be observed. When both Orowan and prismatic loops are present (e.g., Figure 15.7c) the question as to whether the prismatic loops are inside or outside the Orowan loops will depend on the magnitudes of the mutual repulsion between the rows of prismatic loops, and the stand-off distances of the Orowan loops from each other and from the particle. Simple calculations indicate that while the inner Orowan loop may shrink onto the particle, a second or third one will stand off a distance of the order of the particle diameter, thus justifying the geometry of Figure 15.7c.

15.2.4. Experimental evidence

Humphreys and Martin (1967) have examined in detail the microstructures of deformed single crystals of copper containing cobalt or silica particles, and Humphreys (to be published) has also studied single crystals of copper containing alumina particles, and of Cu–Zn crystals of various

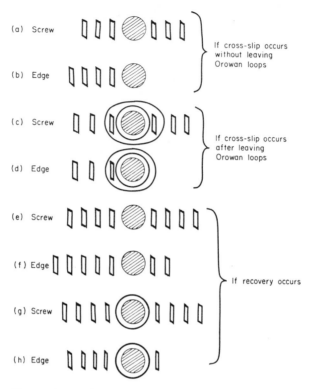

FIGURE 15.7. The microsctructures expected after cross-slip.

compositions containing alumina particles. Figure 15.8 shows a typical structure in a single crystal of copper containing alumina particles deformed into stage I at room temperature and sectioned parallel to the primary slip plane. Rows of loops predominantly on one side of the particle are observed, and this suggests that the deformation occurs mainly either by the movement of edge dislocations (Figure 15.7b) or by the formation and movement of super edge double kinks. It is difficult to tell from the experiments whether one or perhaps two loops are left at the particle on the side opposite to that of the row of loops, and therefore it is not possible to decide whether one or perhaps two Orowan loops are left around the particle before cross-slip occurs.

If the temperature or stacking fault energy is lowered sufficiently, it might be expected that cross-slip becomes more difficult, and that the structures change from those of Figure 15.7b, to those of Figures 15.7c, d, e, or f. However, examination of copper crystals containing silica or alumina particles, deformed into stage I at 77°K, shows essentially similar features to the room-temperature structure of Figure 15.8. A Cu–20 wt percent Zn crystal containing alumina particles again shows a similar dislocation

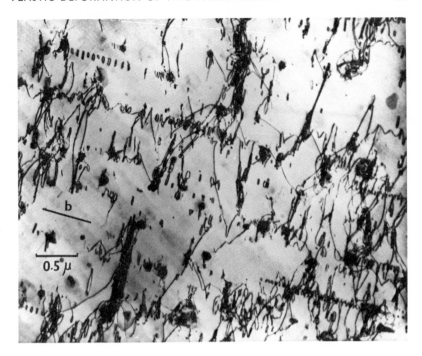

FIGURE 15.8. Section parallel to the primary slip plane (111) of a single crystal of copper containing alumina particles ($f = 2.2 \times 10^{-3}$) deformed at room temperature into stage I to a shear strain (γ) of 0.15.

structure at the precipitate (Figure 15.9), even though the stacking fault energy (γ) of the matrix is ~ 20 ergs/cm^2 (Christian and Swann (1963). However, in a Cu-30wt percent Zn crystal ($\gamma \sim 10$ ergs/cm^2), cross-slip is inhibited, and particles which have been bypassed by up to about three dislocations show Orowan loops around them (Figure 15.10). The passage of further dislocations causes the inner loops to cross-slip, and forms structures such as in Figure 15.11 showing both Orowan and prismatic loops similar to Figures 15.7d or h.

We conclude from these observations that

1. Cross-slip occurs either without leaving an Orowan loop, or after leaving a small number of Orowan loops—the number depending on the stacking fault energy and presumably on the temperature and strain rate.

2. The deformation takes place predominantly by the movement of edges or by the generation and movement of edge super double kinks.

15.2.5. *The Yield Stress*

If the dislocations do not cross-slip but leave Orowan loops the yield stress is that characteristic for the formation of Orowan loops. Combining the effect of anisotropy on the yield stress for edges and screws for a regular

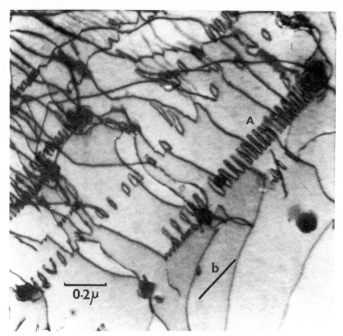

FIGURE 15.9. Cu-20% Zn, containing Al_2O_3 particles ($f = 2.2 \times 10^{-3}$) $\gamma = 2 \times 10^{-2}$, (111) section. Note the formation of helices at A.

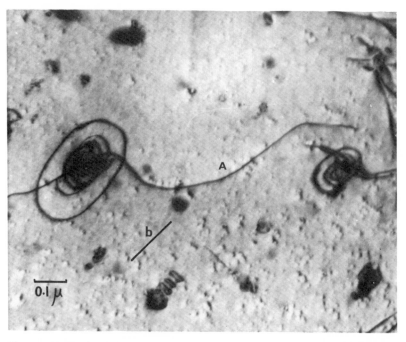

FIGURE 15.10. Cu-30% Zn containing Al_2O_3 particles ($f = 2.2 \times 10^{-3}$) $\gamma = 6 \times 10^{-2}$, (111) section, showing the formation of Orowan loops (the dislocation A is not associated with the structure).

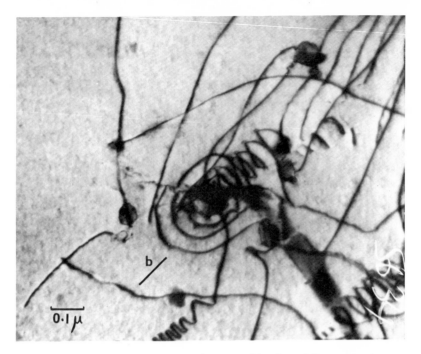

FIGURE 15.11. As Figure 15.10, showing a combination of Orowan and prismatic loops as described in Figure 15.7d.

lattice with the results of the statistical theory based on a model of constant line tension, Ashby (1968) suggests that the yield stress for screws exceeds that for edges according to Equations 15.3 and 15.4. He has also pointed out that it is not sufficient to exceed τ_e, for then ribbons of slip elongated along the screw direction are formed, but dislocation sources cannot operate. The screw parts of the loop must also advance. It is therefore usually assumed that in order to exceed the macroscopic yield stress, the Orowan stress for screws must be exceeded (Ashby 1968, Kocks 1969). However, this conclusion does not necessarily follow for the following reasons.

First, for a random particle lattice it is important to consider not only the relative magnitudes of the line tension for screws and edges, but also the equilibrium shape of the loop, which in the anisotropic case is no longer circular but in many cases roughly elliptical (de Wit and Koehler 1959). The importance of the shape of the loop can be appreciated from the following simple argument. When the dislocation loop bows out into its critical configuration, it must not touch another particle before the critical configuration is reached (Figure 15.1). For a square lattice this condition is more than satisfied; for a random array of obstacles, and a semicircular loop this condition is satisfied if $D_0^{-1} > (\pi/8)^{1/2}N_s^{1/2} = 0.63N_s^{1/2}$, compared with $D_0^{-1} = 0.81N_s^{1/2}$ on the Foreman and Makin (1966) calculation (see Section 15.1). This simple condition is expected to be necessary but of course not sufficient because the process of yielding for a random array is considerably more complex. However, it does indicate that the yield stress for a random array is expected to be smaller than for a square lattice. Extending this argument to the anisotropic case, the condition now becomes that the expanding elliptical loop must not touch any other particle as it expands to the critical configuration. For both edge and screw the areas of the half loops are equal to half the area of the complete elliptical loop, but clearly the effective values of D_0 for screws and edges will now be in the ratio of the major to minor axis of the ellipse (Figure 15.1c and d). We expect therefore that for this reason the mean spacing between obstacles along the screw will be larger than along the edge, and this effect compensates for the difference in the line tension. The ratio of the major to minor axis of the ellipse is $E(\pi/2)/E(0)$, where $E(\pi/2)$ and $E(0)$ are the energies pet unit length of edge and screw dislocations (DeWit and Koehler 1959), which is exactly equal to the ratio τ_s/τ_e for a regular lattice; this ratio has the value $(1 - v)^{-1}$ for isotropic elasticity; see Equations 15.3 and 15.4. Hence on this argument we would expect the yield stress for edges and screws to be exactly equal and no ribbon slip would occur before general yielding. Assuming the shape of the loop to be an ellipse, the yield stress is found to be

$$\tau = 0.81 \sqrt{E_D(\pi/2)E_D(0)N_s}\,b^{-1} \tag{15.8a}$$

where $E_D(\pi/2)$, $E_D(0)$ are the energies per unit length of edge and screw dipoles of width $\sim 2R$; for isotropic elasticity

$$\tau = \frac{0.81GbN_s^{1/2}\ln(2R/r_0)}{2\pi(1 - v)^{1/2}}. \tag{15.8b}$$

The agreement of the experimental results with these equations is just as good as with Equations 15.3 and 15.4.

Second, suppose that the yield stress for a screw is larger than for an edge. Then the screw can advance by forming an edge super double kink (Figure 15.12), which can then travel along the screw direction. For a random distribution of particles, we expect that the super double kinks will be generated at those places where the spacing between the particles is particularly large. The stress to generate the super double kink should therefore be less than that to move the screw along the whole of its length, and the stress to move it will be approximately that needed to pull out a dipole of width D_0, i.e. it will be given by τ_e. However, in order to enable the edge kink to pass through some of the narrower gaps, and thereby bypass the more difficult hard spots, the stress required for the movement of the edge kink will be somewhat larger than that predicted on the statistical theory for edges. This suggests that the yield stress should have a value lying between those of Equations 15.3 and 15.4, as is the case for Equations 15.8.

Foreman (private communication) has used the following ingenious argument to show that the yield stress for screws and edges will be equal. If the shape of the loop is elliptical, the random lattice can be sheared appropriately to change a circle into an ellipse with the required eccentricity; the shear deformation transforms the random lattice into another random lattice. The critical dislocation configuration for circular loops for a random lattice is therefore transformed into the critical configuration for elliptical loops. The yield stress for screws and edges will be equal, and it is easily shown that the yield stress is given by Equations 15.8.

Consider now the effect of cross-slip on the yield stress. In Section 15.2.1 we noted that for edges the Orowan stress must be exceeded in order for the dislocations to turn into the screw orientation prior to cross-slip, and for screws we saw that even if the dislocation cross-slips, the stress to bypass by the Orowan mechanism is less than that required to pull out dipoles. Therefore, even though cross-slip takes place, the yield stress will be given by the Orowan bypassing stress discussed in the previous paragraph. This conclusion is contrary to that of Kocks (1969), who suggested that the occurrence of cross-slip should lead to a yield stress dependent on temperature through the thermally activated cross-slip process. Ashby (private communication) has come to the same conclusion by considering the case of screw dislocations under conditions in which the jogs are not trapped at the particles, and therefore dipoles are not pulled out there. Brown (private communication) has used a somewhat similar argument to reach the same conclusion.

15.3. Work Hardening

There are two models of work hardening of alloys containing a dispersed second phase. In the first of these, due to Fisher, Hart, and Pry (1953), the dislocations are assumed to leave Orowan loops around the particles, exerting a back stress which opposes further slip. The stress-strain curve is found to be linear and is given by

$$\tau - \tau_0 = 6Gf^{3/2}\varepsilon, \tag{15.9}$$

where f is the volume fraction and ε the plastic strain. Ashby (1968) has reviewed the experimental evidence, and he concludes that this model does not explain the observed work-hardening curves (roughly parabolic) except at very low strains (<2 percent) where there is some evidence that the work hardening depends only on f.

In the second theory, due to Ashby (1968) the stresses due to the Orowan

loops are relieved by secondary slip nucleated at or near the particle interface, either in the form of glide loops or punched out prismatic loops.
Assuming maximum relaxation the number of prismatic loops, N is of the
same order as the number of Orowan loops for the same strain (ε), i.e.,

$$N = 2R\varepsilon/b. \tag{15.10}$$

Ashby then assumes that the loops are randomly distributed in orientation
and space in the matrix, and calculates the flow stress using a forest interaction model derived for quenched metals (Kroupa 1962, Kroupa and
Hirsch 1964). He finds a parabolic stress-strain relation

$$\tau - \tau_0 = CG\sqrt{\frac{bf\varepsilon}{2R}}, \tag{15.11}$$

where C (for prismatic glide) is thought to lie between 0.1 and 0.4. This
equation appears to account reasonably well for the limited experimental
data available. For the two systems investigated in detail, i.e., Cu containing
SiO_2 particles (Ebeling and Ashby 1966) and Cu with BeO particles (Jones
and Kelly 1968), C is found experimentally to have the values 0.24 and 0.15
respectively. However, Jones and Kelly (1968) and Humphreys (to be
published) have observed marked recovery for crystals containing BeO and
Al_2O_3 particles respectively, during deformation at room temperature, and
the Ashby theory can be expected to apply only at low temperatures, where
this effect is negligible.

While the Ashby equation describes reasonably well the work-hardening
characteristics of dispersion-hardened alloys, certain assumptions made in
the model are open to question. Ashby assumes that (1) the loops are
prismatic and randomly distributed in the matrix; (2) the Burgers vectors
are random; (3) the loops are fixed in position and not glissile. On the
other hand, electron-microscope observations on single crystals of copper
containing Co, SiO_2, or Al_2O_3 particles, or Cu–Zn crystals containing
Al_2O_3 particles, deformed into stage I, all show that most of the loops are
prismatic, have the primary Burgers vector, and are arranged in rows on
one side of the particle, in agreement with the cross-slip model. The loops
appear to be formed by dislocations on the same slip plane, and the average
number of loops behind a particle increases with increasing strain, showing
that the distribution of loops is far from random. Furthermore, in the cases
where Orowan loops are formed (Cu–Zn crystals) the stresses at the particles
are relieved by cross-slip of the inner Orowan loops, and not by formation
of secondary dislocations.

It is well known that the yield stress of a work-hardened material depends
not only on the density of dislocations but also on their nature and distribution. We shall now consider the work hardening due to rows of prismatic
loops with primary Burgers vector left behind the particles during deformation.

The problem is in two parts. First, we calculate the average number of loops, N, left behind the particles as a function of strain. If Orowan loops are formed around the particles then N is given by Equation 15.10, i.e., one Orowan loop is formed per passing dislocation. For the cross-slip process and deformation by edge dislocations, again one prismatic loop is formed per dislocation, so that Equation 15.10 is again applicable. Preliminary results indicate that after room temperature deformation the number of loops formed is less than that predicted by Equation 15.10, but it is not yet known whether this is an experimental error or a real recovery effect.

Next we consider the interaction between the dislocations and the rows of loops, and we shall treat first the case where the number of loops is small. The loops repel each other, and in the absence of any frictional force on the dislocation loops, they would move as far apart as possible. Adapting the results for circular loops (Bullough and Newman 1960, Hirth and

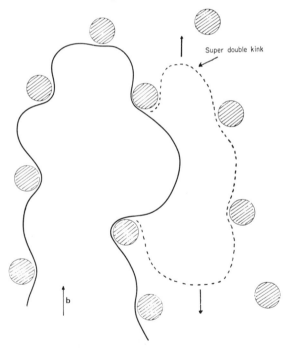

FIGURE 15.12. The propagation of slip by the movement of an edge super double kink.

Lothe 1968), for separations (x_s) large compared with their mean radius (R), the force decreases sharply with increasing x_s, and in equilibrium is balanced by the frictional force (F per unit length), i.e.,

$$F \sim \frac{3Gb^2R^3}{(1-v)x_s^4}. \tag{15.12}$$

Clearly we would expect the equilibrium separation to be smaller for solution hardened alloy such as Cu–Zn, although x_s varies only as $F^{-1/4}$. On the whole, the micrographs of Cu and Cu–Zn alloys containing Al_2O_3 particles confirm that in the latter the mean spacing tends to be smaller.

Suppose now that a screw interacts with the loops (Figure 15.13). There

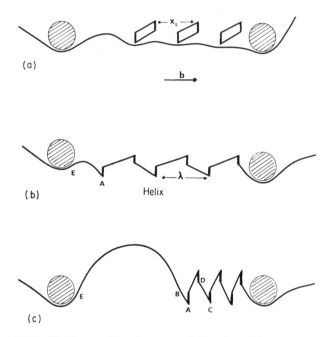

FIGURE 15.13. The interaction of a screw dislocation with a row of prismatic loops to form a helix.

is no net force between the loops and the dislocation, but a couple which causes the screw to combine with loops to form a helix (Figure 15.13b). This interaction to form a helix is often observed in deformed crystals containing particles, and can be seen at A in Figure 15.9.

The repulsion between the turns of the helix and the line tension of the screw will tend to pull out the helix. The pitch λ of the helix will be greater than the equilibrium separation x_s between loops, and for $\lambda \gtrsim 3R$ the line tension becomes the predominant term; and for small friction stresses the equilibrium pitch may be determined by the free length of screw available (Figure 15.13b). If a stress is now applied, and if the free length EA of the screw in Figure 15.13b is greater than the pitch, the helix will shrink (Figure 15.13c) until the distance between the turns is of the same order as the radius of the helix. At this stage the repulsion between the helix turns, which is given approximately by Equation 15.12, increases sharply with decreasing helix pitch, and the helix, acting in many ways as a nonlinear spring, cannot be compressed further. The repulsion between AB and CD on the same plane is strong and it seems likely that the screw must penetrate the gap BE by the Orowan looping mechanism. *We conclude therefore that the loops have the effect of converting the particle into a linear obstacle whose length increases with increasing strain.*

We must now consider the problem of how a dislocation moves through an array of parallel linear obstacles. The mean distance along an approximate screw (dislocation) between neighboring obstacles will no longer be given by the statistical theory for point obstacles. It seems reasonable to use the criterion that when the loop bows out to the appropriate critical configuration it touches one more obstacle (Figure 15.14a), i.e.,

$$k(D + S)D = N_s^{-1}, \tag{15.13}$$

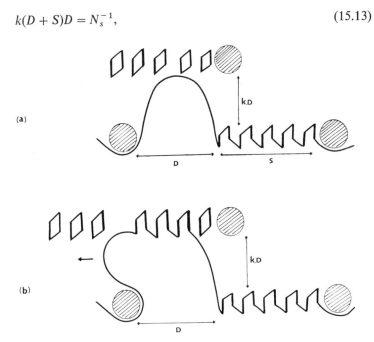

(a)

(b)

FIGURE 15.14. Work hardening by helix formation.

where D is the gap length, S is the obstacle length, and k is a constant depending on statistical considerations and on the need for the "double kink" to propagate between the obstacles. When $S = 0$, i.e., at the initial yield, $D^{-1} = 0.81 \, N_s^{1/2}$ and thus $k = (0.81)^2$. We shall assume that k takes this value for all values of S. We therefore write

$$(D + S)D = D_0^2, \tag{15.14}$$

where D_0 is the effective spacing to be inserted into the equations for the yield stress of the undeformed crystal. The yielding of the crystal requires both the formation and propagation of the double kink, but it is likely that the path of least difficulty for the dislocation will be that for which these two stresses are equal. For our immediate purpose we shall use Equations 15.8. The solution of Equation 15.14 is

$$\frac{D}{D_0} = \sqrt{1 + \left(\frac{S}{2D_0}\right)^2} - \frac{S}{2D_0}.$$

(15.15)

Inserting in Equation 15.8, for $S/D_0 < 1$ we find

$$\tau = \tau_0 + \frac{Gb \ln (2R/r_0)}{2\pi(1 - v)^{1/2}D_0} \left(\frac{S}{2D_0}\right).$$

(15.16)

Assuming that all the loops are pressed against the particles, and that their spacing is $\sim R$, we find

$$\tau = \tau_0 + \frac{3(0.81)^2 G \ln (2R/r_0)}{(2\pi)^2(1 - v)^{1/2}} f\varepsilon,$$

(15.17)

i.e., a linear work-hardening rate, proportional to volume fraction, and only weakly dependent on R. Ashby (1968) has considered the experimental evidence at low strains; the data suggest that the hardening rate is approximately independent of R, and proportional to f^m, where m lies between 1 and 1.3.

Replotting the hardening rates θ/G of Ashby's Figure 12b (Ashby 1968) against f, we find a slope of between 0.3 and 0.7 compared with a value expected from Equation 15.17 of 0.35, using $v = 1/3$ and $2R/r_0 \sim 300$; the agreement is satisfactory.

At larger strains we can no longer expect all the loops to be pressed against the particles. Figure 15.14b illustrates a possible situation where some of the loops will be swept away from the particle by the bowing-out screw dislocation. In addition, some of the loops will be swept away by dislocations moving on neighboring planes. Consequently an increasing fraction of the number of loops will be distributed randomly throughout the matrix; each loop can glide freely over distances of the order of the mean free path between particles, which, for small-volume fractions, is considerably larger than the mean planar particle spacing. The "free" loops are unlikely to cause as much hardening as the loops piled up against the particles, since they are almost equivalent to jogs, glissile along the Burgers vector direction. We expect therefore that the hardening rate will decrease with increasing strain. However, at present, the processes have not been considered in sufficient detail to predict the precise form of the hardening curve.

15.4. Recovery Processes

15.4.1. General

Jones (1969) and Humphreys (to be published) have found that deformed single crystals of copper containing BeO and Al_2O_3 or SiO_2, respectively, recover markedly at room temperature, and a large fraction of the increased flow stress due to deformation may be lost on standing for a few hours. The work-hardening rates in stage I are also very dependent on strain rate, at room temperature, and for a given strain rate the work-hardening rate decreases markedly with increasing temperature. Figure 15.15 summarizes these effects in $Cu-Al_2O_3$, where the work-hardening

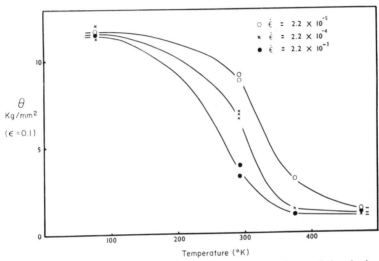

FIGURE 15.15. The effect of temperature and strain rate on the work-hardening (θ) in stage I of copper single crystals containing Al_2O_3 particles ($f = 2.2 \times 10^{-3}$), taken at a strain of 0.1.

rate at an arbitrary strain in stage I is plotted against temperature for three different strain rates.

Electron-microscope observations show that the reduction in hardening rate at high temperature is accompanied by a large reduction in loop density. Figures 15.16 and 15.17 show electron micrographs of specimens deformed at 77°K and 200°C, respectively, and they may be compared with Figure 15.8, which is the room-temperature structure at the same strain.

There are a number of possible recovery mechanisms which may be operating. First, there is the possibility that the glide of loops under their mutual repulsion (see Section 15.3) is thermally activated. In that case the spacing of the loops may increase with increasing time, leading to a reduction in flow stress since the proportion of "free" loops increases. Second,

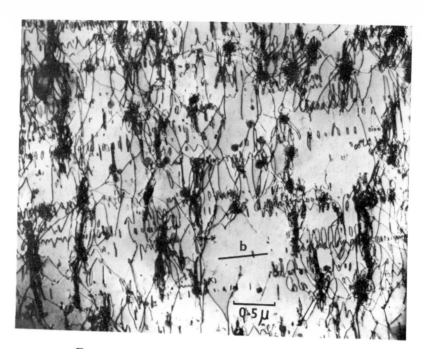

FIGURE 15.16. As Figure 15.8, but deformed at 77°K.

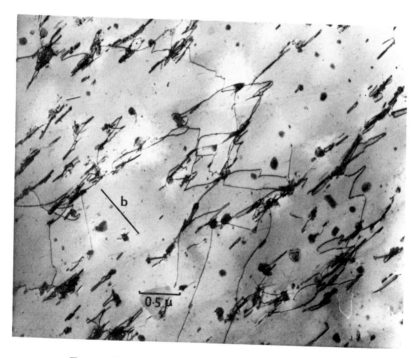

FIGURE 15.17. As Figure 15.8, but deformed at 200°C.

part of the flow stress may be due to the interaction of screws trapped by the loops to form helices. These could gradually annihilate each other by cross-slip. Third, the loops may be swept along by gliding dislocations, and may be annihilated. In general this mechanism will require some climb of the loops, and climb in the range of temperatures considered could be either due to the point defects produced during the deformation, or possibly caused by pipe diffusion or diffusion across the particle interface. Although it is not clear at this stage which of these mechanisms is the most important at room temperature, the large reduction in loop density at higher temperatures (100°C and above) shows that under those conditions loops are either not formed or are annihilated, and the relevant processes will require pipe or interface diffusion. We shall consider first the rate of climb of dislocations around particles during deformation, by the process of pipe diffusion.

15.4.2. Climb of Dislocations around Particles

Suppose that a dislocation leaves an Orowan loop around a particle, and subsequently cross-slip occurs (Figure 15.18a). The loop can now anneal out by climb by vacancy transfer from jogs AA' to BB'. Assuming the rate of climb to be diffusion controlled, the flux of vacancies is

$$\phi = -\frac{D_p}{V c_0} \, \text{grad } c, \tag{15.18}$$

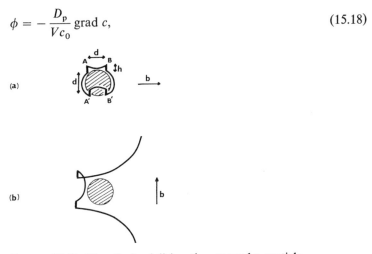

(a)

(b)

FIGURE 15.18. The climb of dislocation around a particle.

where D_p is the diffusion coefficient for pipe diffusion, c_0 is the equilibrium concentration of vacancies, c is the actual vacancy concentration, and V is the atomic volume $= b^3/\sqrt{2}$. The climb force F_{cl} on jogs A and B is equal and opposite, so that, using Friedel's climb theory (1964),

$$\phi = -\frac{z A v_0 \, b^2 \exp - U_p/kT}{V d} \left\{ \exp \frac{F_{cl} b^2}{kT} - \exp \frac{-F_{cl} b^2}{kT} \right\} \tag{15.19}$$

where h, d are height and separation of jogs (Figure 15.18a), z is coordination number (~ 11), A is an entropy factor (~ 10), v_0 is the Debye frequency ($\sim 10^{13}$ sec^{-1}), U_p is the activation energy for pipe diffusion, k is the Boltzmann factor, and T is the temperature.

The total flux of vacancies $= 2\phi a$, where a is the cross-section of the pipe; the number of vacancies to be transported is $\sim dhb/V$, so that the time (t) taken for a loop to disappear is given by

$$t^{-1} = \frac{2\phi a V}{dhb} = \frac{2az A v_0 \, b \, \exp(-U_p/kT)}{d^2 h} \left\{ \exp \frac{F_{cl} b^2}{kT} - \exp - \frac{F_{cl} b^2}{kT} \right\}.$$

$$(15.20)$$

For small $F_{cl} b^2/kT$, which is a reasonable assumption,

$$t^{-1} = \frac{4az A v_0 \, b}{d^2 h} \left(\frac{F_{cl} b^2}{kT} \right) \exp(-U_p/kT).$$

$$(15.21)$$

The number of loops produced per second per particle is obtained by differentiating Equation 15.10,

$$\frac{dN}{dt} = \frac{2R\dot{\varepsilon}}{b}.$$

$$(15.22)$$

Thus if $t^{-1} \lesssim dN/dt$ the hardening effect from loops disappears, since no loops are formed. The critical temperature T_c, at which the hardening disappears is given by

$$T_c = \frac{U_p}{k \ln \left[\dfrac{2az A v_0 \, b^2}{d^2 h R \dot{\varepsilon}} \left(\dfrac{F_{cl} b^2}{kT_c} \right) \right]}$$

$$(15.23)$$

We now have to determine F_{cl}. There are several contributions to the climb force. First there is the attraction between the jogs, e.g., at A, A'; if $h < d$, this contribution is

$$F_{cl} \sim \frac{Gb^2 h}{8\pi(1 - v) \, d^2};$$

$$(15.24)$$

if $h > d$,

$$F_{cl} \sim \frac{Gb^2}{2\pi(1 - v) \, d}.$$

$$(15.25)$$

During the climb process d changes, and Equation 15.25 is probably a better estimate. Second, there is the attraction of the screw segments AB, $A'B'$; this gives a contribution of similar magnitude, i.e.,

$$F_{cl} \sim \frac{Gb^2}{4\pi h}.$$

$$(15.26)$$

The applied stress will make another contribution due to bowing out of the screw segment; this is

$$F_{cl} \sim \frac{Gb^2 d}{2Dh},$$ (15.27)

which again is of a similar order of magnitude. However, if the loop climbs one elementary jog at a time, we must take into account the effect of the line tension, for as the loop climbs it shrinks around the particle. This effect gives

$$F_{cl} \sim \frac{\alpha Gb^2 \log d/r_0}{2\pi d},$$ (15.28)

where α depends on the height of the loop relative to the center plane of the particle. For an average position of the loop we shall take $\alpha \sim 1/2$. This last force appears to be the largest, and the loop is likely to climb in this way. Using Equations 15.23 and 15.28, and writing $a = f^*\pi b^2$, we find

$$T_c = \frac{U_p}{k \ln \left[\dfrac{f^* z A v_0 b^5 \alpha}{d^3 h R \dot{\varepsilon}} \left(\dfrac{Gb^3 \ln d/r_0}{kT_c} \right) \right]}.$$ (15.29)

The activation energy for pipe diffusion is not well known, although recent estimates suggest values of about half the activation energy for self-diffusion; the latter is ~ 2 eV for Cu. We might expect therefore $U_p \sim 1$ eV. Experiments suggest that the hardening rate due to the formation of loops disappears at about 100°C. Taking $T_c \sim 370°$K, for a strain rate of 10^{-4}, putting $d = h = R$, $R/b \sim 10^2$, $f^* = 10$ (corresponding to a pipe radius of $\sim 3b$) we find $U_p \sim 0.9$ eV, in reasonable accord with expectation. The results of the dependence of T_c on the strain rate are not yet sufficiently precise to enable an accurate value of U_p to be obtained.

It should be emphasized that this particular type of climb process can also occur as the dislocation bows around the particle, e.g., as shown in Figure 15.18b. The climb forces involved are similar to those discussed above, and the rate at which the dislocation climbs around the particle is comparable to the rate at which Orowan loops anneal out by climb.

Once a prismatic loop is formed, pipe diffusion has a less drastic effect. Although as pointed out below, helix annihilation can occur and loops can also migrate by conservative climb under their mutual interactions. These interactions fall off rapidly with distance, and this may be the reason that loops produced by deformation at room temperature do not anneal out even if the specimen is held at 300°C for an hour.

It should be noted that in the case when the Orowan loop is situated very close to the particle interface, climb can take place by surface diffusion over the interface. This type of process can become important at these relatively low temperatures if the activation energy for boundary diffusion is of the same order as that for pipe diffusion, and if the dislocation pipe touches or overlaps the interface.

The question arises as to whether the yield stress can fall below the "Orowan" stress as a result of this process. Edge dislocations still have to bow around the particles in order to produce screw segments before jogs can be formed by cross-slip. On the other hand, screw dislocations can now circumvent the particles by combined cross-slip and climb by pipe diffusion, and the yield stress can fall from the value intermediate between that for screws and edges (see Equation 15.8, Section 15.2.5.) possibly to that expected for edges.

15.4.3. *Other recovery processes*

As a result of the pipe-diffusion mechanism several recovery processes are possible. For example, loops of the opposite sign at different points on the slip plane could annihilate or partially annihilate by this process if they are connected by a dislocation pipe (Figure 15.19). Similarly, smaller loops of

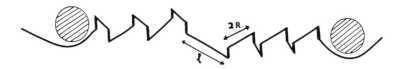

FIGURE 15.19. The mutual annihilation of helices of opposite sign connected by a dislocation "pipe."

the same sign may shrink while larger loops grow. As an example, it can be shown that two loops of opposite sign, of radius R, connected by a pipe of length l, will anneal out in a time t given by

$$t \sim \frac{4\pi l R^3 \exp\left(U_{\mathrm{p}}/kT\right)}{3f^* z A v_0 b^4 \left(\dfrac{Gb^3 \ln R/r_0}{kT}\right)}, \tag{15.30}$$

which, if $l \sim R$, is of the same order as the time taken for an Orowan loop to climb around a particle (Equation 15.20).

This type of process can occur at an appreciable rate at temperatures a little below T_c, since the time available during the test ($= \varepsilon/\dot{\varepsilon}$, typically 10^3 sec for 10 percent strain at $\dot{\varepsilon} = 10^{-4}$) is greater than the time interval between formation of loops at a given particle ($b/2R\dot{\varepsilon} \sim 50$ sec). At room temperature the time necessary for the annihilation to be complete, using the same parameters as before, comes out to be about 30 hours, which is a reasonable value. Loops of the same sign, at distances of the order of R, can coalesce in about the same time.

It is of course possible that point defects produced during the deformation contribute to the recovery effect. Some loops or helices will shrink, others expand, but although this process is likely to contribute to recovery at room temperature, the almost total disappearance of loops above $100°C$ is more likely due to the pipe-diffusion mechanism.

The recovery processes which we have discussed above will take place at temperatures dependent on the value of the activation energy of pipe diffusion, which is at present uncertain. In particular it should be emphasized that the pipe-diffusion mechanism of the recovery effect around room temperature is tentative. The possibility that this effect could be due to the glide and cross-slip processes discussed in Section 15.4.1 cannot be ruled out, and further experiments are in progress.

15.5. Conclusions

1. It is shown in this paper that cross-slip plays an important part in the formation of the microstructure of alloys containing small particles. The microstructure observed in Cu and Cu alloys containing small particles can be explained in terms of this mechanism.

2. The yield stress for such alloys at low temperatures is determined by the Orowan criterion; the yield stress for a random array of obstacles, taking into account anisotropy, lies between that for screws and edges for a regular lattice.

3. Work hardening is considered to be due to the interaction between screws and loops to form helices, and the flow stress of the deformed material is determined by the need to reverse this reaction. The reaction results in the formation of parallel linear obstacles whose lengths increase with increasing strain. An approximate theory for low strains is developed which gives reasonable agreement with the existing data.

4. Recovery effects at room temperature in two-phase copper crystals are considerable. The absence of loops after deformation above about $100°C$ is thought to be due to pipe or interface diffusion; while this process may make a contribution also at room temperature, glide and cross-slip processes may also be important.

REFERENCES

Ashby, M. F., 1964, *Z. Metallk.* **55**, 5.

Ashby, M. F., 1968, in *Proc. Second Bolton Landing Conference on Oxide Dispersion Strengthening*, New York: Gordon and Breach.

Ashby, M. F., private communication.

Brown, L. M., private communication.

Bullough, R., and R. C. Newman, 1960, *Phil. Mag.*, **5**, 921.

Christian, J. W., and P. R. Swann, 1963, in *Alloying and Effects in Concentrated Solid Solutions*, New York: AIME.

DeWit, G., and J. S. Koehler, 1959, *Phys. Rev.*, **116**, 1113.

Duesbery, M. S., and P. B. Hirsch, 1969, to be published.

Ebeling, R., and M. F. Ashby, 1966, Phil. Mag., **13**, 805.

Fisher, J. L., F. W. Hart, and R. H. Pry, 1953, *Acta Met.*, **1**, 336.

Fleischer, R. L., 1959, *Acta Met.*, **7**, 134.

Foreman, A. J. E., and M. J. Makin, 1966, *Phil. Mag.*, **14**, 911.

Friedel, J., 1964, *Dislocations*, Reading, Mass: Addison-Wesley.

Gleiter, H., 1967, *Acta Met.*, **15**, 1213.

Hirsch, P. B., 1957, *J. Inst. Metals*, **86**, 7.

Hirth, J. P., and J. Lothe, 1968, *Theory of Dislocations*, New York: McGraw-Hill.

Humphreys, F. J., 1969, to be published.

Humphreys, F. J., and J. W. Martin, 1967, *Phil. Mag.*, **16**, 927.

Jones, R. L., 1969, *Acta Met.*, **17**, 229.

Jones, R. L., and A. Kelly, 1968, in *Proc. Second Bolton Landing Conference on Oxide Dispersion Strengthening*, New York: Gordon and Breach.

Kocks, U. F., 1966, *Phil. Mag.*, **14**, 1629.

Kocks, U. F., 1969, this volume.

Kroupa, F., 1962, *Phil. Mag.*, **7**, 783.

Kroupa, F., and P. B. Hirsch, 1964, *Disc. Faraday Soc.*, **38**, 40.

Orowan, E., 1948, in *Symposium on Internal Stresses in Metals and Alloys*, London: The Institute of Metals, p. 451.

ABSTRACT. The mutual elimination of screw dislocation segments by cross-slip in easy glide produces segmented edge dislocations which can be characterized by segment distribution functions normal to their slip plane. The statistics of the capture of these segmented dislocations is analyzed to calculate the rate of growth of multipoles. The probability of formation of dislocation sources in dislocation multipoles by the partial capture of segmented dislocations, and the probability of inactivation of active sources in multipoles by capture of glide dislocations, have also been calculated to determine the stress and strain dependence of the dislocation density that the crystal can transmit. The stress-strain curve and a number of geometrical parameters such as dislocation multiplication rate and slip line lengths and depths have been calculated and are all in excellent agreement with experimental measurements.

16. A Statistical Theory for Easy Glide II

A. S. ARGON

16.1. Introduction

Ever since the discovery of plastic slip in rock-salt single crystals by Reusch (1867) more than one hundred years ago, the subjects of plastic deformation, strain hardening, and the rate mechanism of plastic deformation have been sources of continuing fascination to a large group of investigators. The experimental and theoretical developments in this very active area have been periodically reviewed (see Joffe 1928, Elam 1935, Schmid and Boas 1935, Maddin and Chen 1954, Seeger 1958, Clarebrough and Hargreaves 1959, Nabarro, Basinski, and Holt 1964, and Mitchell 1964).

Many theories of varying degrees of complexity have been proposed for the flow-stress and strain-hardening rates of single crystals since the pioneering theory of Taylor (1934) for laminar slip in face-centered cubic metal crystals (for a review see Nabarro, Basinski, and Holt 1964). Very few, if any, of these theories have been able to provide a theoretical basis for the increasing dislocation density of a crystal with increasing plastic strain. This problem of "statistical dislocation mechanics" in its general form remains as one of the outstanding challenges. In the more restricted confines of laminar slip, recent experimental studies have revealed a comparative simplicity of dislocation interactions which makes it possible to attempt a statistical theory for this, the simplest form of plastic deformation. Below we will develop such a statistical theory for laminar slip. This theory incorporates a number of important modifications of an earlier development by Argon and East (1968), which will be referred to as A & E in the following sections.

16.2. Summary of Experimental Observations

In A & E the significant experimental observations in the easy-glide region were listed. The most important ones of these are summarized below.

1. The bulk of the grown-in dislocations do not lie on slip planes and are, therefore, sessile (Young 1967).

2. The friction stress to move mobile dislocations is of the order of 1 g/mm^2 — almost negligible in comparison to the yield stress (Tinder and Washburn 1964).

3. Excluding perfect whiskers (Brenner 1958), yielding is nearly homogeneous in bulk single crystals, and no yield phenomenon is observed. The curvatures of micro-stress-strain curves in the preyield region steadily decrease until they become zero in easy glide (Argon and Brydges 1968).

4. The primary dislocations are retained in the crystal almost completely in dipolar clusters called "multipoles" by Gilman (1962) or "braids" by Mader (1963). From this Argon and Brydges concluded that the range of internal stresses in easy glide is of the order of dislocation spacings in dipoles.

5. The retained primary dislocations are of the edge type. Few, if any, screw dislocations are encountered (Mader, Seeger, and Theiringer 1963). The screw dislocations are eliminated during deformation (Essmann 1965).

6. The square root of the forest dislocation density N_f increases proportional to the flow stress τ in easy glide (Basinski and Basinski 1964, Brydges 1967) according to the relation $\tau = \alpha G b \sqrt{N_f}$, where α is 0.65 for the single-slip orientation tested by Argon and Brydges.

7. The often assumed proportionality between the square root of the primary dislocation density and the flow stress is unjustified (Basinski and Basinski 1964). The measured exponent of the primary dislocation density varies between 0.174 and 1.62 (Argon and Brydges 1968).

8. The rate of strain hardening at room temperature is of the order of 2 Kg/mm^2 and decreases with decreasing temperature and decreasing crystal size: it increases with orientation changes away from the ideal single-slip orientation (Mitchell 1964).

9. Slip lines are nearly a millimeter long but only of the order of 10–20 Burgers vectors deep (Mader 1957).

10. The extent of easy glide decreases with increasing temperature, increasing crystal size, and departure from the ideal single-slip orientation (Mitchell 1964).

16.3. Description of the Deformation Process

In a well-annealed crystal, grown-in dislocations inside subgrains are largely sessile. The mobile segments can be displaced by the application of a very low stress (~ 1 g/mm^2), and are trapped by sessile grown-in dislocations after gliding a certain distance. At higher stresses some short mobile segments of the sessile growth network act as initial dislocation sources. These initial sources merely initiate the self-sustaining process of dislocation source generation: as will be described below, new sources are produced and inactivated as the deformation continues. When no sources are present at all in the starting crystal, yielding is accompanied with a very large yield drop and the formation of Lüders bands as fresh slip is nucleated from one site and is spread into the perfect crystal by multiple cross glide. This process, which is observed only in perfect whiskers (Brenner 1958), remains outside the scope of this theory.

During deformation, screw dislocation segments of opposite sign frequently encounter each other gliding in opposite directions on parallel slip planes. Screw dislocations of the same sign, always moving in the same direction, rarely encounter each other. The existing evidence (Steeds 1966) suggests that when screw dislocations of opposite type encounter on parallel slip planes closer than a critical spacing δ apart, their mutual attraction stress produces cross slip and mutual annihilation. Although the details of this cross-slip annihilation process remain obscure, it is likely that it does not involve a constriction, but is of a type discussed by Fleischer (1959) where the stacking fault is draped over a short segment of a stair-rod dislocation.

When a dislocation loop expands from an initial source the screw segments will glide on the primary slip plane, on the average a distance $l/2$, until they encounter opposite type screw segments belonging to other expanding loops on parallel primary planes a distance less than δ away. When this occurs, the screw dislocation segment of one of the loops will cross-slip and annihilate the opposite type screw dislocation segment of the other loop, as shown in Figure 16.1. The result will be a large loop with parts on two parallel primary slip planes, joined by a long glissile jog on each of the two edge segments. Similar encounters of this loop with other loops will systematically eliminate the screw segments and will result only in segmented long-edge dislocations, as shown in Figure 16.1. A configuration of this type has been observed in copper by Essmann (1965). Thus the edge dislocations will be composed of segments of average length l on primary slip planes, connected to each other by long glissile jogs of random

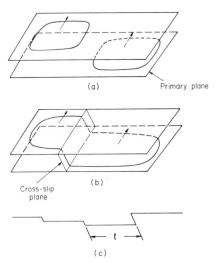

FIGURE 16.1. Formation of segmented edge dislocations by systematic elimination of screw segments from loops.

length between b and $\delta/\cos\psi$ (where $\psi = 70.6$ deg is the angle between the primary and cross-slip planes) and lying on cross-slip planes. If M is the volume density of active sources (to be discussed later), the mean length of edge segments (or the mean free path of screw segments) must be between

$$\frac{1}{\sqrt{\delta M}} < l < \frac{1}{\delta M^{2/3}},$$

depending upon whether a three-dimensional or a two-dimensional model is considered.

If plastic deformation is homogeneous to the extent that mutual encounters of screw dislocation segments can be considered random at all distances of vertical separation, it is easily possible to describe the normalized distribution of segments of the edge dislocation in a direction y, perpendicular to the slip plane, by a function $g(y)$, in such a way that the probability of finding a segment of a long-edge dislocation between y and $y + dy$ is $g(y)dy$. Several such segment-distribution functions for dislocations having between 1 and 5 segments are shown in Figure 16.2.

At a given resolved shear stress τ on the primary slip plane when two segmented edge dislocations of opposite sign encounter each other, the main interaction will be between the opposing segments of the two dislocations immediately facing each other. Side interactions between segments of the two dislocations are not important in the first confrontation, but they become important in further developments between these two dislocations. If a pair of opposing segments of the two segmented edge dislocations encounter each other within the half capture cross-section $\theta = Gb/8\pi(1 - v)\tau$, for edge dislocations, they will mutually trap each other

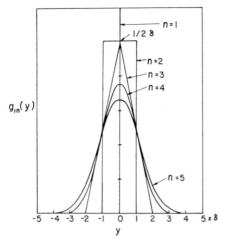

FIGURE 16.2. Segment density dsitribution function $g(y)$ as a function of the distance y/δ out of the slip plane, away from the center of gravity of segments, for dislocations with $n = 1, 2, 3, 4,$ and 5 segments.

by forming a dipole segment. Alternatively, if the two segments encounter each other at a distance greater than θ apart, they will pass each other. If some segment pairs are trapped and some pass, the possibility of unraveling of trapped segments by the line tension of the passing segments, as shown in Figure 16.3, must be considered. If the trapped segments lie within a

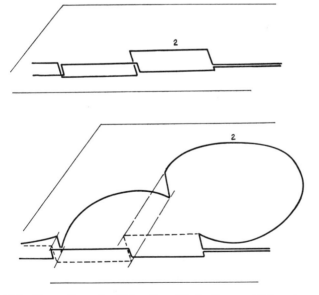

FIGURE 16.3. Unraveling of a dislocation dipole. Segment 2 is not captured; by passing its opposing segment, it tears away the captured segment 1.

critical binding distance θ_b they cannot be torn free by the line tension of any adjacent bowed-out segments.

When a dislocation with a segment distribution $g(y)$ encounters another dislocation with a segment distribution $g(y)$ with its center of gravity on a parallel plane a distance ξ away, the probability $p(\xi)$ of mutual capture of two opposing segments of the dislocations can be readily calculated as the overlap of the two segment distributions. If the dislocation encounters an already developed multipole with $\mu - 1$ dislocations having a segment distribution $g_{\mu-1}(y)$, the probability $p_\mu(\xi)$ of capture of a segment of the new dislocation can be calculated similarly. To obtain the probability of capture, $Q_{\mu n}(\xi)$, of an entire dislocation with all of its n segments by a multipole with $\mu - 1$ dislocations, it is necessary to consider the various interactions of trapped and passing segments as discussed above.

When a pair of passing segments have bound neighbors on both sides, one or both of the passing segments may act as dislocation sources if they lie on planes separated by a distance larger than θ from *all* other dislocation segments, bound or unbound.[1] Some possibilities that need to be considered are sketched in Figure 16.4. Once a source is produced, it cannot remain

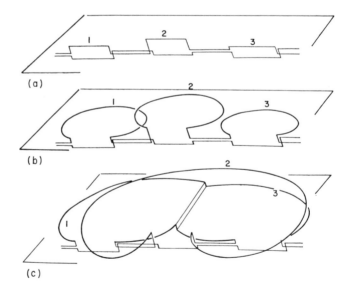

FIGURE 16.4. Possible modes of side interaction of dislocation segments which have to be considered in source formation by the partial capture of a segmented edge dislocation.

[1] It would seem that due to mutual shielding, the capture cross-section of dislocations in multipoles is smaller than that for a free dislocation: examination of capture cross-sections of dipoles, tripoles, and quadrupoles shows, however, no significant decrease (Chen, Gilman, and Head 1964).

active indefinitely but will be inactivated when a new dislocation segment is captured by the active dislocation segment.[2]

From the segment distribution function $g(y)$ it is possible to calculate the probability, $P_{\mu n}(\xi)$, of source production by the partial capture of an n-segmented edge dislocation by a multipole with $\mu - 1$ dislocations. The probability of inactivation, $S_{\mu n}(\xi)$, of active dislocation sources by capture of a new segment by the active segment in a multipole with $\mu - 1$ dislocations is determined along similar lines. The latter probability is proportional to the distribution function of active sources in the multipoles, which in turn is proportional to the spatial distribution of the probability of source formation.

Once these probabilities of dislocation capture $Q_{\mu n}(\xi)$, dislocation source formation $P_{\mu n}(\xi)$, and source inactivation, $S_{\mu n}(\xi)$, are obtained for various resolved shear stresses τ, on the primary slip plane and as a function of μ, the number of dislocations contained in a multipole, the dynamics of glide can be investigated. For this the crystal is idealized as a collection of bundles of parallel multipoles with an average spacing of h, as shown in Figure 16.5, and a dislocation forest with density N_f threading through the

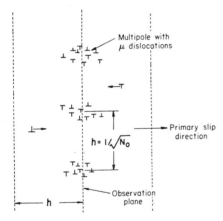

FIGURE 16.5. Idealization of the crystal into a collection of multipole bundles.

primary slip plane. Since only grown-in dislocations act as multipole nuclei (Argon and Brydges 1968), the spacing h is taken as constant. It is assumed that the average velocity of the primary dislocations is governed by thermally activated intersections with forest dislocations. As will be discussed

---

[2] The description of the source production mechanism implies that the segment lengths l should steadily increase with strain. In addition, alignment of the segment ends sketched in Figures 16.3. and 16.4 would seem unrealistic. In reality, the segment ends will not align, but randomly overlap, with many jogs crossing over free segments of other dislocations, and producing junction reactions between some jogs and free segments. Such junction reactions would serve to refine the segment lengths and counteract the lengthening effect described above.

further below, the experiments of Argon and Brydges suggest that through the hardening of secondary-slip systems this velocity remains approximately constant during easy glide even in the face of arising flow stress by strain hardening.

In tensile straining with a homogeneous strain distribution the dislocation flux ϕ_T through any part of the crystal must be constant and equal to the strain rate divided by the magnitude of the Burgers vector. This dislocation flux through any arbitrary observation plane is made up of the rate of emission of dislocations from sources in multipoles on both sides of the observation plane—subject to filtering by multipoles between the sources and the observation plane and the dislocations of opposite type moving in the other direction.

If new dislocations are produced from a source in a multipole as soon as the previously emitted dislocation loop has moved away from the source by a distance equal to the mean distance, $1/\sqrt{N_T}$, between mobile dislocations having a density N_T, the dislocation flux going through a portion h of the observation plane becomes

$$hN_T v = h\varphi_T = 2Kv\sqrt{N_T}\, e^{-h/\Lambda}/(1 - e^{-h/\Lambda}), \qquad (16.1)$$

where K is the number of active sources in a multipole, v the average velocity of dislocations, and Λ the mean free path of dislocations—all functions of the resolved shear stress, and all to be discussed in detail below. Since K and Λ are also dependent on the number μ of dislocations in a multipole, Equation 16.2 is an implicit equation for the dislocation density N_T, which can be transmitted through the crystal having multipoles containing μ dislocations, under a given stress τ. At a constant stress as the multipoles grow in size by accumulating more and more dislocations, the transmitted dislocation density will steadily decrease, resulting in a decreasing external strain rate. Thus in this model the hardening under constant stress results directly from the decreased flux transmittance of the crystal.

Below we will now develop in detail the processes which were described in words above.

16.4. Theoretical Development

16.4.1. Production of Segmented Dislocations

Let y be the position of the dislocation segment relative to the slip plane, measured normal to the slip plane. The density distribution of segments as a function of y can be represented statistically by a function $g(y)$ such that $g(y)\, dy$ gives the probability of finding a primary edge segment of a long-edge dislocation on a parallel primary plane a distance y away from the center of gravity of the long dislocation. The density distribution of segments will depend on the number n of segments of length l along a long dislocation which, in turn, is governed by the subgrain size or width of the crystal.

To illustrate how the segment distribution functions can be obtained, consider a simple case where the dislocation contains three segments (i.e., $n = 3$), separated by two jogs. The segments of the dislocation are confined into a range of 4δ, and the segment distribution will be triangular with a peak value of $1/2\delta$, as shown in Figure 16.2. This distribution is obtained by first dividing the vertical range of 4δ into four cells of size δ. Given one segment of the dislocation, the probability of finding the second segment in a strip of width dy in the two adjoining cells is $dy/2\delta$. For each possible position of the second segment the third segment will again have an equal probability of lying at any distance within plus or minus δ from this position. If y_3 is designated the coordinate of the point of interest for the third segment, the density distribution $g_3(y_3)$ of the third segment is found by integration over all possible positions of the second segment, i.e.,

$$g_3(y_3) = \frac{1}{2\delta} \int_{y_2 = y_3 - \delta}^{\delta} \frac{1}{2\delta} \, dy_2 = \frac{2\delta - y_3}{4\delta^2}. \tag{16.2}$$

The distribution in Equation 16.2 represents the positive half of the symmetric total distribution shown in Figure 16.2. In a similar manner, the segment distribution functions for dislocations with different numbers of segments can be calculated. Examples of segment distributions up to $n = 5$ are shown in Figure 16.2. As the number of segments becomes larger the distribution tends to a normal distribution.

The process of elimination of screw dislocations as described above, in the form of a cascade starting out near one surface and working its way across the width of the crystal, is a limiting form which gives the widest possible range for the distribution. In reality the cascade can start anywhere in the interior of the crystal and advance outward toward the two surfaces — producing segment distributions of narrower range. For simplicity we will ignore any such changes of the segment distribution function along the dislocation line, and will adopt for the whole dislocation the simple limiting scheme of calculation of segment distributions discussed above.

The development given above represents an idealized set of circumstances. For example, it is assumed that the stress levels are always high enough where the distance δ is larger than the half capture cross sections θ_s of a screw dislocation, so that screw dislocation dipoles do not form. In addition it is assumed that while the screw dislocations are eliminated, they do not interact with multipoles. This restriction is imposed for the convenience to deal with only one segment distribution function.

16.4.2. Growth of Multipoles

When two segmented edge dislocations of opposite type encounter each other on parallel planes a distance ξ apart, the interaction will be primarily between opposing segments. If the segments lie on planes less than the capture distance $\theta = Gb(8\pi(1 - v)\tau$ apart, for the prevailing shear stress τ, they will mutually capture. If all opposing segments of the two encountering dislocations capture each other, a dipole is formed which could be moved around by other mobile dislocations until captured by a grown-in dislocation, where it will become part of a growing multipole.

Experiments of Argon and Brydges (1968) show that only a certain fraction of the grown-in dislocations act as effective centers for the growth of multipoles by capturing glide dislocations and dipoles. Call the initial density of such nuclei N_0. Although some increase in the multipole density is observed with strain, for simplicity, we assume here that N_0 remains constant during easy glide.

As shown in Figure 16.5, we idealize the crystal as a collection of multipole nuclei at an average spacing $h = 1/\sqrt{N_0}$ lying on the primary slip planes parallel to the primary edge dislocations, and a uniform forest of dislocations with density N_f threading through the primary plane. We now pick a plane normal to the primary Burgers vector as an observation plane. Schematically we represent all dislocation arrivals from the left and all departures to the right of this observation plane. If the dislocation density in motion, N_T, moves with a velocity v, the observation plane is subjected to a dislocation flux $\phi_T = vN_T$. We assume that the segment distribution $g_n(y_n)$ of the moving dislocations is known and that the frequency of arrival of dislocations is independent of the y coordinates.

Consider a single dislocation with a segment distribution $g_{1n}(y)$ (the first subscript indicates that the dislocation is single and the second subscript stands for the number of segments in the dislocation) situated at the origin. acting as a nucleus of a possible multipole. The probability of capture $p_{2n}(\xi)$ of an opposing segment of a glide dislocation with center of gravity of segments on a plane a distance ξ away from the origin by the dislocation at the origin is

$$p_{2n}(\xi) = \int_{-(n-1)\delta}^{(n-1)\delta} g_{1n}(y)\, dy \int_{y-\theta}^{y+\theta} g_{1n}(\eta - \xi)\, d\eta.$$

The above expression is correct only when the probability of capture is small. When the multipole grows and $g_{\mu n}(y)$ (μ indicates the number of dislocations in the multipole) becomes large, the correct expression for the probability of capture of opposing segments when they are considered alone becomes an exponential which, for example, for $p_{2n}(\xi)$ is

$$p_{2n}(\xi) = 1 - \exp\left(-\int_{-(n-1)\delta}^{(n-1)\delta} g_{1n}(y)\, dy \int_{y-\theta}^{y+\theta} g_{1n}(\eta - \xi)\, d\eta\right). \quad (16.3)$$

When considering the probability of capture of all n segments of a glide dislocation by a multipole we must consider all possible modes of capture. Thus, as discussed in A & E in addition to the simultaneous capture of all

n segments of the glide dislocation outright, it is possible that $(n - m)$ segments pass during the first encounter while only m segments are caught, while the passing $(n - m)$ segments could then be caught by other segments of the glide dislocation or multipole on their return trip, as sketched in Figure 16.4.

As we will discuss shortly below, however, when a segment passes, its line tension exerts a strong force on the adjacent captured segments which tends to unravel these segments in the majority of cases. The captured segments can resist the line tension force of a passing segment only if they lie within a small binding distance θ_b. Since mutual encounter of opposing segments of two edge dislocations within this critical binding distance θ_b is a rare event, it can be safely assumed that in the large majority of cases whenever a pair of opposite segments pass each other, they will unravel all captured segments and the dislocation will manage to slip by intact. For this reason the probability of capture $Q_{\mu n}(\xi)$ of a dislocation with n segments by a multipole with $\mu - 1$ dislocations will be very nearly equal to the probability of capture of all if its n segments outright, i.e.,[3]

$$Q_{\mu n}(\xi) = (p_{\mu n}(\xi))^n. \tag{16.4}$$

The rate R_g of dislocation capture at a multipole would then become

$$R_g = \int_r Q_{\mu n}(\xi)\phi_T \, d\xi, \tag{16.5}$$

where the integration is over the entire range over which the integrand exists.

16.4.3. Rate of Dislocation Source Production

As shown in Figure 16.3, when all segments but one of a dislocation are arrested by a multipole the line tension of the passing segment tends to unravel the adjacent bound segments.[4] It is shown in Appendix 16.1 that the tearing action of the line tension can be resisted only if the bound segments on both sides lie within a critical binding distance θ_b of their corresponding segments belonging to the multipole where

$$\theta_b \cong 5.5 \times 10^{-6} \text{ cm} \tag{16.6}$$

in copper.

---

[3] For more complete treatments of the derivation of the expressions for the various probabilities, see Feller (1966).
[4] Inspection of Figure 16.3 would indicate that the arrest of jogs of a dislocation by segments of other dislocations would also serve to anchor the ends of free segments and influence the rate of source production. Such arrests, however, are effective only in one direction and can therefore not be important in stable source production.

For an operable source to develop in the manner sketched out in Figure 16.4, it is not sufficient that one segment of the trapped dislocation pass, but it is necessary that at least two segments on either side of it be bound to their corresponding trapping segments of the multipole within the binding distance θ_b.

Consider a dislocation with n segments interacting with a multipole of n segmented $\mu - 1$ dislocations. Out of the n segments of the dislocation m may be captured within the binding distance θ_b. This leaves $m + 1$ spaces between the bound elements containing a total of $n - m$ segments in $m + 1$ segment groupings, where a segment grouping may contain from zero to a maximum number of $n - m$ segments.

First we are interested in the frequency distribution of the lengths of segment groupings that lie between the m segments that have been captured within the critical binding distance. This is done by first enumerating all possible capture modes of the m segments out of a total of n segments, to determine the number of segment groups that have j adjacent segments lying outside the critical binding distance θ_b, while the m segments are captured within this binding distance. This number of j segment groupings is then divided by the total number of unbound segments of all lengths lying between the m bound segments. The resulting distribution of lengths j of segment groupings produced by m binding captures of a total of n segments will be referred to as $_{mn}p(j)$, and is discussed in detail in Appendix 16.2. Tables 16A2.1 and 16A2.2 in Appendix 16.2 give the distribution of the function $_{mn}p(j)$ for a dislocation with $n = 10$, and $n = 20$ segments, respectively.

The probability p_a that m segments become caught within the binding distance θ_b while the remaining $n - m$ segments lie outside the binding distance is (note that the $n - m$ segments not caught within the binding distance θ_b may still be caught within the half-capture cross-section θ of an individual dislocation segment),

$$p_a = \frac{n!}{m!(n-m)!} \left[\left(\frac{\theta_b}{\theta} p_{\mu n}(\xi) \right)^m \right] \left[\left(1 - \frac{\theta_b}{\theta} p_{\mu n}(\xi) \right)^{n-m} \right]. \qquad (16.7)$$

If m segments become bound with $m + 1$ gaps in between them (forming $m + 1$ groupings of unbound segments between them) there are then $m + 1$ potential sources possible. We now assume that the first requirement for the formation of a potential source is that at least one segment in a segment grouping of j adjacent unbound segments lies outside the half-capture cross section θ of an individual segment. This segment, which can pass, can then tear out all other $j - 1$ adjacent segments between the two bound segments at the ends. The probability, p_b, that at least one segment among j segments in a segment grouping lies outside the capture distance θ is equal to

$$p_b = 1 - \begin{Bmatrix} \text{probability that all segments lie within} \\ \text{the capture distance } \theta \end{Bmatrix},$$

i.e.,

$$p_b = (1 - (p_{un}(\xi))^j). \tag{16.8}$$

Consider for example a segment grouping of four ($j = 4$) adjacent unbound segments A, B, C, and D as shown in Figure 16.6, where at least

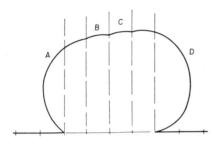

FIGURE 16.6. Source formation by the passage of four segments that have not been captured in a binding mode.

one of the four was outside the capture distance θ and tore out the other unbound but trapped segments. For this source to operate freely either A or D or both must clear all other segments of the dislocation and the multipole. Segments B and C, which lie in between, now become unimportant as they will be carried away by the first loop, with the three long jogs between the four segments being constrained to move on their respective planes, as indicated by the dotted lines in Figure 16.6. The probability p_c that either A or D or both will clear all other segments is given by

$$p_c = 1 - (1 - q)^2 = 2q - q^2, \tag{16.9}$$

where

$$q = (1 - p_{un}(\xi))^n (1 - p_{1n}(\xi = 0))^{n-1}, \tag{16.10}$$

is the probability that a segment (A or D) passes all other segments. In Equation 16.10, $p_{1n}(\xi = 0)$ refers to the segment distribution function of a single dislocation with center of gravity at $\xi = 0$.

When $m + 1$ segment groupings between m bound segments are formed, in principle these could give rise to k sources, where k can be a number between 1 and $m + 1$. Finally then, the probability, $p^{(k)}$ of producing k sources out of $m + 1$ possible segment groupings is

$$p^{(k)}(\xi) = \sum_{j=1}^{n-m} \frac{(m+1)!}{(m+1-k)!k!} (_{mn}p(j))(p_a)(p_b)(p_c). \tag{16.11}$$

On the other hand, k sources can be produced by a variety of different numbers m of bound segments. Hence, a summation over all m is necessary, so that the probability, $P_{\mu n}^{(k)}(\xi)$, of producing k sources by the capture of the μth dislocation by a multipole having $\mu - 1$ dislocations becomes

$$P_{\mu n}^{(k)}(\xi) = \sum_{m=s(k)}^{n} \sum_{j=1}^{n-m} \left(\frac{(m+1)!}{(m+1-k)!k!} \right) \left(\frac{n!}{m!(n-m)!} \right) [L_{mn}p(j)]$$

$$\left[\left(\frac{\theta_b}{\theta} P_{\mu n}(\xi) \right)^m \right] \left[\left(1 - \frac{\theta_b}{\theta} P_{\mu n}(\xi) \right)^{n-m} \right] [1 - (p_{\mu n}(\xi))^j]$$

$$[2(1 - p_{\mu n}(\xi))^n (1 - p_{1n}(\xi = 0))^{n-1}$$

$$- (1 - p_{\mu n}(\xi))^{2n} (1 - p_{1n}(\xi = 0))^{2(n-1)}], \qquad (16.12)$$

where

$s(k) = 1$	for $m = 1$ and
$s(k) = k - 1$	for all other m;
$1 \leqslant k \leqslant m + 1$	in the range $m \leqslant n/2$ and
$k \leqslant n/2 + 1$	in the range $n/2 < m \leqslant n$.

We calculate next the most probable number $\bar{k}$ of sources produced. This is

$$\bar{k} = \frac{\int_r \sum_{k=1}^{n/2} k P_{\mu n}^{(k)}(\xi) \, d\xi}{\int_r \sum_{k=1}^{n/2} P_{\mu n}^{(k)}(\xi) \, d\xi}, \qquad (16.13)$$

where the integration again is performed over the entire range over which the distribution is defined.

The mean time t_s, spent in the capture of a dislocation to produce the most probable number of sources, is

$$t_s \cong \left[\phi_T \int_r \sum_{k=1}^{n/2} P_{\mu n}^{(k)}(\xi) \, d\xi \right]^{-1}, \qquad (16.14)$$

where ϕ_T is the dislocation flux which the bundles of multipoles will transmit.

Once a dislocation is caught in any of the possible modes, it produces the most probable number $\bar{k}$ of sources. The rate R_p of source production, therefore, becomes

$$R_p = \frac{\bar{k}}{t_s} = \bar{k}\phi_T \int_r \sum_{k=1}^{n/2} P_{\mu n}^{(k)}(\xi) \, d\xi, \qquad (16.15)$$

and after substitution of Equation 16.13 into Equation 16.15,

$$R_p = \phi_T \int_r \sum_{k=1}^{n/2} k P_{\mu n}^{(k)}(\xi) \, d\xi. \qquad (16.16)$$

16.4.4. Rate of Dislocation Source Inactivation

To calculate the probability of inactivating a source by the μth dislocation, we obtain first the density distribution for the position of active sources in multipoles. The distribution of active sources along the observation plane away from the center of the multipole must parallel closely the ξ dependence of the probability of source production. Therefore, we write the active source density in the multipole per unit length along the observation plane as

$$\frac{K \sum_{k=1}^{n/2} k P_{\mu n}^{(k)}(\xi)}{\int_r \sum_{k=1}^{n/2} k P_{\mu n}^{(k)}(\eta) \, d\eta}, \qquad (16.17)$$

where K is the number of active sources per multipole, and the rest of the expression is the normalized position distribution of active sources.

The rate of source inactivation R_i by the μth dislocation in a multipole with $\mu - 1$ dislocations now becomes

$$R_i = \phi_T \int_r \left[p_{\mu n}(\xi) \left(\frac{2\theta_b}{\theta} \right) \right] \frac{K \sum_{k=1}^{n/2} k P_{\mu n}^{(k)}(\xi)}{\int_r \sum_{k=1}^{n/2} k P_{\mu n}^{(k)}(\eta) \, d\eta} \, d\xi, \qquad (16.18)$$

where the term in brackets represents the probability that the inactivating segment will be bound to the active segment which was operating as a source. The rate of source inactivation given above is approximate only in that it ignores the specific makeup of the operating source, which may be composed of more than one segment, as illustrated in Figure 16.6.

16.4.5. Average Velocity of Mobile Dislocations

If relativistic effects are excluded, the average velocity of a mobile dislocation in a perfect crystal is governed by its inherent interactions with the lattice, which produces a lattice friction stress, or by its interaction with lattice thermal energy. When other dislocations are present in the crystal, they either influence the motion of mobile dislocation through their internal stress fields (Chen, Gilman and Head 1964, Li 1968, Argon 1968, and others) or slow the mobile dislocations down by acting as forest trees requiring intersection (Cottrell 1953, Seeger 1954, and others).[5]

In the absence of substantial internal stress, in our restricted consideration, the average velocity of a mobile dislocation is governed by its interaction with phonons and by its simple-jog-producing intersections with

---

[5] In impure materials and materials subjected to irradiation or thermal treatment a whole host of other mechanisms may be present checking the velocity of dislocations. As mentioned above we restrict here our attention only to pure close-packed metals. For a more complete discussion of dislocation drag mechanisms, see Gilman (1968), and Nabarro (1967).

forest dislocations. The resulting average velocity can be written as

$$v = c \left[\frac{c}{v_p(\tau, T)} + \exp\left(\frac{H_0}{kT} \left(1 - \frac{F}{F_m} \right) \right) \right]^{-1}, \tag{16.19}$$

where c is the velocity of a shear wave, and v_p the velocity of a glide dislocation under a stress τ in a perfect crystal where only phonon interactions are important. Since both Basinski's (1959) and Mitchell's (1962) measurements of the stress dependence of the activation energy were made in stage II, where a long-range internal stress may be present, the constant H_0 can be evaluated only indirectly, as will be discussed below. The quantity F/F_m at room temperature is equal to the ratio of the flow stress at room temperature to the flow stress at absolute zero divided by the ratio of the shear moduli at these two temperatures. If Diehl and Berner's (1960) measurements of the temperature dependence of the flow stress in easy glide are used in conjunction with Köster's (1948) measurements of the temperature dependence of the Young's modulus, the ratio F/F_m is found to be near 0.65, which is the value of the coefficient between the flow stress and the forest spacing experimentally measured by Brydges (1967). Therefore, we write the ratio

$$\frac{F}{F_m} \cong \frac{\tau l_f}{Gb} = \alpha, \tag{16.20}$$

where $\alpha = 0.65$ in the easy glide orientation tested. Since Brydges found that Equation 16.20 is obeyed for all flow stresses tested, the ratio of F/F_m and therefore the intersection controlled velocity remains constant during deformation, as a first approximation.

16.4.6. Dynamics of Glide

The dislocation flux, actually made up of two oppositely directed streams, loses dislocations not only to the bundles of parallel multipoles but also in streaming through itself. The multipoles extract a certain fraction of dislocations per each multipole spacing h while the fraction lost by mutual capture will be more uniform with distance. When a dislocation moves through a distance dx of crystal, the increase in the probability of its capture dp_a from these two causes is

$$dp_a = dx \left[N_0 \int_r \left[Q_{\mu n}(\xi) + \sum_{k=1}^{n/2} k P_{\mu n}^{(k)}(\xi) \right] d\xi + 2N_T \int_r Q_{2n}(\xi) \, d\xi \right], \tag{16.21}$$

where the first integral represents the probability of capture of a mobile dislocation at a multipole (including the capture modes which result in source production), and the second integral represents the probability of capture of a mobile dislocation by another mobile dislocation to form a dipole. In Equation 16.21, N_0 is the multipole bundle density and N_T the transmitted dislocation density. The fraction f of flux extracted by the multipoles and by mutual capture of mobile dislocations per distance x moved is then

$$f = 1 - \exp\left(-x\left[N_0 \int_r Q_{\mu n}(\xi)\, d\xi + 2N_T \int_r Q_{2n}(\xi)\, d\xi\right]\right), \quad (16.22)$$

where the very small contribution of capture modes by source production has been neglected.

Evidently the mean free path Λ of a mobile dislocation is

$$\Lambda = \left[N_0 \int_r Q_{\mu n}(\xi)\, d\xi + 2N_T \int_r Q_{2n}(\xi)\, d\xi\right]^{-1}, \quad (16.23)$$

from which the dislocation accumulation rate is obtained as

$$\frac{dN}{d\gamma} = N_0 \frac{d\mu}{d\gamma} = \frac{1}{b\Lambda} = \frac{N_0}{b}\int_r Q_{\mu n}(\xi)\, d\xi + \frac{2N_T}{b}\int_r Q_{2n}(\xi)\, d\xi. \quad (16.24)$$

We now focus our attention on an observation plane. There are K active sources in a multipole, each of which emit a dislocation pair under the applied stress as soon as the previously emitted dislocation pair has moved away from the source in the multipole by a distance equal to the average spacing $l_T = 1/\sqrt{N_T}$ of mobile dislocations.

The number of mobile dislocations $(hN_T v)$ arriving at the observation plane per vertical distance h is made up of a series of terms

$$hN_T v = 2Kv\sqrt{N_T}\, e^{-h/\Lambda} + 2Kv\sqrt{N_T}\, e^{-2h/\Lambda} \\ + \cdots + 2Kv\sqrt{N_T}\, e^{-(s+1)h/\Lambda} + \cdots, \quad (16.25)$$

where the first term on the right-hand side gives the number of dislocations emitted by the two immediately adjacent multipoles and screened by the opposed dislocation flux stream, the second term gives the dislocations emitted from the next-nearest-neighbor multipoles and screened by both the opposed flux stream and the nearest neighbor-multipoles, and so on. From the description of the process it is evident that the sum is a geometric progression which in a closed form is

$$hN_T v = 2Kv\sqrt{N_T}\, e^{-h/\Lambda} \frac{1 - e^{-sh/\Lambda}}{1 - e^{-h/\Lambda}}. \quad (16.26)$$

In a very large crystal where s is substantial, the transmitted density assumes the limiting form of

$$(e^{h/\Lambda} - 1)\sqrt{N_T} = \frac{2K}{h}, \quad (16.27)$$

where it is recalled that Λ is a function of N_T.

The number of active sources K in a multipole, on the other hand, obeys a differential equation

$$\frac{1}{\phi_T}\frac{dK}{dt} = \int_r \sum kP_{\mu n}^{(k)}(\xi)\, d\xi - K\int_r S_{\mu n}'(\xi)\, d\xi, \tag{16.28}$$

where

$$S_{\mu n}'(\xi) = \left(\frac{2\theta_b}{\theta}\right)P_{\mu n}(\xi)\frac{\sum kP_{\mu n}^{(k)}(\xi)}{\int_r \sum kP_{\mu n}^{(k)}(\eta)\, d\eta}, \tag{16.29}$$

and the second integral in Equation 16.28 represents the rate of source inactivation per unit mobile dislocation flux.

Both from intuition and from examination of the values of the integrals in Equation 16.28, it can be concluded that the number of sources K must vary only slowly with increasing μ in easy glide. Therefore, the number of active sources per multipole is, in the steady-state region of deformation,

$$K(\mu) = \frac{\int_r \sum kP_{\mu n}^{(k)}(\xi)\, d\xi}{\int_r S_{\mu n}'(\xi)\, d\xi}. \tag{16.30}$$

If the expression for the number of active sources (Equation 16.30) is substituted into Equation 16.27, an expression is obtained for the transmitted density as a function of stress and multipole size (strain)

$$(e^{h/\Lambda} - 1)\sqrt{N_T} = \frac{2}{h}\frac{\int_r \sum kP_{\mu n}^{(k)}(\xi)\, d\xi}{\int_r S_{\mu n}'(\xi)\, d\xi}. \tag{16.31}$$

Multiplication of the transmitted density in Equation 16.31 by the velocity and Burgers vector gives the obtainable strain rate as a function of stress at a certain strain history (size of multipoles). If the stress is held constant, as will be discussed below, there is a gradual decrease of the velocity due to a decrease in the forest spacing accompanying the increase of the size of multipoles, both of which produce a decrease in the transmitted density, and therefore, decrease of strain rate with increasing strain, i.e., transient creep.

To obtain strain at a steady rate the stress has to be systematically increased to keep the transmitted dislocation flux, $\phi_T = N_T v$, equal to the geometrically demanded dislocation flux, $N_m v$, i.e.,

$$\frac{\dot{\gamma}}{b} = vN_m = vN_T. \tag{16.32}$$

Equation 16.32 is an implicit statement of the stress-strain law from the very beginning of deformation through easy glide.

16.4.7. Solution of Process Equations

A computer program was written for dislocations with ten segments, $n = 10$, to obtain the step-by-step change of the main process variables, i.e., the capture probability distributions (Equation 16.4), source production probability distributions (16.12), and source inactivation probability distributions (16.29) of growing multipoles, From these, other important quantities such as number of active sources in multipoles (16.30), transmitted density (16.31), and mean free path (16.23) were calculated for multipole sizes containing up to 500 dislocations. In the probability cross sections the length dimension is given in units of the critical cross slip distance, δ, while the stress is represented as the capture cross section θ of individual segments which is also measured in units of δ.

In Figure 16.7 the change of probability of capture of a glide dislocation

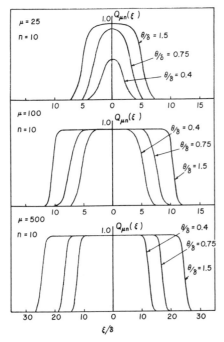

FIGURE 16.7. Change of probability of capture of a segmented dislocation by a multipole having 25, 100, and 500 dislocations, at three capture cross sections ($n = 10$).

across a multipole is plotted for three multipole sizes ($\mu = 25$, 100, and 500) each at three stresses (capture cross sections). Evidently with straining the center of the multipole becomes opaque to glide dislocations; the stress affects the capture probability only along the active margins.

Figure 16.8 shows the change in the source production probability by a

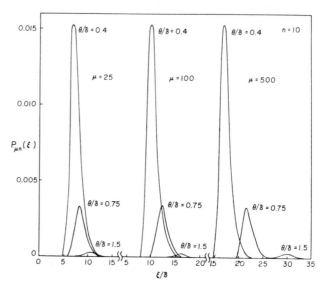

FIGURE 16.8. Change of probability of source formation by a segmented dislocation in a multipole with 25, 100, and 500 dislocations, at three capture cross sections ($n = 10$).

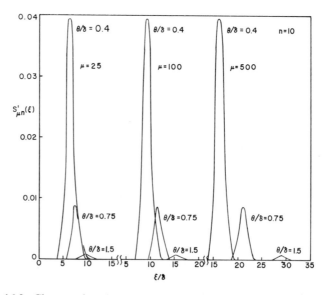

FIGURE 16.9. Change of probability of source inactivation by a segmented dislocation in a multipole with 25, 100, and 500 dislocations, at three capture cross sections ($n = 10$).

glide dislocation across a multipole, while Figure 16.9 shows the corresponding source inactivation probability by a glide dislocation across a multipole for the same multipole size and stress values given in Figure 16.7. It is evident that the active sources are located along the margins of the multipoles where partial capture of a glide dislocation is possible. Figure 16.10 is a summary of several of the calculated quantities, showing the

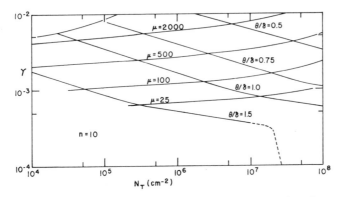

FIGURE 16.10. Transmitted dislocation density as a function of strain at a constant capture cross section, and constant multipole size.

change in the transmitted density with strain both at constant stress as the size of multipoles change, and at constant multipole size with increasing stress. In the determination of the mean free path, and strain accounting in Figure 16.10, a value of 3×10^3 was used for the multipole density N_0, a figure obtained by direct measurement of the average spacing of etch pit clusters in the experiments of Argon and Brydges (1968).

16.4.8. The Stress-Strain Curve

The stress-strain curve can be obtained from the results of Figure 16.10 once the strain rate is prescribed. In the experiments of Argon and Brydges (1968) a strain rate of 10^{-4} per second was used. If we take the mobile dislocation density as 10^5 per cm^2 — or equal to the dislocation density relaxing upon stress release — observed in etching experiments by Argon and Brydges, we can calculate an average dislocation velocity of 8×10^{-3} cm/sec, and a constant $H_0 = 1.27$ electron volts for Equation 16.19. In order to determine the stress-strain curve, plots of the type shown in Figure 16.10 were obtained for several assumed values of δ. For each such set of curves a transmitted density of 10^5 per cm^2 gave a stress $(1/\theta)$ − strain curve from which a value of δ was obtained by equating the yield stress τ_0 to 35 g/mm². This value of δ was checked with the assumed δ. The curves in Figure 16.10 are for the self-consistent scheme in which $\delta = 3 \times 10^{-4}$ cm, while Figure 16.11 shows the stress-strain curve obtained from the curves of Figure 16.10.

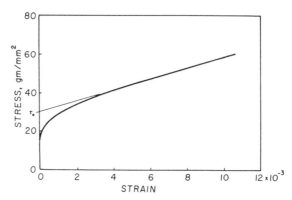

FIGURE 16.11. Stress-strain curve obtained from Figure 16.10 for a constant
dislocation flux to give $\dot{\gamma} = 0.0001$ per second.

16.5. Discussion of Results

Although the statistical theory discussed above can correctly
account for the development of preyield deformation, the stress-strain curve
given in Figure 16.11 is not accurate in the preyield range because the assum
ption of steady-state conditions, made to obtain the transmitted dislocation density at constant stress, is inappropriate for the nonsteady-state
preyield region. For this reason no comparison with experimental microstrain curves is warranted for the preyield region.

On the other hand the agreement of calculated quantities with experimental measurements is good in the steady-state deformation region of easy
glide. The hardening rate, either scaled off directly from Figure 16.11 or
obtained from Figure 16.10 according to the formula

$$\theta = \frac{d\tau}{d\gamma} = \frac{\tau}{\gamma} \frac{(\partial \ln N_T)/(\partial \ln \gamma)_{\theta/\delta}}{(\partial \ln N_T)/(\partial \ln \theta/\delta)_\mu}, \tag{16.33}$$

gives for $\theta_I = d\tau/d\gamma = 2.95$ Kg/mm^2, or very nearly the lowest hardening
rates observed by Argon and Brydges (1968) and quoted under item 8 in
Section 16.2. In the earlier theory (A & E) a much higher hardening rate of
10.3 Kg/mm^2 was reported, for which two reasons were suspected — (a) that
some of the dislocations captured at low stresses in multipoles may be
extracted and reused at higher stresses, and (b) that the dislocations in
multipoles capturing glide dislocations are shielded by the other dislocations of the multipole resulting in cancellation in stress fields and smaller
capture cross sections. Both of these possibilities had been ignored in
A & E. Detailed analysis showed that when these effects are properly
accounted for, the reduction of the hardening rate is insignificantly small.
This is because (a) the opaque centers of the multipoles are highly elongated
in the slip direction as is clear from the experimental observations of Argon
and Brydges, so that if a dislocation is torn out of any interior part of

the multipole it will be trapped again by other dislocations in the multipole before moving an appreciable distance. Only weakly trapped dislocations on the margins of the multipoles could be extracted and reused at high stresses; these, however, make up only a small portion of the trapped dislocations. (b) Examination of dipole and quadrupole capture cross sections calculated by Chen, Gilman, and Head (1964) show that although they are not as symmetrical as that for a monopole dislocation, they are on the average no smaller than the capture cross section of a monopole. As the analysis above shows, the real reason for the decreased hardening rate is due to the unraveling of trapped but unbound dislocation segments by untrapped segments. Unless all segments of a dislocation are trapped outright, they will, in most instances, be torn away by an untrapped segment, decreasing very markedly the rate of stable dislocation trapping.

The rate of dislocation multiplication dN/dy calculated by the theory is about 10^9 lines per square centimeter per unit strain. The rate is considerably higher than what the etch-pit measurements would indicate (Young 1962, Hordon 1962, Argon and Brydges 1968), but leads to a mean free path Λ of 0.4 mm, which is very close to the slip-line lengths measured by Mader (1957). Since it is possible that the etch-pit method may give an underestimated count when dislocations lie close together, the agreement with the slip-line measurements is considered to be more significant. The present modification of the theory has produced no important changes in the calculated slip-line depths λ of about 1.5×10^{-6} cm given in the earlier theory (A & E).

In the present form of the theory, easy glide is assumed to terminate when the multipole capture cross section has become equal to the multipole spacing, so that the entire crystal is filled with primary dipoles and multipoles (see Kuhlmann-Wilsdorf 1962). This occurs in our model of an infinite crystal at a strain of 0.9 percent — a value low but not unreasonable.

The present theory introduces no new understanding to the hardening of secondary slip systems, i.e., whether their hardening responds to the resolved shear stress on the secondary slip planes or to the hardening of the primary slip planes. Since the hardening of the secondary slip systems is an important ingredient of the orientation sensitivity, and the temperature sensitivity of the flow stress and the hardening rate, no discussion of these effects will be given here. Although the phenomenological approach in the earlier theory (A & E) to these effects still gives a very good account for them, there is very little experimental information (Basinski and Basinski 1964) to provide a critical comparison. In view of this it becomes premature to develop the theory further in this direction.

Apart from the observation by Essman (1965) reproduced in A & E there has been no experimental evidence for the assumed mode of source production. This requires further experimental work with transmission electron microscopy.

This work was supported by the National Science Foundation under Grant GK–596. I am grateful to Mr. George East, who performed the computations and in so doing discovered a number of inaccuracies in the formulas. I am also indebted to Professor M. F. Ashby, whose several queries prompted useful clarifications in the text.

REFERENCES

Argon, A. S., 1968, *Materials Science and Engineering*, **3**, 24.

Argon, A. S., and W. T. Brydges, 1968 *Phil. Mag.*, **16**, 817.

Argon, A. S., and G. East 1968, *Proc. Int'l Conf. on the Strength of Metals and Alloys Trans. Jap. Inst. Met.*, **9** (suppl.) 756.

Basinski, Z. S., 1959, *Phil. Mag.*, **4**, 393.

Basinski, Z. S., and S. J. Basinski, 1964 *Phil. Mag.*, **9**, 51.

Brenner, S. S., 1958 in R. H. Doremus et al. (eds.) *Growth and Perfection of Crystals*, International Conference on Crystal Growth, Cooperstown, N. Y., New York: Wiley, p. 157.

Brydges, W. T., 1967, *Phil. Mag.*, **15**, 1079.

Chen, H. S., J. J. Gilman, and A. K. Head, 1964, *J. Appl. Phys.*, **35**, 2502.

Clarebrough, L. M., and M. E. Hargreaves, 1959, in B. Chalmers and R. King (eds.), *Progress in Metal Physics*, New York: Pergamon, vol. 8, p. 1.

Cottrell, A. H., 1953, *Dislocations and Plastic Flow in Crystals*, Oxford: Clarendon Press, p. 174.

Diehl, J., and R. Berner, 1960, *Z. Metallk.*, **51**, 522.

Elam, C. F., 1935, *The Distortion of Metal Crystals*, Oxford: Clarendon Press.

Essmann, U., 1965, *Phys. Stat. Sol.*, **12**, 707.

Feller, W., 1966, *An Introduction to Probability Theory and Its Applications*, New York: Wiley, vol. 1.

Fleischer, R. L., 1959, *Acta Met.*, **7**, 134.

Gilman, J. J., 1962, *J. Appl. Phys.*, **33**, 2703.

Gilman, J. J., 1968 in U. S. Lindholm (ed.), *Mechanical Behavior of Materials under Dynamic Loads*, New York: Springer, p. 152.

Hordon, M. J., 1962, *Acta Met.*, **10**, 999.

Joffee, A. F., 1928, *The Physics of Crystals*, New York: McGraw-Hill.

Köster, W., 1948, *Z. Metallk.*, **39**, 1.

Kuhlmann-Wilsdorf, D., 1962, *Trans. AIME*, **224**, 1047.

Li, J. C. M., 1968, in A. R. Rosenfield et al. (eds.), *Dislocation Dynamics*, McGraw-Hill, New York, p. 87.

Maddin, R., and N. K. Chen, 1954, in B. Chalmers and R. King (eds.), *Progress in Metal Physics*, New York: Pergamon, vol. 5, p. 53.

Mader, S., 1957, *Z. Physik*, **149** 73.

Mader, S., 1963, in G. Thomas and J. Washburn (eds.), *Electron Microscopy and Strength of Crystals*, New York: Wiley-Interscience, p. 183.

Mader, S., A. Seeger, and H. Theiringer, 1963, *J. Appl. Phys.* **34**, 3376.

Mitchell, T. E., 1962, *Ph.D. thesis*, Cambridge University.

Mitchell, T. E., 1964, in E. G. Stanford et al. (eds.), *Progress in Applied Materials Research*, New York: Gordon and Breach, vol. 65, p. 117.

Nabarro, F. R. N., 1967, *Theory of Crystal Dislocations*, Oxford: Clarendon Press, pp. 482–559

Nabarro, F. R. N., Z. S. Basinski, and D. B. Holt, 1964, *Advances in Physics.*, **13**, 193.

Reusch, E., 1867, *Ann. Physik*, Lpz., **132**, 441.

Schmid, E., and W. Boas, 1935, *Kristallplastizität*, Springer, Berlin; (1950) (English translation) *Plasticity of Crystals*, London: Hughes.

Seeger A., 1954, *Phil. Mag.*, **45**, 771.

Seeger, A., 1958, in S. Flügge (ed.), *Encyclopedia of Physics*, Berlin-Heidelberg-New York: Springer, p. 1.

Steeds, I. W., 1966, *Proc. Roy. Soc.*, **A292**, 343.

Taylor, G. I., 1934, *Proc. Roy. Soc.*, **A145**, 362.

Tinder, R. F., and J. Washburn, 1964, *Acta Met.*, **12**, 129.

Young, F. W., 1962, *J. Appl. Phys.* **33**, 963.

Young, F. W., 1967, in H. Steffen Peiser (ed.), *Crystal Growth, Proceedings*, International Conference on Crystal Growth — Boston — 1966, New York: Pergamon, p. 789.

APPENDIX 16.1: UNRAVELING OF IMPERFECTLY BOUND DISLOCATIONS

Consider two-edge dislocations of opposite type on parallel planes a distance t apart, as shown in Figure 16A. 1. When a stress is applied to the

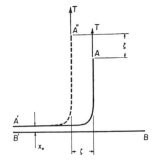

FIGURE 16A.1. The process of tearing away of a dislocation from a multipole.

crystal to exert a force on part of the dislocation line AA', the latter may unravel itself from the dislocation BB' if it is not bound too closely to BB'. We will assume that dislocation BB' is bound to a multipole and is immobile. At the critical binding distance $t = \theta_b$, the work done by the line tension force acting on dislocation AA' at A as it is displaced through a distance ζ must be equal to the change in energy of the bound dislocation of a length ζ, as illustrated in Figure 16A.1.

The energy E per unit length of a pair of dislocations, as shown in the figure, is

$$E = \frac{Gb^2}{2\pi(1-v)}\left[\ln\frac{t\sqrt{1+(x_0/t)^2}}{r_0} - \frac{x_0^2}{x_0^2 + t^2} + 1\right], \qquad (16A1.1)$$

where x_0 is the projected distance between the dislocations under an applied stress. For tear-away under the above assumptions, we have

$$\left[2E_0 - E\left(\frac{x_0}{t}\right)\right]\xi = T\xi = E_0\,\xi,\tag{16A1.2}$$

where E_0 is the energy of a free dislocation in a crystal. Simplification of Equation 16A1.2 gives

$$\ln\frac{Rr_0}{t^2}\frac{1}{[1+(x_0/t)^2]} = 2\frac{1}{[1+(x_0/t)^2]},\tag{16A1.3}$$

where R is the average dislocation spacing in the crystal. For at least one solution to exist, the derivatives of both sides of Equation 16A1.3 must also be equal. This establishes the condition,

$$t = \theta_b \le \sqrt{\frac{Rr_0}{2e}},\tag{16A1.4}$$

for the critical separation between the two dislocations within which they can be considered stably bound. When $R = 1/\sqrt{N_T} = 3.16 \times 10^{-3}$ cm, $\theta_b \le 5.5 \times 10^{-6}$ cm.

APPENDIX 16.2: FREQUENCY DISTRIBUTION OF LENGTHS OF SEGMENT GROUPINGS

When m segments out of n interacting segments of a glide dislocation are captured within the binding distance θ_b, $m + 1$ spaces between the m bound segments are created, which contain a total of $n - m$ segments in $m + 1$ segment groupings, where a segment grouping may contain from zero to a maximum number $n - m$ segments.

Of interest is the frequency distribution of the lengths of segment groupings. When all the possible capture arrangements are enumerated we find that the total number of segment groups containing (j) segments is

$$(m + 1)u_m[k_m(j)],\tag{16A2.1}$$

where $k_m(j) = n - (m - 1) - j$, and $u_m[k_m(j)]$ is given by the recursion formula

$$u_m[k_m(j)] = \sum_{h=1}^{k_m(j)} u_{m-1}(h),$$

with

$$u_1(k_m) = 1$$
$$u_2(k_m) = u_2(n - 1 - j) = [n - (m - 1) - j]_{m=2} = n - 1 - j = h.$$

To find the fraction that this number represents of the total number of segments, we need only count the total number of segments of all lengths (including zero length, i.e. adjacent captures). This number v_m is

$$v_m = (m + 1)\left(\frac{n!}{m!(n-m)!}\right).$$
(16A2.2)

From here, then, the frequency distribution of segment groupings, with (j) elements arising from m binding captures out of a total of n segments becomes

$$_{mn}P(j) = \frac{m!(n-m)!}{n!}\,u_m[k_m(j)].$$
(16A2.3)

Two such distributions for dislocations with 10 and 20 segments are listed in Tables 16A2.1 and 16A2.2 below.

TABLE 16A2.1. Distribution of lengths, j, of segment groupings when m out of n segments suffer binding capture ($n = 10$).

m \ j	1	2	3	4	5	6	7	8	9	10
1	0.100	0.100	0.100	0.100	0.100	0.100	0.100	0.100	0.100	0
2	0.178	0.155	0.133	0.111	0.089	0.067	0.045	0.022	0	
3	0.233	0.175	0.125	0.083	0.050	0.025	0.008	0		
4	0.265	0.167	0.095	0.048	0.019	0.005	0			
5	0.278	0.139	0.060	0.020	0.004	0				
6	0.266	0.100	0.029	0.005	0					
7	0.234	0.058	0.008	0						
8	0.178	0.022	0							
9	0.100	0								
10	0									

TABLE 16A2.2. Distribution of lengths j, of segment groupings when m out of n segments suffer binding capture ($n = 20$).

m \ j	1	2	3	4	5	6	7	8	9	10	11	12	13	14	15	16	17	18	19	20
1	.050	.050	.050	.050	.050	.050	.050	.050	.050	.050	.050	.050	.050	.050	.050	.050	.050	.050	.050	0
2	.095	.089	.084	.079	.074	.068	.063	.058	.053	.047	.042	.037	.032	.026	.021	.016	.011	.005	0	
3	.134	.119	.105	.092	.080	.069	.058	.048	.039	.032	.025	.018	.013	.009	.005	.003	.001	0		
4	.169	.140	.116	.094	.073	.059	.045	.034	.025	.017	.012	.007	.004	.002	.001	.000	0			
5	.197	.154	.117	.088	.065	.046	.032	.021	.013	.008	.005	.002	.001	.000	.000	0				
6	.221	.160	.113	.077	.052	.033	.020	.012	.006	.003	.001	.001	.000	.000	0					
7	.239	.159	.103	.065	.039	.022	.012	.006	.003	.001	.000	.000	.000	0						
8	.253	.154	.091	.051	.027	.014	.006	.003	.001	.000	.000	.000	0							
9	.261	.145	.077	.038	.018	.008	.003	.001	.000	.000	.000	0								
10	.263	.131	.062	.027	.011	.004	.001	.000	.000	.000	0									
11	.261	.116	.048	.018	.006	.002	.000	.000	.000	0										
12	.253	.098	.035	.011	.003	.001	.000	.000	0											
13	.239	.080	.023	.006	.001	.000	.000	0												
14	.221	.061	.014	.003	.000	.000	0													
15	.197	.044	.008	.001	.000	0														
16	.169	.028	.003	.001	0															
17	.134	.015	.001	0																
18	.095	.009	0																	
19	.050	0																		
20	0																			

ABSTRACT. A certain distribution of dislocation loops is examined to determine the stress dependence of strain rate, $\dot{\varepsilon}$, of the density of mobile dislocations, ρ, and of the average velocity, $\bar{v}$, of these dislocations in the light of the Orowan equation, $\dot{\varepsilon} = \phi\rho b\bar{v}$, where ϕ is a geometric factor and b is the magnitude of the Burgers vector. It is found that the stress dependence of the average dislocation velocity approaches that of a single dislocation only if all the dislocations are mobile. Otherwise, the application of the Orowan equation depends strongly on the fraction of mobile dislocations. In the limit of very small mobile fraction, the stress dependence of $\ln \bar{v}$ is twice that of $\ln v$ where v is the velocity of a single dislocation.

17. Dynamic Behavior of a Distribution of Dislocation Loops

J. C. M. LI

17.1. Introduction

In 1940, Orowan proposed the following relation between the strain rate $\dot{\varepsilon}$, the density of mobile dislocations ρ, and the average velocity $\bar{v}$ of these dislocations

$$\dot{\varepsilon} = \phi\rho b\bar{v}, \qquad (17.1)$$

where ϕ is a geometric factor. While this equation is fundamentally exact, its usefulness is limited because of the difficulty of separating the individual effects of ρ and $\bar{v}$.

Then in 1959, Johnston and Gilman reported direct measurements of single dislocation velocities as a function of stress, τ, and suggested the following relation

$$v = B\tau^{m^*}, \qquad (17.2)$$

where B is velocity at unit stress and m^* is a quantity now known as the velocity-stress exponent. Although these direct measurements gave the velocity of only individual dislocations and were conducted under the condition of negligible internal stress, Equation 17.2 has been widely used in conjunction with Equation 17.1 to relate macroscopic plasticity with the microdynamic behavior of dislocations.

245

The effect of internal stress, τ_i, in a work-hardened crystal is usually taken into account by modifying Equation 17.2,

$$\bar{v} = B(\tau - \tau_i)^{m_e^*}. \tag{17.3}$$

Equations 17.1 and 17.3 are then applied to experimental information from which τ_i and m_e^* are usually determined. The value of m_e^* is then compared with m^* from direct etch-pit measurements. Except for a discussion by Argon and East (1968) no attempt has been made concerning the variation of the density of mobile dislocations. Either its constancy is assumed or its constancy is questioned by examining the consequences of each.

The applicability of Equations 17.1 and 17.3 to a complex work-hardened state has never been carefully examined. Of course if such a state consists of straight dislocations moving in a homogeneous internal stress field, Equations 17.1 and 17.3 are clearly applicable. A more complicated work-hardened state will be examined in this paper to determine the limits of applicability of Equations 17.1 and 17.3.

17.2. Distribution of Dislocation Loops

A work-hardened state will be described by a distribution of dislocation loops. The distribution function is chosen to be simply[1]

$$N(r)\,dr = Ar\,\exp\left(-r/r_m\right)dr \tag{17.4}$$

where $N(r)\,dr$ is the number of loops between r and $r + dr$ and A and r_m are constants. Equation 17.4 is shown in Figure 17.1. The distribution satisfies the condition that very small and very large loops are rare.

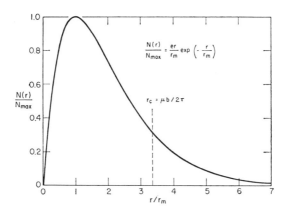

FIGURE 17.1. A distribution of dislocation loop radii.

[1] Kocks (1966) dicsussed other distribution functions. A random distribution of pinning points on a line would give the same exponential dependence as Equation 17.4 (see Koehler 1950).

Although both dislocation density and plastic strain will be calculated by assuming complete circular loops, the conclusion will be similar if, instead of complete loops, they are only arcs of different radii of curvature bulged between obstacles.

According to Equation 17.4, $N(r)$ has a maximum of Ar_m/e at $r = r_m$. The average radius is $2r_m$ and the root mean square radius is $\sqrt{6}r_m$. The total number of loops is Ar_m^2, and the total dislocation density is

$$\rho_t = 2\pi \int_0^\infty rN(r)\, dr = 4\pi Ar_m^3 \tag{17.5}$$

so that $A = \rho_t/4\pi r_m^3$.

The internal stress exerted at each loop is taken as $\mu b/2r$, which includes the interaction from all other loops. A loop is mobile only if the external stress τ exceeds $\mu b/2r$ or the loop radius exceeds $r_c = \mu b/2\tau$. Hence under the applied stress τ, the density of mobile dislocations is

$$\rho = 2\pi \int_{r_c}^\infty rN(r)\, dr = 2\pi Ar_m(r_c^2 + 2r_c r_m + 2r_m^2)\exp(-r_c/r_m) \tag{17.6}$$

or the fraction of mobile dislocations is

$$\frac{\rho}{\rho_t} = \left(1 + X_c + \frac{X_c^2}{2}\right)\exp(-X_c), \tag{17.7}$$

where $X_c = r_c/r_m$. Equation 17.7 is plotted in Figure 17.2, showing how this fraction decreases with increasing X_c or decreasing applied stress.

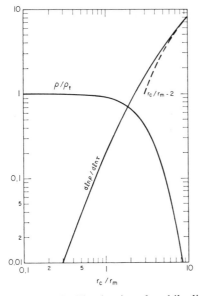

FIGURE 17.2. The density of mobile dislocations.

17.3. Variation of the Density of Mobile Dislocations with Stress

Equation 17.7 can be used to calculate a quantity usually ignored in treating deformation data, namely, the change of the density of mobile dislocations with stress,

$$\frac{d \ln \rho}{d \ln \tau} = -\frac{d \ln \rho}{d \ln X_c} = \frac{X_c^3}{2 + 2X_c + X_c^2}, \qquad (17.8)$$

which is plotted also in Figure 17.2. It is seen that this quantity is small for small X_c or large stress but becomes large and approaches $X_c - 2$ for large X_c or small stress. The variation of the density of mobile dislocations with stress is then more important in microstrain deformation, low-stress creep, and very low strain rate. It would be less important in high-speed deformation, in a state of uniform distribution of dislocations before cell formation, and in deformation under large stress.

17.4. Variation of Strain Rate with Stress

By assuming that the velocity of each loop is determined by Equation 17.3, the strain rate at an applied stress τ is

$$\dot{\varepsilon} = \phi b (2\pi) B \int_{\tau_c}^{\infty} r N(r)(\tau - \tau_i)^{m^*} \, dr$$

$$= \phi \pi A B \mu^{m^*} b^{(m^*+1)} 2^{(1-m^*)} \int_{r_c}^{\infty} r^2 \left(\frac{1}{r_c} - \frac{1}{r} \right)^{m^*} \exp\left(-r/r_m\right) dr. \quad (17.9)$$

Equation 17.9 can be used to calculate the change of strain rate with stress by means of a quantity usually denoted as m:

$$m = \frac{d \ln \dot{\varepsilon}}{d \ln \tau} = -\frac{d \ln \dot{\varepsilon}}{d \ln r_c} = \frac{m^*}{r_c} \frac{\int_{r_c}^{\infty} r^2 (1/r_c - 1/r)^{m^*-1} \exp(-r/r_m) \, dr}{\int_{r_c}^{\infty} r^2 (1/r_c - 1/r)^{m^*} \exp(-r/r_m) \, dr}.$$

$$(17.10)$$

Simple expressions for m can be obtained for small integers of m^*. For example, when $m^* = 1$ and 2

$$m_1 = 1 + \frac{X_c(1 + X_c)}{2 + X_c} \qquad (17.11)$$

$$m_2 = 2 + X_c. \qquad (17.12)$$

For large m^* the expressions are more complicated such as, for $m^* = 3$, 4, and 5,

$$m_3 = 6/[X_c^2 - X_c + 2 - X_c^3 E_1(X_c)\exp X_c] \tag{17.13}$$

$$m_4 = \frac{24/m_3}{X_c^3 + 3X_c^2 - 2X_c + 2 - X_c^3(X_c + 4)E_1(X_c)\exp X_c} \tag{17.14}$$

$$m_5 = \frac{240/m_3 m_4}{X_c^4 + 9X_c^3 + 12X_c^2 - 6X_c + 4 - X_c^3(X_c^2 + 10X_c + 20)E_1(X_c)\exp(X_c)} \tag{17.15}$$

where E_1 is the exponential integral (Abramowitz and Stegun, 1965)

$$E_1(X_c) = \int_{X_c}^{\infty} \exp(-t)\, dt/t \tag{17.16}$$

whose values are tabulated.

Equations 17.11–17.15 are plotted in Figure 17.3. It is seen that m could

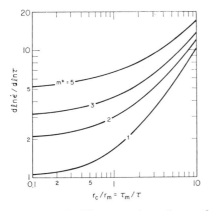

FIGURE 17.3. The stress dependence of strain rate.

be much larger than m^* and approaches m^* only at small X_c or when all dislocations are mobile. However, the percentage change of m with X_c decreases with increasing m^*. For example, between $X_c = 0.1$ and 10, m increases by a factor of 10 for $m^* = 1$, 6 for $m^* = 2$, 4.5 for $m^* = 3$, and 3.3 for $m^* = 5$. Such phenomena seem to have been observed experimentally. For example, between 0 and 10 percent plastic strain, m changes from 5 to 20 (a factor 4) in Fe (Michalak 1965), 15 to 45 (a factor of 3) in LiF; and 45 to 80 (a factor of less than 2) in Si-Fe (Johnston and Stein, 1963).

If such increase of m arises from an increase of X_c, an inspection of Figure 17.2 suggests that the fraction of mobile dislocations decreases with plastic strain. Some experimental support is shown by Argon and Brydges (1968). This seems to question the validity of some treatment of plastic deformation by assuming a constant fraction of mobile dislocations.

17.5. Stress Dependence of Average Dislocation Velocity

In view of Equation 17.1, m can be decomposed into two parts:

$$m = \frac{d \ln \dot{\varepsilon}}{d \ln \tau} = \frac{d \ln \rho}{d \ln \tau} + \frac{d \ln \bar{v}}{d \ln \tau}. \qquad (17.17)$$

Experimentally it is difficult to determine the change of the density of mobile dislocations with stress. However, in this analysis, it is given by Equation 17.8. Hence it is possible to examine the second term, which will be denoted by $m_v = d \ln \bar{v}/d \ln \tau$. For example, when $m^* = 1$,

$$m_v = 1 + \frac{X_c(2 + 4X_c + X_c^2)}{(2 + X_c)(2 + 2X_c + X_c^2)}, \qquad (17.18)$$

which is seen to approach 2 as a limit when $X_c \to \infty$. Similarly, when $m^* = 2$,

$$m_v = 2\left[1 + \frac{X_c(1 + X_c)}{2 + 2X_c + X_c^2}\right], \qquad (17.19)$$

which approaches 4 as a limit when $X_c \to \infty$. It can be shown for any m^*, m_v approaches $2m^*$ as a limit.

The foregoing indicates that, at least for the distribution of Equation 17.4, the average velocity-stress exponent, m_v, is greater than the actual velocity-stress exponent, m^*, by at most a factor of 2. However, experimental verification for such behavior requires a knowledge of $d \ln \rho/d \ln \tau$ which is difficult to obtain.

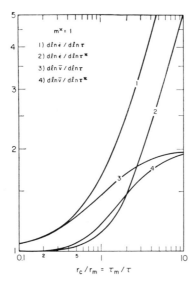

FIGURE 17.4. Stress dependence of strain rate and of average dislocation velocity for $m^* = 1$.

Equations 17.18 and 17.19 are plotted in Figures 17.4 and 17.5, together with other quantities to be discussed later.

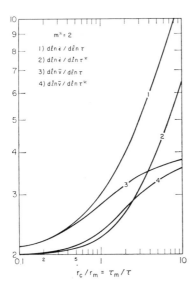

FIGURE 17.5. Stress dependence of strain rate and of average dislocation velocity for $m^* = 2$.

17.6. Effect of Internal Stress

In evaluating m and m_v, applied stress is used without correction of the internal stress. Since each loop has an internal stress of $\mu b/2r$, such internal stress should be taken into account. An average internal stress is defined by Equation 17.3 for the average dislocation velocity. Based on this definition, it is given by

$$\tau_i = \tau(1 - m_e^*/m_v). \tag{17.20}$$

The difference between τ and τ_i is defined as the effective stress:

$$\tau^* = \tau - \tau_i = \tau m_e^*/m_v. \tag{17.21}$$

The quantity m_e^* can be obtained from the variation of m_v with stress:

$$m_e^* = \frac{d \ln \bar{v}}{d \ln \tau^*} = m_v \Big/ \left(1 - \frac{d \ln m_v}{d \ln \tau}\right). \tag{17.22}$$

Equation 17.22 is plotted in Figures 17.4 and 17.5 for $m^* = 1$ and 2, respectively. Since m_v approaches $2m^*$ when $X_c \to \infty$, $d \ln m_v/d \ln \tau$ approaches zero, and consequently m_e^* approaches $2m^*$ also. This fact is clearly demonstrated for $m^* = 1$ and 2 in Figures 17.4 and 17.5. It is seen that, although a consideration of the effect of internal stress does bring m_e^* much closer to m^* than m_v, such correction is important only if X_c is small (less than 1). As mentioned earlier, experimental verification is difficult because of the unavailability of m_v.

A usual practice is to use m instead of m_v. This involves the assumption that $d \ln \rho/d \ln \tau$ is zero. The quantity so obtained will be denoted by m_e,

$$m_e = \frac{d \ln \dot{\varepsilon}}{d \ln \tau^*} = m \Big/ \left(1 - \frac{d \ln m}{d \ln \tau}\right), \tag{17.23}$$

which is plotted also in Figures 17.4 and 17.5 for $m^* = 1$ and 2, respectively. Here again, although a consideration of the effect of internal stress does bring m_e much closer to m^* than m, such correction is useful only for small X_c (less than 1). Unlike m_e^*, m_e does not approach a finite limit (it approaches $m/2$ or $(X_c/2) - 1 + m^*$) when $X_c \to \infty$.

17.7. Conclusions

By accepting Equation 17.2 as an experimental finding for the behavior of individual dislocations, the application of the Orowan equation, Equation 17.1, together with Equation 17.3, which is assumed to describe the behavior of the average velocity of dislocations, is limited to the case in which nearly all dislocations are mobile. In this case, m_e^* approcahes m^*, $d \ln \rho/d \ln \tau$ is nearly zero, and the internal stress is negligible.

When only a fraction of dislocations is mobile, the applicability of Equations 17.1 and 17.3 depends strongly on this fraction. First of all, for the distribution of dislocation loops of Equation 17.4, $d \ln \rho/d \ln \tau$ is less than unity if this fraction is more than 60 percent, but increases rapidly with decreasing fraction of mobile dislocations, as shown in Figure 17.2. Consequently, when the mobile fraction is small, $d \ln \rho/d \ln \tau$ must be taken into account, and in the limit of very small mobile fraction $d \ln \bar{v}/d \ln \tau$ approaches $2m^*$.

When the mobile fraction is not too small, $d \ln \dot{\varepsilon}/d \ln \tau$ can be used to compute m^* by taking into account the internal stress, or $d \ln \rho/d \ln \tau$, or both. As shown in Figures 17.4 and 17.5, the internal stress is more important at large mobile fraction (small r_c) and $d \ln \rho/d \ln \tau$ is more important at small mobile fraction (large r_c). Hence when the mobile fraction is not too small (50 percent for $m^* = 1$, 35 percent for $m^* = 2$, and smaller for larger m^*), $d \ln \dot{\varepsilon}/d \ln \tau$ can be used to compute m^* by neglecting $d \ln \rho/d \ln \tau$, but taking into account the internal stress.

REFERENCES

Abramowitz, M., and I. A. Stegun, 1965, *Handbook of Mathematical Functions* New York: Dover, p. 228.

Argon, A. S., and W. T. Brydges, 1968, *Phil. Mag.*, **18**, 817.

Argon, A. S., and G. East, 1968, *Trans. Japan Inst. Metals*, **9** Suppl, p. 756.

Johnston, W. G., and J. J. Gilman, 1959, *J. Appl. Phys.* **30**, 129.

Johnston, W. G., and D. F. Stein, 1963, *Acta Met.* **11**, 317.

Kocks, U. F., 1966, *Phil. Mag.* **13**, 541.

Koehler, J. S., 1950, in W. Shockley et al. (eds.), *Imperfections in Nearly Perfect Crystals* New York: Wiley, p. 197.

Michalak, J. T., 1965, *Acta Met.* **13**, 213.

Orowan, E., 1940, *Proc. Phys. Soc.* (London) **52**, 8.

II. Physics of Strength

3. Cracks and Fracture

ABSTRACT. The concept of a classical elastic crack is generalized in a way that enables the law of force between the faces of the fracture to be included explicitly in the formalism. The Griffith-Orowan-Irwin equation is then derived in terms of a specific law of force, and the application of the theory to brittle and semibrittle fractures is briefly discussed.

18. The Structure of a Crack

A. H. COTTRELL

18.1. Introduction

An important advance in the theory of fracture was made by Orowan (1948–1949) and Irwin (1948) when they interpreted the surface energy term in Griffith's crack equation as the total work, including plastic work, done in the zone of fracture during the propagation of the crack. This extension of Griffith's theory was made by general arguments, without recourse to any specific structure of the crack. A theoretical analysis of crack structure has now confirmed the Orowan-Irwin interpretation, but this has so far been limited to models of cracks in which the fracture faces are held together by a somewhat artificial rectilinear law of force (Bilby, Cottrell, and Swinden 1963). The purpose of this note is to present an alternative approach which enables more natural laws of force to be examined. Earlier investigations of this problem have of course been made, e.g. by Elliot (1947), and Rosenfield, Dai, and Hahn (1966).

18.2 The Distributed Crack

A classical crack in an elastic solid of Young's modulus E and Poisson's ratio v is represented by the deformation

$$
\begin{aligned}
y &= y_0[1 - (x^2/c_0^2)]^{1/2} && \text{for } x^2 \leq c_0^2, \\
y &= 0 && \text{for } x^2 \geq c_0^2,
\end{aligned}
$$

(18.1)

of the lower face of an infinite ha lf-body occupying the region $y \geq 0$ together with the stress distribution

$$
\begin{aligned}
\sigma(= \sigma_{yy}) &= \sigma_0[1 - (c_0^2/x^2)]^{-1/2} && \text{for } x^2 \geq c_0^2, \\
\sigma &= 0 && \text{for } x^2 \leq c_0^2,
\end{aligned}
\tag{18.2}
$$

with

$$
\sigma_0 = Ey_0/[2(1 - v^2)c_0] \tag{18.3}
$$

as the uniform (applied) stress at large distances from the crack, which maintains (unstable) equilibrium.

This crack could be regarded as formed by the superposition of an infinite number of coincident cracks each with an infinitesimal displacement dy_0 and correspondingly infinitesimal stress $d\sigma_0$. We can thus obtain a more general result by distributing these over a range of crack lengths. Let $dy_0(c)$ be the contribution from those infinitesimal cracks of semilength in the range c to $c + dc$. Thus

$$
y_0 = \int_0^\infty \frac{dy_0(c)}{dc}\, dc. \tag{18.4}
$$

The displacement is then

$$
y(x) = \int_{c=x}^\infty \frac{dy_0(c)}{dc}\, \frac{(c^2 - x^2)^{1/2}}{c}\, dc \tag{18.5}
$$

and the stress is

$$
\sigma(x) = \frac{Ex}{2(1 - v^2)} \int_0^x \frac{dy_0(c)}{dc}\, \frac{dc}{(x^2 - c^2)^{1/2}\, c}, \tag{18.6}
$$

with

$$
\sigma_0 = \frac{E}{2(1 - v^2)} \int_0^\infty \frac{dy_0(c)}{dc}\, \frac{dc}{c}. \tag{18.7}
$$

We can choose various functions for $dy_0(c)/dc$, determine $y(x)$ and $\sigma(x)$ for each of these, and so find the σ, y relation, i.e., the law of force across the plane of fracture $y = 0$. Many realistic laws of force can readily be found in this way, but most have to be evaluated numerically.

18.3. A Simple Example

A simple analytical solution can be obtained by assuming

$$
\begin{aligned}
dy_0(c)/dc &= Ac^2 && \text{for } c_1 \leq c \leq c_2 \\
&= 0 && \text{outside these limits.}
\end{aligned}
\tag{18.8}
$$

The assumption that $dy_0(c)/dc = 0$ at a finite upper limit c_2 implies that the material has a finite elastic limit before it begins to fail. With this distribution function in Equation 18.6 we see that $\sigma(x)$ increases as x is increased toward c. Hence the stress is a maximum at c_2, i.e.,

$$\sigma = \sigma_f \qquad \text{at } c = c_2, \tag{18.9}$$

where σ_f is a property of the material, its *failure stress*. The lower limit c_1 is the point at which all cohesion across the plane of fracture is lost. Thus

$$y = u_f \qquad \text{at } c = c_1 \tag{18.10}$$

where u_f is a property of the material, its *failure displacement*.

We obtain from Equation 18.4

$$y_0 = \tfrac{1}{3}A(c_2^3 - c_1^3) \tag{18.11}$$

and from Equation 18.7

$$\sigma_0 = \frac{E}{2(1 - v^2)} \tfrac{1}{2}A(c_2^2 - c_1^2). \tag{18.12}$$

These give the classical σ_0, y_0 relation in the limit $c_2 \to c_1 \to c_0$. We also obtain, in the range $c_1 < x < c_2$

$$y(x) = \tfrac{1}{3}A(c_2^2 - x^2)^{3/2} \tag{18.13}$$

and

$$\sigma(x) = \frac{E}{2(1 - v^2)} Ax(x^2 - c_1^2)^{1/2}. \tag{18.14}$$

Let

$$p(x) = \sigma(x)\frac{2(1 - v^2)}{EA} = x(x^2 - c_1^2)^{1/2} \tag{18.15}$$

and

$$q(x) = y(x)\frac{3}{A} = (c_2^2 - x^2)^{3/2}. \tag{18.16}$$

We then obtain

$$p = \tfrac{1}{2}[(2c_2^2 - c_1^2 - 2q^{2/3})^2 - c_1^4]^{1/2}, \tag{18.17}$$

which is the law of force, $p = f(q)$, across the plane of fracture. The stress attains its greatest value at $q = 0$ and then falls, at first gradually and then more steeply with increasing q, reaching zero at $q = (c_2^2 - c_1^2)^{3/2}$, which corresponds to the displacement u_f. A linear component can be added to this law of force, as required, to take account of reversible Hookeian deformation between the bounding faces of the fracture zone.

We further obtain, with $E/2(1 - v^2)$ as the unit of stress

$$u_f = \tfrac{1}{3}A(c_2^2 - c_1^2)^{3/2}, \tag{18.18}$$

$$\sigma_f = Ac_2(c_2^2 - c_1^2)^{1/2}, \tag{18.19}$$

$$A = 8\sigma_0^3/9u_f^2, \tag{18.20}$$

$$\sigma_0 = (3u_f\sigma_f/4c_2)^{1/2}, \tag{18.21}$$

$$c_1 = c_2[1 - (4\sigma_0^2/\sigma_f^2)]^{1/2}. \tag{18.22}$$

Since $\sigma_f u_f$ is work per unit area, Equation 18.21 is the Orowan-Irwin form of the Griffith equation, in the limit $(c_2 - c_1) \ll c_2$, i.e., $\sigma_0 \ll \sigma_f$, for which c_2 can be regarded as the half-length of the crack. In this same limit, Equation 18.22 becomes

$$(c_2 - c_1)/c_2 = 2\sigma_0^2/\sigma_f^2 \tag{18.23}$$

and so confirms the "plastic particle" concept of Neuber (1937) and Felbeck and Orowan (1955).

Combining this formula with Equation 18.21, we obtain

$$c_2 - c_1 = 3u_f/2\sigma_f \tag{18.24}$$

and see that, in this limit of long cracks and low applied stresses, the size of the failure zone $c_2 - c_1$ is a property of the material, independent of the crack length. Let $c_2 - c_1 = \delta c$. Then, when $\delta c \ll c_2$, Equations 18.18 and 18.19 reduce to

$$u_f = \tfrac{1}{3}Ac_2^{3/2}(2\delta c)^{3/2}, \tag{18.25}$$

$$\sigma_f = Ac_2^{3/2}(2\delta c)^{1/2}, \tag{18.26}$$

so that, since u_f, σ_f, and δc are properties of the material, then

$$A = \lambda c_2^{-3/2} \qquad (\lambda = \text{constant}) \tag{18.27}$$

and this gives, with Equation 18.21, a constant "stress-intensity factor"

$$K = \sigma_0(\pi c_2)^{1/2} \tag{18.28}$$

for a long crack of any length that is on the point of growing, in the material (Barenblatt 1962).[1] A short calculation then gives the law of force across the plane of fracture as

$$\frac{\sigma}{\sigma_f} = \left[1 - \left(\frac{y}{u_f}\right)^{2/3}\right]^{1/2}. \tag{18.29}$$

[1] I am grateful to Dr. J. D. Eshelby and Professor B. A. Bilby for pointing this out.

18.4. Physical Interpretation

The most direct application of the above results is to brittle cleavage, in which case Equation 18.29, together with a reversible linear term, represents the law of atomic force across the plane of fracture and $(1/2)u_f\,\sigma_f$ is approximately the specific surface energy of this plane. The assumption of a finite elastic limit is an unreal feature of the theory, when applied to this case, since in practice there is always some deviation from linearity, however small, when atomic bonds are strained. In fact no relation of the Griffith type can ever be exact when applied to elastic failure in a discrete atomic system. The error is of course very small when $\sigma_0 \ll \sigma_f$.

Orowan (1948–1949) emphasized that in many macroscopically brittle fractures in, for example, steel, the work of fracture is vastly greater than the specific surface energy, and he explained this in terms of localized plastic deformation in the failure zone. For such fractures, then, the failure displacement u_f is correspondingly large and is determined by features of the microstructure, e.g., grain size or spacing of foreign inclusions, rather than by atomic properties. Under the plane-strain conditions which are typical of these semibrittle fractures in steel, the mode of failure must be heterogeneous, consisting of alternate regions of brittle cracking and ductile tearing; and it must also be discontinuous, i.e., with small separate cracks in the failure zone ahead of the main crack (Tipper 1957), Cottrell 1963). Furthermore, rather more than about two-thirds of the fracture area must consist of brittle cracking, since the plastic failure strength of the bridging regions, which is enhanced approximately threefold by the plastic constraint factor of Orowan (1946), will otherwise exceed the general yield strength of the material and this kind of macroscopically brittle fracture will then no longer be possible (Cottrell 1963). In fact, Hodgson and Boyd (1958) observed that long cracks in broken steel ships stopped when more than one-third of the fracture area became ductile.

REFERENCES

Barenblatt, G. I., 1962, in *Adv. Appl. Mech.*, **7**, 55.

Bilby, B. A., A. H. Cottrell, and T. Swinden 1963, *Proc. Roy. Soc.*, **A272**, 304.

Cottrell, A. H., 1963, *Proc. Roy. Soc.*, **A276**, 1.

Elliot, H. A., 1947, *Proc. Phys. Soc.*, **59**, 208.

Felbeck, D. K., and E. Orowan, 1955, *The Welding Journal*, **34**, 570.

Hodgson, J., and G. M. Boyd, 1958, *Trans. Inst. Nav. Arch.*, London, **100**, 141.

Irwin, G. R., 1948, *Trans. Amer. Soc. Metals*, **40**, 147.

Neuber, H., 1937, *Kerbspannungslehre*, Berlin, J. Springer.

Orowan, E., 1946, *Trans. Inst. Engr. Shipbuilders in Scotland*, **89**, 165.

Orowan, E., 1948–1949, in *Reports on Progress in Physics*. **12**, 214.

Rosenfield, A. R., P. K. Dai, and G. T. Hahn, 1966, *Proc. First Int'l Conf. on Fracture* (Sendai), **1**, 253.

Tipper, C. F., 1957, *J. Iron Steel Inst.*, **185**, 4.

ABSTRACT. A theorem of Bateman's is used to find the elastic field near the tip of an antiplane crack which starts moving in an arbitrary manner. The energy release rate is calculated and found to confirm a general expression previously proposed.

19. The Starting of a Crack

J. D. ESHELBY

19.1. Introduction

An intuitive feeling for the running of cracks, useful in their work, must have been acquired by both the earlier (Baden-Powell 1949) and later (Skertchley 1879) flint-knappers. Modern fracture mechanics is concerned with the prevention of crack propagation rather than with its encouragement. It is the object of this paper to obtain some insight into what goes on near a moving crack tip according to the linear isotropic theory of elasticity. We limit ourselves to states of antiplane strain. It is perhaps worth pointing out that certain interesting effects related to crack branching studied by Orowan and Yoffe (Yoffe 1951) still survive in antiplane strain (McClintock and Sukhatme 1960). It is to be hoped that our type of analysis can also be used for plane strain at the expense of the usual complications consequent on the existence of two velocities of wave propagation.

Elegant mathematical discussions have been given (Broberg 1960, Craggs 1963) of cracks which, starting from zero length, spread with constant velocity. Physically they are perhaps a little unrealistic, not so much because of the limitation to constant velocity, but because according to any reasonable criterion it needs an infinite stress to get a crack of zero length started.

We consider the case where the crack tip starts from rest and moves arbitrarily, the crack being initially of finite length. The analysis is, however, limited to the initial motion in the following sense. In a region close enough to the tip of a static crack, the elastic field, which is independent of the crack length and the details of the loading, is completely characterized by a single parameter, the stress intensity factor. Our results begin to be inaccurate when the disturbance due to the motion of the tip moves out of this region. Also, our analysis ignores the fact that after a finite interval of time the field at the tip will be upset by reflections from the surface of the specimen, and, loosely speaking, reflections from the other end of the crack, and also by disturbances generated at the other tip if it too is in motion. Which of these effects first invalidates our analysis, and after how long, will depend on the details of tip motion, geometry, and loading. Despite these limitations it seems worth working out the problem in detail. It can be solved (Section 19.2) with the help of a result of Bateman's (1915) or by fitting together simple solutions of the wave equation.

The ideas of Griffith (1920), Orowan (1949), and others on the energetics of crack propagation have, in fracture mechanics, become formalized into the concept of crack extension force or energy release rate.

In the static case we may, as Orowan (1955) has emphasized, distinguish two extreme cases. If a specimen containing a crack is deformed by rigid grips, an extension of the crack will obviously reduce the elastic energy. On the other hand, if the specimen is loaded by hanging a weight on it (dead loading), then when the crack extends the specimen becomes more flexible, and the weight descends and does work on the specimen, whose elastic energy consequently increases. However, the weight has lost an amount of potential energy which, as it turns out, is twice the increase of the elastic energy, so that there is again an overall reduction in energy. It is also true that when the rigid and dead loadings happen to produce the same elastic field near the crack tip, the energy reduction per unit extension of the crack is the same. Hence, other things being equal, the energy release rate is a quantity independent of the way in which the load is applied.

For the initial dynamic motion of a crack tip the nature of the loading mechanism is irrelevant until the disturbance produced by the motion of the tip has reached the surface of the specimen. Of the potential energy originally in the region now disturbed, some remains, some is converted into kinetic energy, and some flows out at the crack tip. The outflow, reckoned per unit length of crack front and per unit advance of the tip, is the energy release rate. In Section 19.3 we use the results of Section 19.2 to calculate it, and show that a general expression which has been proposed for the energy release rate gives the correct result.

In Section 19.4 we compare and contrast the elastic field near a moving crack tip with the fields around two other moving elastic singularities which have been studied, the kink and the screw dislocation.

19.2. The Moving Antiplane Strain Crack Tip

In a state of antiplane strain the displacement w is everywhere perpendicular to the xy plane and independent of z. The nonzero stress components are

$$p_{zx} = \mu \, \partial w/\partial x, \qquad p_{zy} = \mu \, \partial w/\partial y, \tag{19.1}$$

where μ is the shear modulus. The displacement must satisfy the wave equation

$$\nabla^2 w - \frac{1}{c^2} \frac{\partial^2 w}{\partial t^2} = 0 \tag{19.2}$$

where

$$c = (\mu/\rho)^{1/2}$$

and ρ is the density of the medium. If w is independent of time, Equation 19.2 becomes $\nabla^2 w = 0$ and solutions can be found by taking w to be the real part of some analytic function of the complex variable $z = x + iy$, which, of course has nothing to do with the z which figures as a suffix in p_{zx} and p_{zy}. If we take $w = \mathrm{Re}\, f(z)$, then the stresses are given by

$$p_{zx} - i p_{zy} = \mu f'(z).$$

The simplest crack-like solution is obtained by taking w to be a multiple of $z^{1/2}$. In this way we get the basic solution

$$w = W(x, y),$$

where

$$\left. \begin{aligned} W(x, y) &= B \, \mathrm{Re}(-iz^{1/2}) = B \, \mathrm{Im}\, z^{1/2} \\ &= 2^{-1/2} B\{(x^2 + y^2)^{1/2} - x\}^{1/2} \, \mathrm{sgn}\, y \\ &= Br^{1/2} \sin \tfrac{1}{2}\theta \end{aligned} \right\} \tag{19.3}$$

with the notation of the figure. The stresses are

$$p_{zx} = -\tfrac{1}{2}\mu Br^{-1/2} \sin \tfrac{1}{2}\theta, \quad p_{zy} = \tfrac{1}{2}\mu Br^{-1/2} \cos \tfrac{1}{2}\theta. \tag{19.4}$$

The displacement w is continuous across the x axis ahead of the crack ($\theta = 0$) but it has a discontinuity $\Delta w = 2B(-x)^{1/2}$ behind the tip. The stress is zero on the faces of the crack ($\theta = \pm \pi$). If we put

$$B = (2/\mu^2 \pi)^{1/2} K,$$

then

$$p_{zy} = (2\pi)^{-1/2} K x^{-1/2}, \, x > 0 \tag{19.5}$$

is the stress ahead of the crack in the plane of the crack. The constant K defined by Equation 19.5 is the stress intensity factor of fracture mechanics: strictly it should be written K_{III} to show that we are dealing with mode III deformation, that is, antiplane strain.

Across a cylinder $r = $ const. the traction is

$$p_{zr} = \mu \, \partial w/\partial r,$$

so that the displacement Equation 19.3 could be produced in a cylinder with a crack along the radius $\theta = \pi$ by forces proportional to $\sin (1/2)\theta$ distributed around its circumference and directed parallel to the generators. However, Equations 19.3 and 19.4 describe the elastic field near the tip of any crack in antiplane strain provided K is given a suitable value. In particular, if by forces remote from the origin we impose an applied stress p_{zx}^A, p_{zy}^A and then make a straight crack from the origin to $x = -l$, $y = 0$ the displacement is

$$w = (2/\mu^2\pi)^{1/2} K r^{1/2}[\sin \tfrac{1}{2}\theta + a(r/l)\sin \tfrac{3}{2}\theta$$
$$+ b(r/l)^2\sin \tfrac{5}{2}\theta + \cdots], \tag{19.6}$$

where

$$K = \frac{1}{(\tfrac{1}{2}\pi l)^{1/2}} \int_{-l}^{0} \left(\frac{x' + l}{-x'}\right)^{1/2} p_{zy}^A(x', 0) \, dx'.$$

Consequently for any type of loading, the elastic field near enough to the tip is the same as that of an elementary crack with a suitably chosen stress-intensity factor K.

In order to set the crack tip in motion we use the following result which is a special case of a somewhat more general theorem due to Bateman (1915). If $W(x, y)$ satisfies $\nabla^2 w = 0$ and is homogeneous of degree 1/2 in x, y, then

$$W[x - \xi(\tau), y]$$

with

$$c\tau = ct - \{[x - \xi(\tau)]^2 + y^2\}^{1/2} \tag{19.7}$$

and

$$\xi = \xi(\tau) \tag{19.8}$$

satisfies the wave equation 19.2. The theorem may be verified by substituting in Equation 19.2, and using the implicit relation Equation 19.7 together with Euler's theorem for homogeneous functions.

The relations 19.7 and 19.8, familiar in electromagnetic theory, can be given the following intuitive interpretation. A point moves along the x axis according to the law $x = \xi(t)$, emitting signals which propagate with velocity c. A signal received at time t and position (x, y) must have been emitted at time τ when the point was at $(\xi(\tau), 0)$.

For W we take the expression 19.3 for the static crack, so that

$$w = B \operatorname{Im}\{x - \xi(\tau) + iy\}^{1/2}. \tag{19.9}$$

We now try to show that the solution 19.9 of the wave equation actually represents a crack whose tip is moving according to the law $x = \xi(t)$. When $y = 0$, Equation 19.7 shows that, as the point x approaches the point $\xi(\tau)$ from left or right, $\tau \to t$, $\xi(\tau) \to \xi(t)$. But on the x axis

$$w = \pm B\{|x - \xi(\tau)| - [x - \xi(\tau)]\}^{1/2}$$

(see Equation 19.3) so that w is continuous or discontinuous across the x axis according as x is greater or less than $\xi(t)$. Also, it is not hard to show that for $y = 0$, $p_{zy} \neq 0$ or $p_{zy} = 0$ according as x is greater or less than $\xi(t)$. The main point is that in calculating $p_{zy} = \mu \, \partial w / \partial y$ from Equation 19.1 we have to take into account the variation of $\xi(\tau)$ with y, but that Equation 19.7 shows that $\partial \xi(\tau) / \partial y = 0$ when $y = 0$. Thus Equation 19.7 has the basic properties required of a crack extending from $-\infty$ to $\xi(t)$. It is not hard to convince oneself that for x, y close to the tip $\xi(\tau) \sim \xi(t)$ so that the elastic field near the tip resembles that for a stationary crack. This point is considered more carefully later.

These conclusions are confirmed by considering a special case. Suppose that the tip of a stationary crack starts to move with uniform velocity v at time $t = 0$. Then

$$\left.\begin{array}{l} \xi(\tau) = 0, \, \tau < 0 \\ \quad\quad = v\tau, \, \tau > 0 \end{array}\right\} \tag{19.10}$$

and Equation 19.7 can be solved to give

$$w = W(x, y), \qquad x^2 + y^2 > c^2 t^2 \tag{19.11}$$

$$= A(v)W\left[\frac{x - vt}{\beta(v)}, y\right], \qquad x^2 + y^2 < c^2 t^2 \tag{19.12}$$

where

$$\beta(v) = (1 - v^2/c^2)^{1/2}$$

and

$$A(v) = \left(\frac{1 - v/c}{1 + v/c}\right)^{1/4}. \tag{19.13}$$

The elastic field is divided into two regions separated by a circle of radius $r = ct$ expanding from the original position of the tip. Outside the circle the static solution remains unaltered. Inside the circle the solution

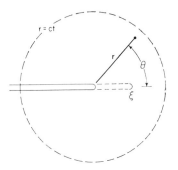

FIGURE 19.1.

takes a form which can be described as the static solution subjected to the Lorentz transformation

$$x \to (x - vt)/\beta(v), \qquad y \to y \tag{19.14}$$

and multiplied by the velocity-dependent constant $A(v)$.

It is worth considering how we might have derived Equations 19.11 and 19.12 if we were unaware of Bateman's theorem. It is obvious that the elastic field will be unaltered outside the circle $r = ct$ and, since the transformation 19.14 is a well-known way of generating moving solutions of the wave equation from stationary ones, we might perhaps have guessed the general form of Equation 19.12, though not the magnitude of A. To fix A we must impose some continuity condition across $r = ct$. If affixed "in," "out" refer to adjacent points just inside and outside the circle, we must obviously have

$$w_{in} = w_{out}. \tag{19.15}$$

According to Love (1927), the relation

$$c(\partial w/\partial n)_{in} + (\partial w/\partial t)_{in} = c(\partial w/\partial n)_{out} + (\partial w/\partial t)_{out} \tag{19.16}$$

must be satisfied across a surface of discontinuity which is advancing with velocity c in the direction of its normal n. Although it is not obvious without some algebra, Equations 19.11 and 19.12 in fact have the same angular dependence round $r = ct$, and Equation 19.15 leads to the value 19.16 for A. Fortunately, Equation 19.16 can also be satisfied all around the circle, and with the same value of A. Actually this is not a coincidence: if Equation 19.15 is valid along a surface of discontinuity which is advancing with velocity c, then 19.16 is also valid. (For a careful discussion, see Hill 1961).

Suppose that after moving for a time t_1 the crack tip increases its velocity to v_1. A new circle of discontinuity will begin to spread with velocity c from the point $(vt_1, 0)$ and outside it (but inside $r = ct$) the displacement will still be Equation 19.12. The previous argument suggests that inside the new circle the solution will be derived from Equation 19.3 by making the substitution

$$x \to [x - vt - v_1(t - t_1)]/\beta(v_1), \qquad y \to y$$

and multiplying by a suitable constant. Calculation shows that if we choose the constant to be $A(v_1)$, that is, Equation 19.13 with v replaced by v_1, then the new solution indeed satisfies Equations 19.15 or 19.16 at its junction with Equation 19.12 across the new circle of discontinuity.

We can go on to the case where the crack tip suffers a succession of changes of velocity. The displacement will evidently be

$$w = A[\dot{\xi}(t)]W\left[\frac{x - \xi(t)}{\beta[\dot{\xi}(t)]}, y\right] + O(R^{3/2}), \tag{19.17}$$

where

$$A[\dot{\xi}(t)] = [c - \dot{\xi}(t)]^{1/4}/[c + \dot{\xi}(t)]^{1/4}.$$

If the point of observation lies within the last discontinuity circle, the first term of Equation 19.17 is exact. If not there is an error, represented by the second term, which increases with distance R from the current position of the tip.

The following argument is supposed to show that, as indicated, the second term in Equation 19.17 is actually of the order $R^{3/2}$. If the reader finds it unacceptable it does not matter, since for the important case of continuous motion the result follows from the discussion of Equation 19.18 below.

Suppose the tip suffers a succession of roughly equal velocity increments at roughly equal intervals of time. If we move a distance R away from the crack tip we cross a number of discontinuity circles proportional to R and arrive at a point where the elastic field is that appropriate to a fictitious crack with a velocity which falls short of that of the actual crack by an amount proportional to R. If there had been no intervening changes of velocity, the fictitious crack would have kept up with the actual one. As things are, it has lagged behind the real crack with a relative velocity of order R for a time about R/c, and so it has fallen behind the real crack by a distance of order R^2. Since w behaves like $R^{1/2}$ near the tip (see Equation 19.3) the first term of Equation 19.17 is in error by an amount of order $R^2 \, \partial R^{1/2}/\partial R = \frac{1}{2}R^{3/2}$.

We now return to Bateman's theorem. The problem of a crack moving with successive discontinuous changes of velocity can be solved by an obvious extension of the method which gave Equation 19.12. It is harder to derive Equation 19.17 directly for continuous motion. The first term of Equation 19.17 for unrestricted R (the distance from the current position of the tip) is (see Equation 19.12) the displacement associated with a crack which has been moving uniformly for all time and which at time t happens to coincide in position and velocity with the actual crack, and we have to show that close enough to the crack (Equation 19.9) leads to this expression with an error of order $R^{3/2}$. Equation 19.9 gives the displacement as a function $w[x, y, \tau]$ of position and the retarded time τ defined implicitly by Equation 19.7. What we want is the displacement $w[x, y, \tau(x, y, t)]$ as a function of x, y, t. As in similar electromagnetic problems it can be found with the help of Lagrange's expansion (Whittaker and Watson 1927). The result is

$$w[x, y, \tau(x, y, t)] = w[x, y, t]$$
$$- \frac{1}{c} \sum_{n=1}^{\infty} \frac{1}{n!} \left(-\frac{1}{c} \frac{\partial}{\partial t} \right)^{n-1} [\dot{\xi}(t)Q(t)\{R(t)\}^n],$$

(19.18)

where

$$R(t) = \{[x - \xi(t)]^2 + y^2\}^{1/2}$$

is the distance from the current position of the crack tip, and

$$Q(t) = [\dot{\xi}(t)]^{-1} \partial w[x, y, t]/\partial t = O(R^{-1/2}).$$

If we were to pick out the terms of order $R^{1/2}$ in Equation 19.18 we could not easily see that they add up to the first term of Equation 19.17, (supposing that they do); since Equation 19.18 is a power series in c^{-1}, the square roots in Equation 19.17 would appear in an unidentifiable expanded form. However, we can complete the proof by an artifice. In expanding the nth term of Equation 19.18 we have to apply $n - 1$ operators $\partial/\partial t$ to the $n + 2$ factors $\dot{\xi}(t)$, Q, R, R. ... A term of order $R^{1/2}$ results when we leave $\dot{\xi}(t)$ undifferentiated and apply $\partial/\partial t$ to all but one of the remaining factors. Any other arrangement (e.g. with $\dot{\xi}(t)$ differentiated or $\partial/\partial t$ applied twice to Q or one of the factors R) leads to terms of order $R^{3/2}$ or higher, and, further, such terms will contain one or more of $\ddot{\xi}(t)$, $\dddot{\xi}(t)$ ... as a factor. In addition there may be a term of order $R^{1/2}$ which only involves $\dot{\xi}(t)$. So the terms of order $R^{1/2}$ may be picked out by expanding each term of Equation 19.18 and, in the result, throwing out all terms which contain $\ddot{\xi}(t)$, $\dddot{\xi}(t)$ .... But what is left is the exact expansion for a fictitious crack which has been moving uniformly for all time and which, at time t, happens to coincide in position and velocity with the real crack, and so its sum must be the first term of Equation 19.17.

19.3. The Energy Release Rate

Atkinson and Eshelby (1968) have argued that the energy release rate for a moving crack may be calculated from

$$vG = \lim_{S \to 0} \int_S (\mathbf{T} \cdot \mathbf{u} + vEn_x)\, dS,$$

(19.19)

where S is a circuit surrounding the crack tip, $\mathbf{u}$ is the elastic displacement, $\mathbf{T}$ is the traction across S, (n_x, n_y) is the outward normal to S, E is the total energy density (elastic plus kinetic), and v is the instantaneous velocity of the tip. The limit sign must be taken to mean that S is allowed to shrink on to the tip, and that, for a meaningful result, the integral must ultimately be independent of the shape as well as the size of S. In Equation 19.19, vG is the energy release per unit time and G the energy release per unit advance of the tip.

In the present case, Equation 19.19 reduces to

$$vG = \lim_{S \to 0} \int_S (\mu \dot{w} \, \partial w/\partial n + vEn_x) \, dS \tag{19.20}$$

with

$$E = \tfrac{1}{2}\mu(\partial w/\partial r)^2 + \tfrac{1}{2}\mu(\partial w/r \, \partial\theta)^2 + \tfrac{1}{2}\rho\dot{w}^2 \tag{19.21}$$

With Equation 19.12 the recipe 19.20 gives

$$G = A^2 K^2/2\mu, \tag{19.22}$$

which reduces to the accepted static value when $v = 0$.

It is possible to verify directly that, at least in the case of a crack starting off with uniform velocity, Equation 19.19 gives the correct result. Since the elastic state outside $r = ct$ is unchanged and the surface of discontinuity does not carry a finite amount of energy along with it, at time t the energy which has disappeared from the elastic field since the crack started to move is given by the integral

$$\int (E_s - E_m) \, dx \, dy \tag{19.23}$$

extended over the interior of the circle $r = ct$. Here E_s and E_m are the energy densities calculated from Equation 19.21 and the static solution 19.3 or the dynamic solution 19.12, respectively. Since energy conservation is taken care of everywhere else, the energy 19.23 must have disappeared via the tip. If Equation 19.19 is correct, 19.23 should be t times 19.20. A tedious calculation shows that this is so. The following less direct method is instructive and also useful in other connections.

One can verify that for any solution derived from a static solution by the transformation 19.14 the integral 19.19 or 19.20 is independent of the shape and size of S, so that we may ignore the limit sign and, for convenience, take S to be the circle $r = ct$. (Of course we must use the displacement 19.12 rather than 19.11). It is easy to see that E_m is homogeneous of degree -1 in $x - vt$ and y. Consequently,

$$\left[(x - vt)\frac{\partial}{\partial x} + y \frac{\partial}{\partial y}\right] E_m = -E_m$$

or

$$E_m = -\frac{\partial}{\partial x}[(x - vt)E_m] - \frac{\partial}{\partial y}[yE_m],$$

and similarly,

$$E_s = -\frac{\partial}{\partial x}[xE_s] - \frac{\partial}{\partial y}[yE_s].$$

Thus, with the help of Gauss's theorem, 19.23 becomes

$$\int (E_s - E_m)\, dx\, dy = \int_{r=ct} [vtE_m\, n_x + r(E_s - E_m)]\, dS.$$

But the static and moving solutions confront one another across the circle $r = ct$ and the difference in the energy densities (in the form 19.21) can be found by squaring and rearranging 19.16 and adding $(\partial w/r\, \partial\theta)^2_{in} = (\partial w/r\, \partial\theta)^2_{out}$ which follows from 19.15. The result is

$$E_s - E_m = (\mu/c)[\dot{w}\, \partial w/\partial r]_{in}$$

so that

$$\int (E_s - E_m)\, dx\, dy = t\int_{r=ct} (\mu\dot{w}\, \partial w/\partial n + vEn_x)\, dS,$$

as required.

So far we have simply shown that for a uniformly moving crack, Equation 19.20 gives the correct result. However, near enough to the tip of a crack in arbitrary motion the elastic field, Equation 19.17, is the same, apart from notation, as Equation 19.12, and so the energy release rate must be the same, namely,

$$G = \left[\frac{1 - \dot{\xi}(t)/c}{1 + \dot{\xi}(t)/c}\right]^{1/2} \frac{K^2}{2\mu} \tag{19.24}$$

for a crack tip in arbitrary motion. This expression depends only on the initial K value and the instantaneous velocity of the crack and not, for example, on its acceleration. The fears expressed by Atkinson and Eshelby about the applicability of Equation 19.19 to an accelerating crack thus seem to be unfounded. Of course the acceleration is not entirely without effect. The argument following 19.18 shows that the remainder term in 19.17 contains terms proportional to the acceleration and higher derivatives of the velocity. The larger they are the smaller must be the circuit S in 19.20 before the integral approaches a limiting value. But unless the size of S turns out to be of atomic dimensions, this causes no trouble.

19.4. Discussion

In the ideal case of a solid with surface energy γ a crack tip when released will initially move with a constant velocity found by equating 2γ with the energy release rate G given by Equation 19.22. Note that if K is not large enough A may have to be greater than unity, in which case v is negative and the crack shrinks.

More generally γ may be an equivalent surface energy representing plastic work, or 2γ may be taken to stand for some velocity-dependent dissipative process going on in a small region near the crack tip. Then the equation of motion takes the form

$$G(\dot{\xi}) = 2\gamma(\xi, \dot{\xi})$$

where G is given by Equation 19.24 and γ is a prescribed function of position and velocity.

Since G does not depend on the acceleration, the crack exhibits no inertia. If the tip encounters a region where γ suddenly decreases it immediately speeds up so as to decrease G (see Equation 19.24); there is no question of the tip overshooting, or tripping over its own feet in an attempt to feed more energy into the crack plane than the latter will accept. Closely connected with the absence from G of terms involving the acceleration or higher derivatives of the velocity is the fact that the discontinuity associated with an abrupt change of velocity carries no energy with it.

It is interesting to compare the behavior of a crack tip with that of other moving elastic signularities.

A kink (Eshelby 1962) behaves very much like an electron. It has a mass and is subject to a radiation reaction proportional to the rate of change of acceleration. If it suffers a rapid change of velocity two spheres of discontinuity spread out with the velocities of longitudinal and shear waves, and each carries with it a finite amount of energy.

A screw dislocation (Nabarro 1967) behaves like a moving electrically charged rod. The radiation reaction depends on the whole previous history of the motion so that there is an integral equation of motion. We may say that this is because the problem, though nominally two-dimensional, is really three-dimensional, so that each element of the dislocation is perpetually subjected to disturbances produced by remote elements at earlier times. However, precisely similar effects occur in "really" two-dimensional systems. A smooth heavy particle sliding on a horizontal drumhead has an integral equation of motion similar to that of a charged rod or dislocation. Unfortunately the equations are not quite identical since the particle is coupled to a scalar field whereas the rod or dislocation is coupled to a vector field. As a result the particle does not feel anything analogous to the Lorentz force which acts on the rod, and so this simple model throws no light on the problem (Nabarro 1967) of whether a screw dislocation experiences a Lorentz force or not.

It is not very surprising that a crack tip does not behave much like a kink, but we might have expected that some of the peculiarities of the two-dimensional wave equation which affect the dislocation would also affect the crack. The reason why they do not is that, roughly speaking, the peculiarity of being two dimensional is canceled by the peculiar angular dependence of the elastic field through sines and cosines of half-integral multiples of the angle θ, in the static case, or of the corresponding aberrated (Lorentz-transformed) angle in the moving case. The matter is bound up with the problem of whether the N-dimensional wave equation admits distortionless solutions or not.

A solution of the wave equation is considered to be distortionless if it takes the form

$$w = g(r)f(r \pm ct)\Omega(\theta, \phi \ldots). \tag{19.25}$$

In N dimensions the equation may be written

$$w'' + (N - 1)r^{-1}w' + r^{-2}\Lambda w - c^{-2}\ddot{w} = 0, \tag{19.26}$$

where the dash denotes $\partial/\partial r$ and $r^{-2}\Lambda$ is the angular part of the Laplace operator. To get rid of the term in f' we must take $g = r^{-1/2N+1/2}$. Then if we take Ω to be an eigenfunction of Λ, with $\Lambda\Omega = \lambda\Omega$, Equation 19.26 becomes the one-dimensional wave equation only if

$$\lambda = \tfrac{1}{4}(N - 1)(N - 3).$$

In three dimensions, $g = r^{-1}$, $\lambda = 0$ and there is a distortionless solution, namely,

$$w = r^{-1}f(r \pm ct)$$

only if there is no angular dependence. Otherwise there is no solution of the the form 19.25. For instance, if $\Omega = \cos\theta$, the best we can do is

$$w = [r^{-2}f(r \pm ct) - r^{-1}f'(r \pm ct)]\cos\theta,$$

which is not distortionless in the sense of Equation 19.25; hence the distinction between the induction and radiation zones in electromagnetic and other fields. When $N = 2$,

$$g = r^{-1/2}, \lambda = -\tfrac{1}{4},$$

and

$$w = r^{-1/2}f(r \pm ct)\frac{\sin}{\cos}\tfrac{1}{2}\theta, \tag{19.27}$$

and we have a distortionless solution in which, as stated, the usual peculiarities of the two-dimensional wave equation are offset by the peculiar angular dependence.

We can, in fact, use Equation 19.27 to construct our basic solution 19.9. It can be rewritten as

$$w = (x + iy)^{-1/2} f(r \pm ct).$$

For f take the step function $H(ct - r)$, which is unity for $ct > r$ and zero for $ct < r$. Shift the origin of time and space to τ', $\xi(\tau')$, 0, multiply by $-\frac{1}{2}B\dot{\xi}(\tau')d\tau'$, and integrate over all values of the parameter τ'. The result is

$$w = -\frac{1}{2} \int_{-\infty}^{\infty} \frac{BH\{c(t - \tau') - [(x - \xi(\tau'))^2 + y^2]^{1/2}\}}{\{x - \xi(\tau') + iy\}^{1/2}} \dot{\xi}(\tau') \, d\tau'$$

$$= B\{x - \xi(\tau) + iy\}^{1/2},$$

where τ is defined by Equation 19.7. The imaginary part of this expression is 19.9. (The rather offhand treatment of the lower limit can be justified).

The expression 19.27 is the only distortionless solution in the two-dimensional case. The absence of others for angular dependence $\sin (3/2)\theta$, $\sin (5/2)\theta \ldots$ is what prevents us treating the higher terms in Equation 19.6 in the same simple way as the first.

REFERENCES

Atkinson, C., and J. D. Eshelby, 1968, *Int. J. Fracture Mech.*, **4**, 3.

Baden-Powell, D. F. W., 1949, *Proc. Prehist. Soc.*, **15**, 38.

Bateman, H., 1915, *Electrical and Optical Wave-Motion*, Cambridge: Cambridge University Press, p. 138.

Broberg, K. B., 1960, *Arkiv för Fiz.*, **18**, 159.

Craggs, J. W., 1963, in *Fracture of Solids*, New York: Interscience, p. 51.

Eshelby, J. D., 1962, *Proc. Roy. Soc.*, **A266**, 222.

Griffith, A. A., 1920, *Phil. Trans. Roy. Soc.*, **A221**, 163.

Hill, R., 1961, in I. N. Sneddon and R. Hill (eds.), *Progress in Solid Mechanics*, New York: Wiley-Interscience, vol. 2, p. 247.

Love, A. E. H., 1927, *The Mathematical Theory of Elasticity*, Cambridge: Cambridge University Press, p. 295.

McClintock, F. A., and S. P. Sukhatme, 1960, *J. Mech. Phys. Solids*, **8**, 187.

Nabarro, F. R. N., 1967, *Theory of Crystal Dislocations*, Oxford: Clarendon Press, p. 496.

Orowan, E., 1949, in *Rep. Prog. Physics*, **12**, London: Physical Society, p. 185.

Orowan, E., 1955, *Welding J.*, **34**, 157s.

Skertchley, S. B. J., 1879, *On the Manufacture of Gunflints*, Mem. Geol. Survey, London: H.M. Stationary Office.

Whittaker, E. T., and G. N. Watson, 1927, *Modern Analysis*, Cambridge: Cambridge University Press, p. 132.

Yoffe, E. H., 1951, *Phil. Mag.*, **42**, 739.

ABSTRACT. Temperature elevations in the plastic zone of a crack are calculated. Stress and strain distributions in nonhardening materials are used, and plastic zones are regarded as distributed sources of heat proportional to the plastic work rate. Results are approximate in that temperature-independent mechanical and thermal properties are assumed and thermal stressing is neglected.

For small-scale yielding and moderately rapid rates, it is found that, under increasing load, the temperature rise at a stationary crack tip is proportional to $K^2/E\sqrt{\rho ckT}$. The same form applies for a running crack, with loading time replaced by the ratio of plastic zone size to velocity v, giving a temperature elevation proportional to $K\sigma_0\sqrt{v}/E\sqrt{\rho ck}$.

Numerical results are given for 2024 aluminum alloy, 6Al–4V titanium alloy, and mild steel. Temperature rises predicted for Krafft's experiments seldom exceed 100°C. These may still be large enough to influence fracture toughness at very fast rates. Assuming that crack tip temperature alone governs fracture toughness, the behavior of the titanium alloy can be explained, and a toughness minimum at slower rates can be predicted for some metals.

20. Local Heating by Plastic Deformation at a Crack Tip

J. R. RICE AND N. LEVY

20.1. Introduction

This paper presents calculations of local temperature elevations accompanying plastic deformation near the tips of stationary and running cracks. The plastic zone at the tip is regarded as the site of heat sources varying in intensity with the distribution of plastic work rates. Calculations are based on some of the known solutions for the stress and strain distribution near a crack tip (Rice 1968, Rice and Rosengren 1968, McClintock and Irwin 1965, Rice 1967, Hutchinson 1968) for elastic perfectly plastic materials. Previous attempts at estimating the temperature rise at the crack tip (Krafft and Irwin 1965, Williams 1965) viewed the plastic zone as the site of a uniform heat generation and were directed at determining whether conditions are more nearly adiabatic or isothermal. Our treatment differs in that the plastic zone grows under increasing load, with plastic work rates (and hence heat generation rates) being far from uniform and tending to very large values within the fracture process zone adjacent to the tip.

The motivation for a precise treatment is the suspicion that such a rise in temperature might be large enough to influence fracture, particularly when loading rates made possible by high speed testing machines are applied. Eftis and Krafft (1965) observed a minimum in the toughness of mild steel when employing a combined rate scale of reciprocal loading time for stationary cracks and propagation velocity for running cracks. Even when this minimum occurred in the running crack range, the velocity was a small fraction of the wave speed, so it is difficult to believe that inertia effects are responsible. Moreover, results by Krafft and Irwin (1965) on a high-strength 6Al–4V titanium alloy tested at room temperature and below reveal no drop in toughness with rate but rather a steady increase. However, a minimum occurs when the test is made at a higher temperature.

A conventional explanation of fracture rate sensitivity lies with the elevation of flow stress curve with strain rate. Maximum stresses achievable in a nonhardening material will be limited by the initial yield value multiplied by a factor on the order of 2.6 to 3.0 accounting for plane strain constraint so that high strain rates increase local stresses. Strain hardening nullifies this elevation somewhat, for the maximum stress triaxiallity achievable in plane strain increases with strain hardening exponent (Rice and Rosengren 1968, Drucker and Rice 1968) and postyield flow in mild steel reveals a reduction of strain-hardening rate with strain rate. Still on balance, this view of rate sensitivity leads to decreasing toughness with rate and is alone inadequate to explain the observed toughness minimum. Some ambiguities are removed by assuming variations in toughness proportional to corresponding hardening exponent variations, as Krafft has proposed (Krafft 1964, Eftis and Krafft 1965, Krafft and Irwin 1965). This does an admirable job in correlating the behavior of mild steel. As to the behavior of 6Al–4V titanium alloy, Krafft showed that the strain-hardening exponent increases with strain rate, and this could explain the peculiar behavior under increasing loading rate. However, this does not explain the appearance of a minimum toughness for the 180°F test on this alloy (Krafft and Irwin 1965).

As a result of our calculations, it was found that the temperature rise for the titanium alloy is much higher than the rise for steel (by a factor of 6 for the same stress intensity factor). It is, therefore, logical to examine this temperature rise as a possible factor in explaining the difference in the behavior of the titanium alloy and mild steel. The strain-rate elevation of flow stress and local temperature rise will each, in isolation, have opposite effects with the latter, enhancing toughness. They are, of course, both present and highly coupled. For example, isothermal straining will result in a higher flow curve than adiabatic straining at the same rate. Also, thermal expansion accompanying highly localized heating at the crack tip will tend somewhat to counteract the stress-concentrating function of the crack. The energy dissipated in ductile separation of surfaces would presumably increase with temperature of the fracture process zone. In this paper, we do not take all these possibilities into account. We take, rather, the view that loading conditions are prescribed and, with several approximations, we estimate the associated temperature field. These estimates provide a useful guide for interpreting experimental results, as noted in our concluding sections.

20.2. Plasticity Models and Assumptions for Temperature Calculations

For simplicity, we limit attention to cases of a small plane strain plastic zone near a crack tip. The boundary layer formulation of small-scale yielding (Rice 1967, 1968) may then be employed with the characteristic elastic singularity setting asymptotic boundary conditions. Applied load and crack length thus appear only as combined in Irwin's stress intensity factor K (equal to $\sigma_\infty \sqrt{\pi l/2}$ for the Inglis configuration of a crack of length l in an infinite body subjected to the remote tension σ_∞). Further, we employ nonhardening plasticity models with an effective yield stress σ_0 (or τ_0 in shear) and use the plastic work rate from these models as the heat-generation rate for the temperature calculation. This neglects all thermomechanical coupling in affecting the near tip temperature field, except that a numerical value for the yield stress may later be assigned in cognizance of the average temperature and strain rate in the plastic zone.

For constant conductivity and specific heat the equation governing the temperature rise $u = u(x, y, t)$ due to heating by plastic work is

$$\frac{\partial u}{\partial t} = a^2 \nabla^2 u + \frac{f(x, y, t)}{\rho c}, \tag{20.1}$$

where $f(x, y, t) = \sigma_{ij} \dot{\varepsilon}_{ij}^p$ is the plastic work rate, ρ the mass density, c the specific heat, k the conductivity, and $a^2 = (k/\rho c)$ the diffusivity. For partial conversion of work to heat, f should be scaled down proportionally. Carslaw and Jaeger (1959) present the fundamental two-dimensional solution for an instantaneously delivered quantity of heat at a point in the infinite xy plane. Taking our plastic work rate $f(x, y, t)$ as a continuous distribution of heat sources and superposing, the solution to Equation 20.1 is

$$u(x, y, t) = \int_0^t \left\{ \iint_{A_P(\tau)} \frac{f(\xi, \eta, \tau)}{\rho c} \exp\left[-\frac{(x - \xi)^2 + (y - \eta)^2}{4a^2(t - \tau)} \right] d\xi \, d\eta \right\}$$

$$\times \frac{d\tau}{4\pi a^2(t - \tau)} \tag{20.2}$$

Here $A_P(t)$ is the plastically deforming region at time t. It is appropriate to use the fundamental solution for an uncracked plane here, since the work rate f will be symmetric about the crack line so as to nullify any heat conduction across the crack surface.

This integral is the desired form of solution to the temperature equation for the rapid loading of a stationary crack. It is inconvenient for a running crack. A suitable approximation to the latter case would regard the crack velocity and plastic zone size as constant (for purposes of temperature calculation only) so that a plastic work rate independent of time, say $g(x, y) = \sigma_{ij}\dot{\varepsilon}^p_{ij}$, results referred to a system of coordinates moving with the crack tip. Again the solution is given by superposing heat sources, this time employing the fundamental solution (Carslaw and Jaeger 1959) for a point source moving at constant speed v equal to the crack velocity. There results

$$u(x, y) = \iint\limits_{A_P} \frac{g(\xi, \eta)}{2\pi k} \exp\left[-\frac{v(x - \xi)}{2a^2}\right]$$

$$\times K_0\left\{\frac{v}{2a^2}\left[(x - \xi)^2 + (y - \eta)^2\right]^{1/2}\right\} d\xi\, d\eta. \qquad (20.3)$$

Here A_P is the plastically deforming region in the moving coordinates and K_0 is the modified Bessel function of the second kind.

The functions $f(x, y, t)$ and $g(x, y)$ are determined from the crack-tip plasticity model employed. An approximate nonhardening plane strain analysis for small-scale yielding at a stationary crack tip has been given by Rice (1968), as based on the slip-line theory and his path-independent line integral. Plastically deforming regions result in centered fans above and below the crack tip, so that if a polar coordinate system is chosen with origin at the tip with $\theta = 0$ being the line ahead of the crack, the extent of the plastic region is approximately

$$R_P(\theta) = \omega \cos 2(\theta - \pi/2) \quad \text{where} \quad \omega = \frac{3(1 - v)}{\sqrt{2}(2 + \pi)}\left(\frac{K}{2\tau_0}\right)^2. \quad (20.4)$$

Here ω is the maximum dimension of the plastic zone and $R_P(\theta)$ is the distance from the crack tip to the elastic-plastic boundary along a ray at angle θ, the boundary cutting into the tip along 45-deg boundaries of the centered fans. The associated plastic work rate at points within the plastic zone may be obtained from Rice (1968) as

$$f(r, \theta, t) = \frac{d}{dt}\left[\frac{\tau_0^2}{G}\frac{R_P(\theta)}{r}\right] = \frac{\tau_0^2}{Gr}\cos 2(\theta - \pi/2)\frac{d\omega}{dt} \quad \text{for} \quad r < R_P(\theta)$$

$$(20.5)$$

(G is the elastic shear modulus).

No analogous tensile plasticity model is available for an advancing crack. We therefore employ the simple Dugdale (1960) model which represents yielding by essentially viewing the crack as longer by a length equal to the plastic zone size, with the plastic separation displacements

of the extended crack surface being opposed by the tensile yield stress σ_0.
The small-scale yielding plastic zone size then is

$$\omega = \frac{\pi}{8}\left(\frac{K}{\sigma_0}\right)^2.$$

(20.6)

Separation displacements are the same whether the crack is moving or not.
From the expressions given in Rice (1967) one may compute the Dugdale
plastic work rate in plane strain for a stationary crack as

$$f(x, y, t) = \frac{4(1 - v)\sigma_0^2}{\pi G}\left(1 - \frac{x}{\omega}\right)^{1/2} \delta_D(y)\frac{d\omega}{dt}, \quad \text{for} \quad 0 < x < \omega.$$

(20.7)

Here the coordinate origin is at the tip with the x axis extending directly
ahead, and $\delta_D(\ldots)$ is the Dirac delta function resulting from the displacement discontinuity in the plastic zone. For the running-crack case, we take
the coordinate origin at the moving tip with the x axis again directly
ahead. Presuming the plastic zone to remain fixed in size as the crack
advances, the work rate is computed from Rice (1967) as

$$g(x, y) = \frac{2(1 - v)\sigma_0^2}{\pi G} \log\left[\frac{1 + (1 - x/\omega)^{1/2}}{1 - (1 - x/\omega)^{1/2}}\right] \delta_D(y)v, \quad \text{for} \quad 0 < x < \omega.$$

(20.8)

Were the growth of the plastic zone with crack length taken into account,
the resulting work rate would be the sum of Equations 20.7 and 20.8.
Thus our treatment of the running-crack case as steady-state problem is
justified as long as the crack speed v greatly exceeds the zone growth rate
$d\omega/dt$, which is satisfied by definition for small-scale yielding. The Dugdale
model is most appropriate for plane stress rather than plane strain, and
this poses an important limitation on our running-crack results as noted
later.

20.3. Temperature Rise at the Tip of a Stationary Crack

When the plastic work rate $f(r, \theta, t)$ is represented in polar coordinates with the crack tip at the coordinate origin, Equation 20.2 gives
the crack-tip temperature rise after loading from times 0 to t as

$$u = \int_0^t \left\{ \iint\limits_{A_P(\tau)} \frac{f(r, \theta, \tau)}{\rho c} \exp\left[-\frac{r^2}{4a^2(t - \tau)}\right] r\, d\theta\, dr \right\} \frac{d\tau}{4\pi a^2(t - \tau)}.$$

(20.9)

Note that only the angular average (in θ) of the work rate contributes.
This is the reason why the two plasticity models we consider, involving

very different yielding patterns, lead to similar results. Inserting the work rate for the plane-strain slip-line model (Equation 20.5) and integrating in θ,

$$
u = \frac{\tau_0^2}{2\pi kG} \int_0^t \left\{ \int_0^{\omega(\tau)} \sqrt{1 - [r/\omega(\tau)]^2} \, \exp\left[-\frac{r^2}{4a^2(t - \tau)} \right] dr \right\}
$$

$$
\times \frac{d\omega(\tau)}{d\tau} \frac{d\tau}{t - \tau}. \tag{20.10}
$$

Here the time variation of $\omega(\tau)$ is determined through Equation 20.4 in terms of the imposed time variation in the stress-intensity factor. An identical formula results for the Dugdale model, except that the τ_0^2 in front is replaced by $2(1 - v)\sigma_0^2/\pi$ and the r/ω appearing within the square root sign is then raised only to the first power.

Dimensionless variables are now convenient. Let T denote the time at which the maximum load is reached on a cracked specimen, let K_{max} be the corresponding stress intensity and ω_{max} the plastic zone dimension resulting at K_{max}. Define

$$
\alpha = t/T, \qquad \Omega(\alpha) = \omega(t)/\omega_{max} = [K(t)/K_{max}]^2, \qquad R = r/\omega_{max}
$$
$$
\tag{20.11}
$$

as dimensionless time, plastic zone size, and distance, respectively. Further let

$$
\delta = \frac{2a\sqrt{T}}{\omega_{max}} \tag{20.12}
$$

be a dimensionless measure of the shortness of time to maximum load. Then Equation 20.10 for the crack-tip temperature rise at the time of maximum load may be put in the form

$$
u = u_c h(\delta), \tag{20.13}
$$

where u_c has temperature dimensions and may be expressed in terms of loading time and maximum load, and $h(\delta)$ is a dimensionless function depending on the manner of load variation with time. For the plane-strain slip-line model

$$
u_c = \sqrt{\frac{2}{\pi}} \frac{1}{2 + \pi} \frac{(1 - v^2)K_{max}^2}{E\sqrt{\rho ckT}} = 0.156 \frac{(1 - v^2)K_{max}^2}{E\sqrt{\rho ckT}}, \tag{20.14}
$$

(E is Young's modulus), and for the Dugdale model,

$$
u_c = \frac{2}{3\sqrt{\pi}} \frac{(1 - v^2)K_{max}^2}{E\sqrt{\rho ckT}} = 0.377 \frac{(1 - v^2)K_{max}^2}{E\sqrt{\rho ckT}}. \tag{20.15}
$$

We interpret u_c shortly. The function $h(\delta)$ for the plane-strain slip-line model is given by the double integral

$$h(\delta) = \frac{3}{4\sqrt{\pi}} \int_0^1 \frac{\Omega'(\alpha)}{\sqrt{1-\alpha}} \int_0^{\Omega(\alpha)} \frac{\sqrt{1-[R/\Omega(\alpha)]^2}}{\delta\sqrt{1-\alpha}}$$

$$\times \exp\left[-\frac{R^2}{\delta^2(1-\alpha)}\right] dR\, d\alpha. \qquad (20.16)$$

The same formula applies for the Dugdale model, eccept that the $R/\Omega(\alpha)$ ratio within the square root sign is raised to only the first power.

3.1 Rapid Loading

Consider a very short time to attainment of maximum load so that δ (Equation 20.12) is small compared to unity. The situation is approximated by seeking the limit of $h(\delta)$ as $\delta \to 0$. In that limit the entire contribution to the integral on R in Equation 20.16 is shifted to the origin $R = 0^+$. The integrand becomes infinite there, but not the integral itself which approaches the form of the half-infinite normal error integral of value $\sqrt{\pi/2}$. Hence, in the short loading time limit the crack-tip temperature rise for both models is

$$u = \frac{3}{8} u_c \int_0^1 \frac{\Omega'(\alpha)}{\sqrt{1-\alpha}}\, d\alpha. \quad \text{for } \delta \ll 1. \qquad (20.17)$$

If the applied load rises linearly with time up to maximum load, $\Omega(\alpha) = \alpha^2$ by Equation 20.11 and the integral turns out to have the value 8/3 so that the temperature rise at maximum load is

$$u = u_c \quad \text{(constant loading rate and } \delta \ll 1). \qquad (20.18)$$

Thus our arrangement of Equation 20.13, with the temperature rise referenced to this short time constant-rate limit as given by Equations 20.14 and 20.15. One may also show from Equation 20.17 that the short-time limit temperature elevation is $3u_c/4$ if the load rises as $t^{1/2}$ and $6u_c/5$ if the load rises at $t^{3/2}$.

The function $h(\delta)$ or u/u_c is plotted as a function of δ in Figure 20.1, as obtained through numerical integration of Equation 20.16 for the special case of the applied load rising linearly with time. It is seen that $h(\delta)$ tends very rapidly to unity as δ or $2a\sqrt{T}$ becomes small. In other words, the temperature at the crack tip, for a given material, becomes proportional to u_c which is inversely proportional to loading time, as can be seen from Equation 20.14. It is interesting to note that the thermal properties influencing the temperature at the crack tip appear in the term $1/\sqrt{k\rho c}$.

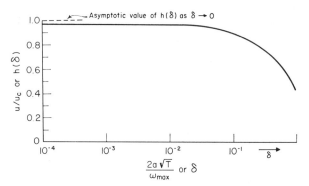

FIGURE 20.1. The value of the function $h(\delta)$ for constant rate of loading (temperature rise at stationary crack tip $= u_c\, h(\delta)$).

3.2 *Temperature Distribution Near the Tip*

It is clear that the maximum temperatures occurring at the tip (Equations 20.13 to 20.18) will hold over a highly localized region. From the presence of the thermal conductivity k in expressions for u_c, one might conclude that it is never meaningful to speak of "adiabatic conditions" at the tip even for very short loading times. Nevertheless, adiabatic conditions may result a short distance away from the tip. The $1/r$ strain singularity of the slip-line model suggests that the temperature gradient will then be quite steep. The gradient is, of course, unrealistically steep along the plastic zone of the Dugdale model due to the concentration of deformation in a band. This dependence on artificial localization of the deformation is alleviated by focusing on the average temperature on a circle of radius r_0:

$$\bar{u}(r_0, t) = \frac{1}{2\pi} \int_{-\pi}^{+\pi} u(r_0, \phi, t)\, d\phi. \tag{20.19}$$

Upon expressing both the source points and response point in polar coordinates, the general solution of Equation 20.2 becomes

$$u(r_0, \phi, t) = \int_0^t \left\{ \iint_{A_P(\tau)} \frac{f(r, \theta, \tau)}{\rho c} \exp\left[-\frac{r^2 - 2rr_0\cos(\phi - \theta) + r_0^2}{4a^2(t - \tau)} \right] r\, d\theta\, dr \right\}$$

$$\times \frac{d\tau}{4\pi a^2(t - \tau)} \tag{20.20}$$

Averaging as in Equation 20.19 converts the ϕ-dependent part of the exponential term to a modified Bessel function of the first kind (McLachlan 1961) so that

$$\bar{u}(r_0, t) = \int_0^t \left\{ \iint_{A_P(\tau)} \frac{f(r, \theta, \tau)}{\rho c} \exp\left[-\frac{r^2 + r_0^2}{4a^2(t - \tau)} \right] I_0\left[\frac{rr_0}{2a^2(t - \tau)} \right] r\, d\theta\, dr \right\}$$

$$\times \frac{d\tau}{4\pi a^2(t - \tau)} \tag{20.21}$$

This equation is identical in form to Equation 20.9 in that only the angular average of the work rate contributes, as desirable in view of uncertainties in the models. The remaining steps are identical to the progression from Equations 20.9 to 20.16 so that the average temperature on a circle of radius r_0 at the time T of loading to a plastic zone dimension ω_{max} is

$$\bar{u} = \frac{3}{4\sqrt{\pi}} u_c \int_0^1 \frac{\Omega'(\alpha)}{\sqrt{1-\alpha}} \int_0^{\Omega(\alpha)} \frac{\sqrt{1 - [R/\Omega(\alpha)]^2}}{\delta\sqrt{1-\alpha}}$$

$$\times \exp\left[-\frac{R^2 + R_0^2}{\delta^2(1-\alpha)}\right] I_0\left[\frac{2RR_0}{\delta^2(1-\alpha)}\right] dR \, d\alpha \qquad (20.22)$$

where $R_0 = r_0/\omega_{max}$ and the remaining notation is as given above. This reduces to the crack-tip form of Equations 20.13 and 20.16 when $R_0 = 0$. Here the square power on $R/\Omega(\alpha)$ applies again for the plane-strain slip-line model, with a power of unity for the Dugdale model.

One need only consider small values of r_0 since temperature rises will be entirely negligible at distances from the tip comparable to the plastic zone dimension. For example, the plane-strain slip-line model estimates the maximum plastic shear strain on the circle $r_0 = \omega_{max}/10$ as only $9\gamma_0$. Further, negligible temperature elevations will occur throughout the zone unless the dimensionless loading time δ is small. When both R_0 and δ are small compared to unity, the major contribution of the integral on R is localized around R_0 and the integrand falls off rapidly to zero due to $\exp\left[-R^2/\delta^2(1-\alpha)\right]$. Thus one introduces negligible error (and zero in the limit as both δ and $R_0 \to 0$) by replacing the term $\sqrt{1 - [R/\Omega]^2}$ by unity and the upper limit $\Omega(\alpha)$ by ∞. Consequently, the temperature elevation then depends on R_0 and δ only through the combination R_0/δ:

$$\bar{u} = u_c \, p\left(\frac{R_0}{\delta}\right) = u_c \, p\left(\frac{r_0}{\delta\omega_{max}}\right) = u_c \, p\left(\frac{r_0}{2a\sqrt{T}}\right), \quad \text{for} \quad \delta \quad \text{and} \quad R \ll 1,$$

$$(20.23)$$

where the function $p(\ldots)$ is defined by

$$p(\lambda) = \frac{3}{4\sqrt{\pi}} \int_0^1 \frac{\Omega'(\alpha)}{1-\alpha} \int_0^\infty \exp\left[-\frac{\lambda^2 + \mu^2}{1-\alpha}\right] I_0\left[\frac{2\lambda\mu}{1-\alpha}\right] d\mu \, d\alpha.$$

$$(20.24)$$

Equation 20.23 reveals the parameter $2a\sqrt{T}$ as the size scale over which temperatures comparable to the tip value will result.

If the loading time is so short that $2a\sqrt{T}$ is small compared to the radius r_0 of interest, the asymptotic formula (McLachlan 1961) $I_0(z) = (2\pi z)^{-1/2}e^z$ may be employed for the Bessel function so that $p(\lambda) \approx 3/8\sqrt{\pi\lambda}$, and

$$\bar{u} \approx \frac{3}{8\sqrt{\pi}} u_c \frac{2a\sqrt{T}}{r_0} \quad \text{for} \quad r_0 \gg 2a\sqrt{T}. \qquad (20.25)$$

This is independent of the loading time (since $u_c \sim 1/\sqrt{T}$) and could have been written down immediately from adiabatic considerations. The function $p(\lambda)$ has been evaluated numerically for the case of a linearly rising load ($\Omega = \alpha^2$), and the angular averaged temperature at r_0 is shown in dimensionless form as $\bar{u}/u_c$ vs. $r_0/2a\sqrt{T}$ (or $r_0/\delta\omega_{max}$) in Figure 20.2.

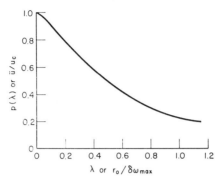

FIGURE 20.2. The value of the function $p(\lambda)$ or $\bar{u}/u_c$. Nomenclature: $\bar{u}$ = average temperature along a circle with radius r_0 from the crack tip; u_c = characteristic temperature as defined; $\delta = 2a\sqrt{T/\omega_{max}}$; $a^2 = k/\rho c$ or diffusivity; T = loading time; ω_{max} = maximum length of plastic zone.

Finite geometry changes occurring through blunting of the crack tip will limit conditions at which sharp crack temperature predictions are considered acceptable. We may approximate the tip temperature in a rough way by taking r_0 as half the tip opening displacement, which is $r_0 = 4(1 - v^2)\varepsilon_0 \omega_{max}/\pi \approx \varepsilon_0 \omega_{max}$ for the Dugdale model ($\varepsilon_0 = \sigma_0/E$ is the initial yield strain). This correction would estimate the tip temperature as

$$u \approx u_c p\left(\frac{\varepsilon_0}{\delta}\right), \tag{20.26}$$

so that our previous formulae apply if δ is comparable to or larger than ε_0. An upper limit to the achievable tip temperature is then given by Equation 20.24, when the loading time is so short that δ is small compared to the initial yield strain. Employing Equations 20.15 and 20.16, this upper limit is

$$u \approx \sigma_0/\pi\rho c. \tag{20.27}$$

The numerical factor of π is probably not very accurate, but otherwise the form is plausible as it represents the adiabatic temperature rise accompanying a strain on the order of unity. With the π factor, this value is 120°C for the titanium alloy 6Al–4V, and 60°C for mild steel and 80°C for aluminum 2024, for values σ_0 as obtained in "static" tensile tests. If, however, the effect of the high strain rates is to be taken into account, these upper bounds should be scaled up in proportion to the increase in the yield strengths. A figure of 3 is usually taken for steel, and the upper limit is 180°C.

20.4. Temperature Rise at a Running Crack Tip

In contrast to the stationary crack for which the tip temperature rise is insensitive to the details of the work rate distribution (Equation 20.9), a strong directionality is suggested by the exponential term of the running crack solution in Equation 20.3. The Dugdale model predicts an unrealistically large concentration of plastic work directly ahead of the crack tip. While the lack of more accurate advancing crack models forces a limitation to this case, it is expected that predicted temperatures will somewhat exceed actual values.

Setting $x = y = 0$ in Equation 20.3 so as to describe the temperature rise at the running crack tip, employing the Dugdale work rate of Equation 20.8 and integrating in η, one obtains

$$u = \frac{(1 - v)\sigma_0^2 v}{\pi^2 Gk} \int_0^\omega \log\left[\frac{1 + (1 - \xi/\omega)^{1/2}}{1 - (1 - \xi/\omega)^{1/2}}\right] \exp\left(\frac{v\xi}{2a^2}\right) K_0\left(\frac{v\xi}{2a^2}\right) d\xi.$$

$$(20.28)$$

Here ω is the plastic zone size as given by Equation 20.6 and v is the crack speed. Let $R = \xi/\omega$ be a dimensionless distance and define

$$m = \frac{v\omega}{2a^2} \tag{20.29}$$

as a dimensionless measure of the rapidity of crack speed. Then Equation 20.19 may be placed in a form analogous to the stationary crack case as

$$\frac{u}{u_c} = h(m) = \frac{\sqrt{m}}{\pi\sqrt{2\pi}} \int_0^1 \log\left[\frac{1 + (1 - R)^{1/2}}{1 - (1 - R)^{1/2}}\right] \exp(mR) K_0(mR) \, dR.$$

The characteristic temperature for the running crack is

$$u_c = \frac{\sqrt{\pi}}{2}\frac{(1 - v^2)K^2}{E\sqrt{\rho ck}}\sqrt{\frac{v}{\omega}} = 0.886\frac{(1 - v^2)K^2}{E\sqrt{\rho ck}}\sqrt{\frac{v}{\omega}}. \tag{20.31}$$

This value of u_c (which will shortly be shown to be the large m limit) is in the same form as Equation 20.15, except that the velocity over plastic zone size replaces reciprocal loading time and the numerical factor is larger. Note that here a dependence on yield stress (through ω) results, in con' ast to the stationary crack case. In terms of σ_0,

$$u_c = \sqrt{2}\frac{(1 - v^2)K\sigma_0\sqrt{v}}{E\sqrt{\rho ck}} = 1.414\frac{(1 - v^2)K\sigma_0\sqrt{v}}{E\sqrt{\rho ck}}. \tag{20.32}$$

The high velocity case is handled by computing the limit of $h(m)$ as $m \to \infty$. For large values of its argument the Bessel function behaves as (McLachlan 1961) $K_0(z) \to (\pi/2z)^{1/2} e^{-z}$. Inserting this form for $K_0(mR)$ as $m \to \infty$ in Equation 20.30, one finds that $h(\infty) = 1$ so that the u_c of Equations 20.31 and 20.32 is the large velocity form for the tip temperature. The ratio u/u_c is plotted in Figure 20.3 as a function of m. The function

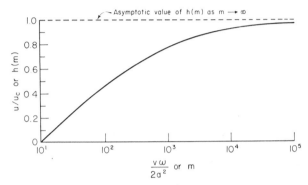

FIGURE 20.3. The value of the function $h(m)$ (temperature rise at the tip of a running crack $= u_c h(m)$).

$h(m)$ tends asymptotically to 1.0 as $m \to \infty$. Therefore, for very high values of m (or $v\omega/2a^2$), the temperature at the tip of the crack is directly proportional to u_c or the square root of the velocity. Again, the thermal properties influence the temperature rise through the term $1/\sqrt{k\rho c}$.

20.5. Numerical Results

The magnitude of the temperature rise at the tip of a stationary or a running crack is shown as a function of the stress-intensity factor for three materials (Figures 20.4 and 20.5). The materials, mechanical, and

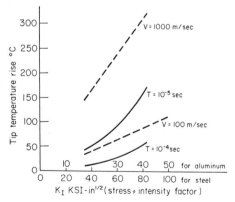

FIGURE 20.4. Temperature dependence on stress intensity factor K_1 for aluminum 2024 alloy with Y–S 30 KSI, and mild steel with Y–S 60 KS$_1$.

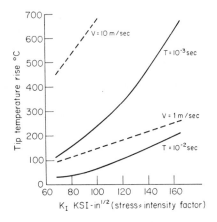

FIGURE 20.5. Dependence of root temperature on stress-intensity factor, for Titanium-6 Al-4V alloy with Y–S 120 KSI.

thermal properties at room temperature relevant to the calculations are taken as shown in the following table.

Yield Strength ksi	Thermal Conductivity cal/sec°C	Specific Heat cal/g°C	Diffusivity cm²/sec	$\sqrt{k\rho c}$ cal/cm²°C sec$^{1/2}$
		2024 Aluminum Alloy		
30	0.410	0.185	0.85	0.475
		Mild Steel		
60	0.143	0.10	0.20	0.330
		6Al-4V Titanium Alloy		
120	0.018	0.13	0.032	0.110

Noting that the temperature rise for a given K_I is proportional to $1/\sqrt{T}$ (Equation 20.24) or $\sqrt{v}$ (Equation 20.23) for the stationary crack and the running crack, respectively, it is easy to calculate the rise in temperature for loading times or propagation velocities other than those shown in the figures.

It should be noted that the results for v = constant for the running crack depend on the yield strength chosen, since the rise in temperature is proportional to $\sigma_0 K$. In Figures 20.4 and 20.5, the yield strengths used are those obtained from a standard tensile test. Therefore, if the rise in yield strength due to the very high strain rates such as occur at the tip of a running crack is taken into account, the results in Figures 20.4 and 20.5 for the running crack should be increased proportionately to the increase in the yield strength.

To have an idea on the magnitude of the temperature rise in actual experiments, the data from Krafft and Irwin (1965) was considered. For example, Figure 5 of Krafft and Irwin shows the variation of the critical stress-intensity factor with loading time for 6Al-4V titanium alloy. The critical stress-intensity factor is equal to 70 ksi$\sqrt{\text{in}}$ at loading time $T = 10^{-3}$ sec, for a specimen tested at 180°F. For this point, the temperature rise at the tip of the crack is calculated to be 120°C. The temperature rise at the point of minimum critical-stress intensity factor is only 10°C.

In Figure 6 of Krafft and Irwin (1965) for mild steel the temperature rise at the minimum critical-stress intensity factor for the test run at $-320°$F is less than 1°C. However, at the highest stress-intensity rate in the same test, the rise is estimated at 50°C. As to the running crack, the rise in temperature is calculated for the various velocities of propagation as shown below:

Velocity of Propagation	Temperature Rise
100 ft./sec.	80°C
1000 ft./sec.	400°C
3000 ft./sec.	1400°C

Being calculated from the Dugdale model, these are probably overestimates. Also, we have not assessed the role of tip blunting in providing an effective upper limit (to parallel Equation 20.27) for the running-crack case.

Although no data are available for the propagating crack in 6Al-4V titanium alloy, the temperature rise is twelve times as high as for mild steel for comparable stress-intensity factors and velocities of propagation.

20.6. Influence of Local Heating at the Tip of a Stationary Crack on Toughness

As can be seen from the preceding results, the rise in temperature at the tip of a stationary crack is not huge even at very high loading rates. Nowhere in the data of Krafft and Irwin (1965) does this rise exceed 120°C. However, whether this local temperature increase is to be considered as "small" or "large" depends on its effect on the behavior in fracture. For the moment, let us assume that a local heating at the tip to a temperature u will result in a behavior similar to that of a specimen at a uniform temperature u. This seems to be a plausible assumption, since fracture being a local phenomenon, the mechanical properties at or near the tip govern.

It is generally observed that increasing the temperature of a notched specimen increases the toughness for a given loading rate. Moreover, as the loading rate is increased, the toughness usually decreases for a given specimen temperature. If, therefore, we consider the influence of loading rate and the temperature rise at the tip which increases with loading rate, we will have two opposing factors acting simultaneously. At very high loading rates and, hence, very high average strain rates near the tip, the mechanical properties tend to be little sensitive to further increases in strain rates (Marsh and Campbell 1963). At this stage, the local temperature rise dependence on increased loading rate will perhaps emerge as the sole important factor in affecting toughness. Beyond this stage, therefore, toughness will increase as the loading rate is increased. It follows that materials which show a reduction in toughness versus loading rates, at low loading rates, will exhibit a minimum toughness at critical loading rate beyond which toughness will increase as the loading rate is increased.

Factors such as lower testing temperatures which reduce the sensitivity of mechanical properties to strain rates and at the same time enhance the term $1/\sqrt{k\rho c}$ (which accounts for the temperature rise) will shift the critical loading rate (point of minimum toughness) to lower values of loading rates. Moreover, if the sensitivity to strain rate is small, such as for F.C.C. metals, and if the term $1/\sqrt{k\rho c}$ is large, the critical loading rate can be extremely small and might be suppressed altogether. This is perhaps the case for the titanium alloy 6Al–4V, which shows a rising toughness versus loading rate when tested at room temperature or below. However, as the testing temperature is increased, the sensitivity to strain rates is enhanced simultaneously with a reduction of the term $1/\sqrt{k\rho c}$, and thus a reduction in toughness versus loading rate should be observed up to a critical loading rate, beyond which toughness will increase as the loading rate is increased.

The assumption that a given local temperature rise affects toughness to the same degree as an equal uniform rise in temperature of the whole specimen (at very high loading rates), suggests a method of determining the dependence of toughness on loading rate for a given test temperature when the toughness at various test temperatures at a given loading rate is known. Suppose, as in Figure 20.6, that the toughness for test temperatures $u(1)$ and $u(2)$ at loading time T_1 is known. We assume that at the given loading time, the average strain rate near the tip is high enough so that the effect of further increase in strain rates on toughness is negligible. The temperature rises $\Delta u(1)$ at (1), and $\Delta u(2)$ at (2) can be calculated. To reach point (3), where $K(3) = K(1)$, the temperature at the tip of the crack must be equal to that at 1. In other words,

$$u(2) + \Delta u(3) = u(1) + \Delta u(1) \quad \text{or} \quad \Delta u(3) = u(1) - u(2) + \Delta u(1).$$

Since $\Delta u(3)/\Delta u(1) = \sqrt{T_1}/\sqrt{T_3}$, one can calculate T_3 and the point (3) can be thus located. Thus the toughness versus loading time at a temperature $u(2)$ can be predicted from the variation of toughness versus loading time at $u(1)$ if toughness versus temperature at a given loading time is known. Using the data of Krafft and Irwin (1965) for the 6Al-4V titanium alloy

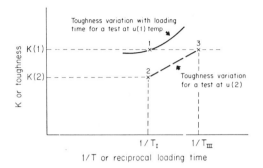

FIGURE 20.6. Method for deriving toughness vs. loading time for a test at $u(2)$ when behavior is known at $u(1)$. Nomenclature: (2): point on the curve for a test at $u(2)$ when $u(2) < u(1)$; $\Delta u(1)$: elevation of temperature at crack tip for point (1); $\Delta u(2)$:elevation of temperature at crack tip for point (2); T_{III} : loading time at which the toughness reaches the value $K(1)$;

$$T_{\text{III}} = \left[\frac{\Delta u(1)}{u(1) - u(2) + \Delta u(1)} \right]^2 T_{\text{I}}.$$

tested at 180°F, the predicted toughness versus loading rate for the test at 80°F was found to be equal to the one reported in the reference. Point (1) is taken as the point of minimum toughness for the 180°F test; point (2) is the point on the 80°F test at the same loading time as point (1), and point (3) is the toughness at the time of loading 10^{-3} seconds on the 80°F test.

Insufficient data are available for a more thorough quantitative test of this method. Nevertheless, a qualitative understanding of experimental data does result as noted above.

This research was supported by the Office of Naval Research under Contracts Nonr 562(20) and N00014–67–A–0191–003.

REFERENCES

Carslaw, H. S., and J. C. Jaeger, 1959, *Conduction of Heat in Solids*, Oxford: Clarendon Press. 2dn ed.

Drucker D. C., and J. R. Rice, 1969, *Eng. Fract. Mech.*, in press.

Dugdale, D. S., 1960, *J. Mech. Phys. Sol.*, **8**, 100.

Eftis, J., and J. M. Krafft, 1965, *J. Basic Eng.*, **87**, 257.

Hutchinson, J. W., 1968, *J. Mech. Phys. Sol.*, **16**, 13.

Krafft, J. M., 1964, *Appl. Mat. Res.*, **3**, 88.

Krafft, J. M., and G. R. Irwin, 1965, *Symposium on Fracture Toughness Testing*, STP No. 381, Philadelphia: ASTM, p. 114.

Marsh, K. J., and J. D. Campbell, 1963, *J. Mech. Phys. Sol.*, **11**, 49.

McClintock, F. A., and G. R. Irwin, 1965, *Symposium on Fracture Toughness Testing*, STP No. 381, Philadelphia: ASTM, p. 84.

McLachlan, N. W., 1961, *Bessel Functions for Engineers*, Oxford: Clarendon Press.

Rice, J. R., 1967, *Symposium on Fatigue Crack Growth*, STP No. 415, Philadelphia: ASTM, p. 247.

Rice, J. R., 1968, *J. Appl. Mech.*, **35**, 379.

Rice, J. R., and G. F. Rosengren, 1968, *J. Mech. Phys. Sol.*, **16**, 1.

Williams, J. G., 1965, *Appl. Mat. Res.*, **4**, 104.

ABSTRACT. Carbides in iron-carbon alloys and mild steels are subject to cracking during tensile testing, and thereby furnish potential Griffith flaws for cleavage fracture. The degree of such carbide cracking increases with plastic strain, presenting the adjacent ferrite with suddenly developed sharp notches. Whether the ferrite then undergoes microcleavage depends on a Griffith-Orowan criterion, involving the acting stress and the carbide-crack length; the latter is usually determined by the thickness dimension of the carbide being cracked.

Based on these ideas, it becomes possible to account for (a) the number of microcleavages produced in tensile testing as a function of temperature, (b) the effect of carbon content and ferritic grain size on the microcleavage frequency, (c) the ductility-transition temperature as influenced by several metallurgical variables, and (d) the unusual experimental scatter in ductility and necking behavior encountered under certain conditions.

The statistical model for cleavage initiation presented here does not apply to those cases where the fracturing mechanism is controlled by sources other than carbides (or brittle inclusions).

21. A Statistical Treatment of Cleavage Initiation in Iron by Cracking of Carbides

MORRIS COHEN AND M. R. VUKCEVICH

21.1. Introduction

This paper deals principally with low-carbon iron-carbon alloys and mild steels in which the carbide phase, particularly that lying along ferritic grain boundaries, undergoes cracking during the early stages of plastic deformation (Bruckner 1950; Allen, Rees, Hopkins, and Tipler 1953; McMahon and Cohen 1965; McMahon 1966). The main question to be considered here is under what conditions such carbide cracks will act as Griffith flaws to initiate cleavage in the adjacent grains of ferrite. This is a problem of microcleavage, rather than long-range cleavage, because local cleavage can often occur without running to complete fracture.

A good material for such studies is iron—0.035 percent carbon (McMahon and Cohen 1965), and the relevant tensile and microcleavage characteristics are summarized in Figure 21.1 as a function of testing temperature. It is seen that the number of microcleavages in the ferritic matrix of fractured tensile specimens first increases with decreasing test temperature, and then decreases, thus generating a bell-shaped curve. Occasional microcleavages appear at temperatures above the ductility transition (T_d = temperature at which fracture occurs just at the onset of necking in a tensile test), but they become noticeably more abundant as

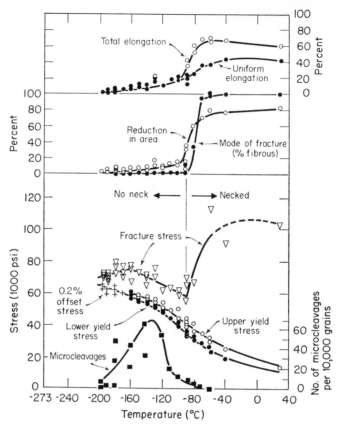

FIGURE 21.1. Tensile properties of vacuum-melted iron (0.035 percent carbon); 0.30 to 0.40 mm grain size

the temperature drops below T_d. For the iron-carbon alloy represented by Figure 21.1, $T_d = -90°C$. Below about $-200°C$, no microcleavages are found because the conditions for initiating a microcrack then satisfy the conditions for propagation to ultimate failure.

At any given temperature, the number of such microcleavages increases during the plastic straining, and they originate predominantly at carbide cracks. However, an important ancillary observation is that each such microcleavage comes into existence just at the instant of formation of the associated carbide crack. If a carbide crack does not happen to initiate microcleavage at once, the adjacent ferrite yields locally and the carbide crack is blunted at the ferrite-carbide interface. As a result, this carbide crack will open up into a void on further loading along the stress-strain curve. In the materials under investigation, decohesion along the carbide ferrite interface rarely occurs. The observed carbide cracks usually correspond to the full thickness dimension of carbide platelets lying along ferritic grain boundaries.

21.2. Cleavage-Initiation Model

We assume that, when a carbide cracks during a tensile test and thereby produces a Griffith flaw of length L, the adjacent ferrite will undergo microcleavage if the tensile stress σ_t on a family of potential cleavage planes satisfies the Griffith-Orowan relationship

$$\sigma_t \geqslant \left[\frac{2E\gamma_{\text{eff}}}{\pi(1 - v^2)L} \right]^{1/2} \tag{21.1}$$

where

$E =$ Young's modulus
$\gamma_{\text{eff}} =$ effective surface energy (including energy expended in plastic straining)
$v =$ Poisson's ratio

γ_{eff} is the order of 5000 ergs/cm^2 in the temperature range where micro-cleavages are first observed (Averbach 1965), and $[2E\gamma_{\text{eff}}/\pi(1 - v^2)]^{1/2}$ then equals 7×10^{15} cgs units for the materials under consideration. L is taken to be the full length of the carbide crack (rather than one-half its length, as is usually done for internal cracks) because this type of Griffith flaw produces microcleavage in a given ferritic grain like an external edge notch. This condition obtains because the intergranular carbides typically crack in bending due to the fact that one of the adjacent ferrite grains yields before the other, causing the crack to travel across the carbide thickness from one interface to the other.

There will be a distribution of carbide thicknesses in the specimen which will range from $L_{\max}$ down to zero. For simplicity, we adopt a parabolic distribution function, as a first approximation to a Gaussian distribution,

$$F(L) = \eta L(L_{\max} - L) \tag{21.2}$$

where $F(L) \, dL$ is the number of carbides per unit volume having thicknesses between L and $L + dL$. This relationship is shown in Figure 21.2. The factor η is determined by the total number of carbides[1] per unit volume (N_v) and the area under the curve in Figure 21.2:

$$N_v = \int_0^{L_{\max}} (F)L \, dL, \quad \text{or}$$

$$\eta = \frac{N_v}{\int_0^{L_{\max}} L(L_{\max} - L) \, dL} = 6N_v/L_{\max}^3 \tag{21.3}$$

[1] We are concerned only with the more massive carbides which, upon cracking, can initiate microcleavage of the ferrite.

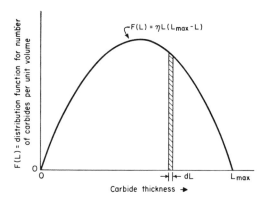

FIGURE 21.2. Assumed parabolic distribution of carbide thicknesses ranging from L_{max} down to zero.

Other things being equal, L_{max} is proportional to the carbon content, and is also proportional to the grain diameter because most of the relevant carbides are spread out along the grain boundaries:

$$L_{max} = K_L CD \tag{21.4}$$

where C is the carbon content in weight percent and D is the ferritic grain size (in the same units as L_{max}). K_L is about unity for the case at hand.

The carbides crack during straining of the ferrite (undoubtedly because of bending of the carbides), and we assume that the increase in number (dN_v^+) of such cracks per unit volume is proportional to the increment of plastic strain $(d\varepsilon_p)$. This will give a distribution of carbide-crack lengths ranging up to L_{max} for each ε_p, but the total number of such cracks in the distribution will increase with plastic straining, as indicated schematically in Figure 21.3 for four positions along the stress-strain curve. At each such point, microcleavage will result only from those carbide cracks whose length L is large enough to satisfy Equation 21.1 at the accompanying stress level.

As we move from one carbide crack-length distribution to the next with increasing plastic strain in Figure 21.3, the acting stress also increases, and carbide cracks of smaller and smaller lengths are able to initiate microcleavage of the ferrite by satisfying Equation 21.1. The L values corresponding to the four stress levels (represented by the four distribution curves) in Figure 21.3 are given along the horizontal axis. At stress σ_1, which might be the yield stress, this illustrative example indicates that the critical Griffith-flaw length $L(\sigma_1)$ is larger than L_{max}, and hence microcleavage does not occur at this level of stress (and strain). At some higher stress σ_2, the critical flaw length is then down to $L(\sigma_2)$, which is below

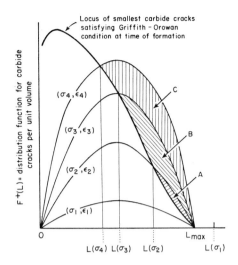

FIGURE 21.3. Distribution of carbide-crack lengths for various positions (σ, ϵ) along the stress-strain curve. Shaded areas signify number of carbide cracks satisfying Griffith-Orowan condition at time of formation. Area A: Number of microcleavages formed up to (σ_2, ϵ_2). Area B: number of microcleavages formed between (σ_2, ϵ_2) and (σ_3, ϵ_3). Area C: number of microcleavages formed between (σ_3, ϵ_3) and (σ_4, ϵ_4).

L_{max}, and so microcleavage can take place. The number of microcleavages per unit volume (N_v^m) thus generated up to (σ_2, ϵ_2) is given by area A, and is equal to the number of supercritical carbide cracks in the carbide crack-length distribution at strain ϵ_2, minus all those in the same size range that formed previously at lower strains but were then subcritical because the stress was too low. The latter carbide cracks are considered to remain inactive at stress σ_2 because of the aforementioned blunting. Ordinarily, carbide cracks either initiate microcleavage at their instant of formation (if they are supercritical) or not at all.

As the tensile loading is further raised, say to (σ_3, ϵ_3) and then to (σ_4, ϵ_4), areas $A + B$ and $A + B + C$ denote the corresponding number of microcleavages per unit volume (N_v^m) generated by the carbide-cracking process. Of course, this is actually a progressive rather than a stepwise phenomenon, and for each successive level of (σ, ϵ), there is a distribution of carbide-crack lengths $(N_v^+$ vs. $L)$ due to the strain ϵ, and a minimum value of L that just satisfies the Griffith-Orowan condition for microcleaving the ferrite at the existing stress σ. The latter relationship forms the lower boundary of the shaded areas in Figure 21.3, thus defining the number of microcleavage events per unit volume (N_v^m) occurring up to any point (σ, ϵ) along the stress-strain curve under the assumptions at hand.

It is now possible to calculate N_v^m up to the point of fracture (σ_f, ϵ_f), knowing the stress-strain curve and the number of carbide cracks (N_v^+) that form as a function of plastic strain. For simplicity, we take

$$dN_v^+ = K_v N_v \, d\varepsilon_p, \tag{21.5}$$

and then from Equation 21.3,

$$N_v^+ = K_v \varepsilon_p \int_0^{L_{max}} F(L) \, dL. \tag{21.6}$$

Here, the proportionality constant K_v turns out to be about unity for the materials under investigation.

N_v^m follows from Equation 21.6 upon subtracting the number of carbide cracks which did not initiate microcleavage (N_n^-). Following the above discussion and employing Equation 21.6, we obtain

$$N_v^- = K_v \varepsilon_f \int_0^{L(\sigma_f)} F(L) \, dL + K_v \int_{L(\sigma_f)}^{L_0} \varepsilon_p F(L) \, dL \tag{21.7}$$

where

> $L_0 = L_{max}$ for $L(\sigma_y) \geqslant L_{max}$
> (This is the case shown in Figure 21.3.)
>
> $L_0 = L(\sigma_y)$ for $L(\sigma_y) \lesssim L_{max}$.
> (This is the case when some of the carbide cracks
> forming at σ_y are immediately supercritical.)

In Equation 21.7, the first term on the right side gives the number of carbide cracks which did not propagate because they were smaller than $L(\sigma_f)$. The second term gives the number of carbide cracks which were larger than $L(\sigma_f)$, but did not propagate because the Griffith-Orowan criterion was not satified at the time of their formation (see the area under the "locus of smallest carbide cracks" in Figure 21.3).

Thus, the total number of microcleavages initiated up to the point of fracture (σ_f, ε_f) becomes

$$N_v^m = N_v^+ - N_v^- = K_v \varepsilon_f \int_{L(\sigma_f)}^{L_{max}} F(L) \, dL - K_v \int_{L(\sigma_f)}^{L_0} \varepsilon_p F(L) \, dL$$

$$= 6\left(\frac{K_v N_v}{L_{max}^3}\right)\left[\varepsilon_f \int_{L(\sigma_f)}^{L_{max}} L(L_{max} - L) \, dL - \int_{L(\sigma_f)}^{L_0} \varepsilon_p L(L_{max} - L) \, dL\right]. \tag{21.8}$$

ε_f is taken as the uniform strain up to the point of fracture because the available quantitative measurements on microcleaving have been made in the uniform portion of the gauge length.

As a first step, we allow $K_v N_v / L_{max}^3$ to be an adjustable parameter in order to show the form of the N_v^m vs. T curve. Equation 21.8 can then be solved graphically for each stress-strain curve (at each temperature), and the results are summarized in Figure 21.4. Here it is evident that the calculated curve successfully matches the bell shape of the experimental curve.

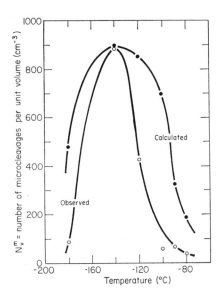

FIGURE 21.4. Comparison of calculated and observed microcleavage frequency in tensile-tested specimens of vacuum-melted iron (0.035 percent carbon); curves matched at $-140°C$.

A $K_v N_v / L_{max}^3$ value of 10^{13} cm^{-6} gives the best fit for this vacuum-melted iron, and checks within a factor of 2 with that obtained from the measured data on K_v, N_v, and L_{max}. The above value of 10^{13} cm^{-6} includes a factor of 1/3 to take into account the angular distribution of the potential cleavage planes relative to the tensile axis. The details of this geometric problem will be published elsewhere.

21.3. Interpretation of Some Metallurgical Variables

In the vacuum-melted iron under discussion, since most of the carbon is in the form of intergranular carbides, the metallographic parameters N_v and L_{max} in Equation 21.8 change accordingly with carbon content and grain size. The effect of these variables are indicated in Figures 21.5 and 21.6, where comparisons are drawn between the observed and calculated microcleavage frequencies. Again, the statistical model accounts for observed trends.

It has been found that, when carbide cracks forming at the yield stress happen to be of sufficient size to initiate microcleavage at that stress,

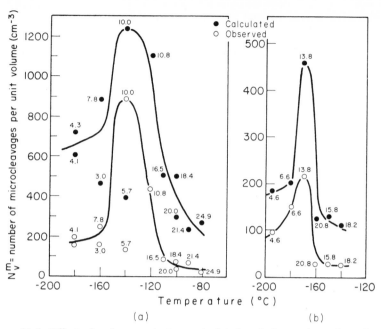

FIGURE 21.5. Effect of carbon content on calculated and observed number of microcleavages in tensile-tested vacuum-melted iron with (a) 0.035 percent carbon, (b) 0.007 percent carbon; grain diameter = 0.25 to 0.35 mm. Numbers indicate percent uniform elongation to fracture.

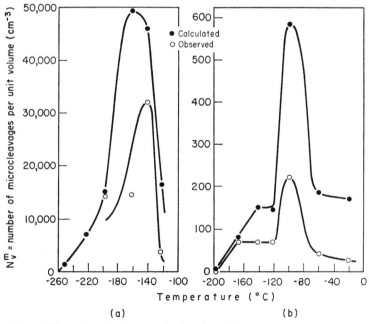

FIGURE 21.6. Effect of grain size on calculated and observed number of microcleavages in tensile-tested vacuum-melted iron (0.039 percent carbon) with (a) grain diameter = 0.11 mm, and (b) grain diameter = 0.41 mm.

302

there is then a good chance that complete fracture will occur before the necking stress is reached. Hence, as an approximation, we shall now identify T_d with the temperature at which L_{max} satisfies the Griffith-Orowan condition at the yield stress (lower yield stress if discontinuous yielding occurs), that is

$$L(\sigma_y) = L_{max} \text{ at } T = T_d. \tag{21.9}$$

As shown in Table 21.1, the transition temperatures calculated on this basis compare nicely with the experimental ductility-transition temperatures for a series of iron-carbon alloys in various conditions.

Another more striking test is given in Table 21.2, in which the ductility and necking behavior of four presumably identical specimens exhibit

TABLE 21.1. Ductility-transition temperatures of a series of vacuum-melted iron-carbon alloys, as measured in tensile tests and calculated according to the statistical carbide-cracking model.

w/o C	Grain Size (mm)	Condition	$L_{max}(\mu)$	$T_d(°C)$ Observed	Calculated
0.007	0.24	Annealed	3.7	−170	−165
0.018	0.11	Recrystallized	2.0	−190	−190
	0.14	Recrystallized	2.6	−140	−170
0.035	0.15	Annealed	5.3	−110	−110
	0.34	Annealed	12.0	−90	−90
	0.32	WQ from 705° C	5.5	−90	−80
0.039	0.11	Annealed	4.3	−130	−120
	0.40	Annealed	15.8	−50	−50

TABLE 21.2. Elucidation of scatter in ductility and necking behavior on tensile testing at $T_d = -90°C$. Vacuum-melted iron with 0.035% carbon.

Grain Size (mm)	σ_y dy/cm^2	Necking	% Reduction in Area	$L(\sigma_y)$ (μ)	L_{max} (μ)	Comparisons
0.34	2.42×10^9	No	7.9	12.0	12.0	$L(\sigma_y) = L_{max}$
0.34	2.45×10^9	No	13.6	11.7	12.0	$L(\sigma_y) < L_{max}$
0.30	2.34×10^9	Yes	31.3	12.8	10.6	$L(\sigma_y) > L_{max}$
0.31	2.47×10^9	Yes	32.3	11.5	11.0	$L(\sigma_y) > L_{max}$

substantial scatter at T_d, as is often noted. For the two specimens that fractured before necking, their parameters show that $L(\sigma_y) \leqslant L_{max}$ which means that the condition for microcleavage was satisfied at the yield stress, whereas for the other two specimens which necked, $L(\sigma_y) > L_{max}$. Thus, relatively small differences in the yield strength and the carbide size can lead to major differences in ductility, merely by controlling whether or not necking sets in before fracture.

With this evidence as justification for the proposed ductility-transition criterion, an explicit formula for T_d can be derived. In Equation 21.9, we substitute Equation 21.1 for $L(\sigma_y)$, and make use of the well-known empirical relation (Petch 1959)

$$\sigma_y = \sigma_0 \exp(-T/\theta) + K_y D^{-1/2} \qquad (21.10)$$

where σ_0 is the extrapolated lower yield point for an iron single crystal at absolute zero, and θ and K_y are constants. We now set $T = T_d$ and rearrange Equation 21.10 to obtain

$$T_d = \theta \ln \sigma_0 \left\{ \left[\frac{2E\gamma_{\text{eff}}}{\pi(1 - v^2)L_{\max}} \right]^{1/2} - K_y D^{-1/2} \right\}^{-1}. \qquad (21.11)$$

Substituting Equation 21.4 for $L_{\max}$ in Equation 21.11 results in the following simple form:

$$T_d = \theta[f(C) + \ln D^{1/2}], \qquad (21.12)$$

where $f(C)$ is a function of the carbon content and does not depend on the grain size.

Differentiating Equation 21.12 with respect to grain size,

$$\frac{\partial T_d}{\partial \ln D} = \frac{\theta}{2}, \qquad (21.13)$$

which can be tested against experiment, as in Table 21.3. The tabulated $(\partial T_d/\partial \ln D)$ values come from the observed dependence of T_d on the grain size, and the $\theta/2$ values are obtained from the temperature dependence of the yield stress. The agreement is reasonably good over the range of carbon contents where the appropriate data are available.

The extent of agreement between the statistical model proposed here and and the experimental observations, as reflected by Tables 21.1 through 21.3, may be better than justified, but it does point up the importance of carbide cracking as a source of cleavage fracture in these materials.

TABLE 21.3. Grain-size dependence of the ductility-transition temperature of vacuum-melted iron-carbon alloys. Test of Equation 21.13.

w/o C	$\partial T_d/\partial \ln D$	$\theta/2$	Source
0.007*	45	48	Tahmoush (1961)
0.018	42	43	Hahn et al. (1959)
0.035	27†	45	McMahon, and Cohen (1965)
0.039	53	46	Hahn et al. (1959)

* Slow-bend test.
† Grain size not under good control.

It should be emphasized, however, that this treatment of the fracture-initiation process cannot be expected to apply to those materials, steels or otherwise, in which the carbide-cracking phenomenon is either absent or superseded by other fracture sources (McMahon and Cohen 1965).

21.4. Conclusions

1. The initiation of microcleavage via carbide cracking in iron-carbon alloys and mild steel can be treated statistically, on the assumption that the number of carbide cracks per unit volume is proportional to the plastic strain, and that the size distribution of carbide cracks at any given strain is parabolic up to the maximum size present.

2. The number of ferrite microcleavages generated at each point along the stress-strain curve has been calculated graphically, based on the number of carbide cracks that are supercritical in length when they come into existence. Any carbide crack that does not happen to satisfy the Griffith-Orowan condition at the time of its formation becomes blunted and is taken to remain inert thereafter relative to microcleavage of the ferrite at subsequently higher stresses.

3. These considerations permit calculations of the number of micro-cleavages as observed in tensile specimens tested at various temperatures, and also the ductility-transition temperature as influenced by several metallurgical variables.

The authors wish to thank the Committee on Ship Steel and the Ship Hull Research Committee for their technical guidance of this research program, which was part of a long-range investigation on the brittle-fracture behavior of low-carbon steel. This work was sponsored by the Ship Structure Committee under Contract Nobs–88279.

REFERENCES

Allen, N. P., W. P. Rees, B. E. Hopkins, and H. R. Tipler, 1953, *J. Iron Steel Inst.*, **174**, 108.

Averbach, B. L., 1965, *Proc. First Int'l Conf. on Fracture*, Sendai, vol. II, p. 779.

Bruckner, W. H., 1950, *Welding J.-Res. Suppl.*, **29**, 467S.

Hahn, G. T., B. L. Averbach, W. S. Owen, and M. Cohen, 1959, in B. L. Averbach et al. (eds.), *Fracture*, Cambridge, Mass: M.I.T. Press, p. 91.

McMahon, C. J., 1966, *Acta Met.*, **14**, 839.

McMahon, C. J., and M. Cohen, 1965, *Acta Met.*, **13**, 591.

Petch, N. J., 1959, in B. L. Averbach et al. (eds.), *Fracture*, Cambridge, Mass.: M.I.T. Press, p. 54.

Tahmoush, F. G., 1961, *S. M. thesis*, M.I.T., Cambridge, Mass.

ABSTRACT. An approximate theory for ductile fracture by the growth and coalescence of holes, combined with the strain distribution in the blunted region of a fully plastic, plane strain, doubly grooved, tensile specimen, predicts that the total crack opening angle varies from 30 degrees to 60 degrees as the initial ratio of hole spacing to diameter varies from 10 to 50. For smaller spacings, interactions between the holes and the tip would have to be taken into account, and there is also evidence that a macroscopically normal fracture often progresses by zigzagging with each step under combined tension and shear (combined modes I and II).

22. Crack Growth in Fully Plastic Grooved Tensile Specimens

F. A. McCLINTOCK

22.1. Introduction

How can a criterion for ductile fracture be found that is analogous to the one for brittle fracture? Should a criterion be based on an energy balance as Griffith (1920) did for tension, or on a local stress as Griffith (1924) did for combined stress, which, as Orowan (1955) showed, provided sufficient conditions for tensile fracture? Should a criterion for initiation be sought, or one for growth?

A criterion should arise naturally out of a simplified model based on experimental observations. Let the model suggest whether an energy or some other principle is appropriate. A criterion for slow growth in equilibrium with the applied loads or deformations seems preferable to one for initiation or for dynamic growth, because it is a growing and not a stationary crack that finally causes fracture, because rate effects can be neglected in most structural alloys, in which changes in rate by a factor of 1000 change the flow stress by only perhaps 10 percent, and because inertia effects can also be neglected, thus excluding some crack arrest problems, since we are not usually concerned with whether a structure fails in seconds or milliseconds.

What is observed experimentally? Aside from cleavage, fracture in ductile metals appears to occur almost exclusively by the growth of holes. The classic example is afforded by Figure 22.1, obtained by Bluhm and Morrissey (1965) on sectioning an ordinary tensile specimen of copper.

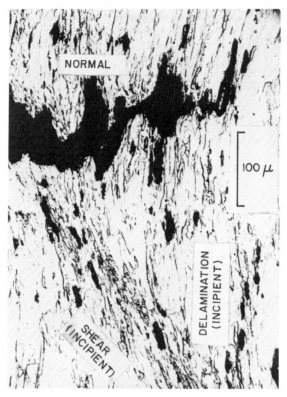

FIGURE 22.1. Growing normal fracture and incipient delaminating and shear fractures in a necked copper tensile specimen. (Bluhm and Morrisey (1965), courtesy of U.S. Army Materials Research Agency. U.S. Army photograph.

The primary fracture is macroscopically normal to the tensile axis, but fracture on a 45-deg shear plane and by delamination are also pending. In the terminology of fracture mechanics, the actual fracture is mode I, and the pending 45-deg fracture is combined modes I and II. 7075–T6 aluminum-zinc alloy lies toward the other extreme of ductility and notch sensitivity in structural alloys. As shown in the series of scanning electron micrographs in Figure 22.2, even this alloy fractures by dimples suggestive of hole growth, although in some cases the dimples are arranged in sheets, as might be expected from numerous sources in grain boundaries. This confirms electron micrographs of replicas, for example by Burghard and Davidson (1965). Dimples are also the dominant mode in fracture of 0.45 Ni–Cr steels with tensile strengths of up to 300,000 psi according to Birkle, Wei, and Pellissier (1966).

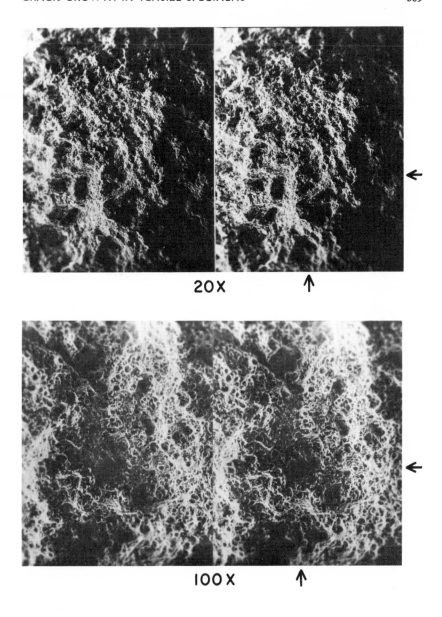

20X ↑

100X ↑

FIGURE 22.2. Stereo scanning micrographs of unnotched 7075–T6 aluminum tensile fractures at increasingly larger magnifications (see also p. 310).

As with ductile fracture in shear, it seems easiest to begin an analysis of tensile fracture with a specimen small enough so that fully plastic stress and strain fields are attained before fracture. It might at first seem easier to consider sharp grooves rather than blunted ones, but sharp grooves have

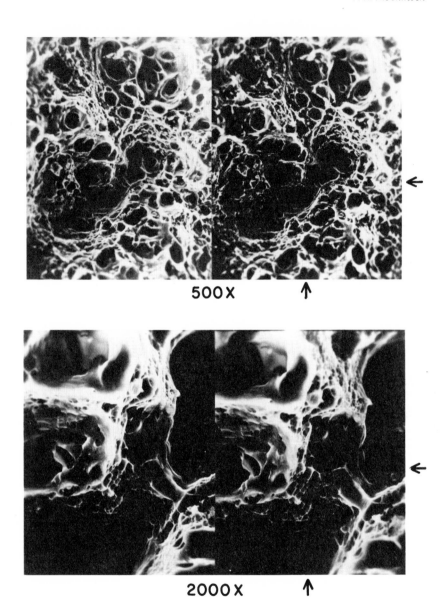

500X ↑

2000X ↑

FIGURE 22.2 (concluded).

no strain concentration in the diamond-shaped region in front of the crack, although there is a concentration varying as $1/r$ in the fan-shaped regions at either side of the tip, as shown in Figure 22.3 (Neimark 1968). With any dependence of fracture on strain, one would expect cracks to branch laterally rather than to run directly ahead. While such delamination has

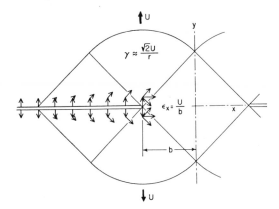

FIGURE 22.3. Flow field and flank displacements for a sharp groove.

been observed (see McClintock 1961), fractures usually propagate ahead where there appears to be no strain concentration. My concern for this paradox was reawakened by a conversation with Professor J. R. Rice at the AIME meeting in New York in March 1966, in which we agreed that the same difficulty was likely to arise in the elastic-plastic case.

The resolution of the paradox of crack propagation with an apparent absence of strain concentration may lie in the effect of crack blunting, as shown by the following discussion of the deformation. The initial flow field around a sharp groove leads to the nearly rectangular blunting shown in Figure 22.3. According to Lee and Wang (1952), a rectangular tip would then become somewhat rounded.[1] On the other hand, consider a groove with a circular tip, as shown in Figure 22.4. Taking the deformation in the flank into account, one would expect from the analysis of Wang (1954) that the arc would be somewhat flattened toward a rectangular shape, suggesting some intermediate steady-state configuration, which has in fact recently been found by Johnson (1968). For convenience, the groove will be assumed to have a circular tip.

For fracture, the important aspect of both the rectangular and circular roots is that the characteristics or slip lines governing the strain in front come in from fan-shaped regions above and below the tip of the groove. The strain concentration in the fans produces a large strain concentration in the region directly in front of the groove. This strain concentration can produce fracture, and the cycle of blunting and fracture can be repeated.

[1] Note that the parts of the Lee (1952) and Lee and Wang (1952) flow fields that extend over the straight flanks of a notch are in error because, contrary to the statement by Lee and Wang, it *is* possible to extend a kinematically admissable slip-line field throughout the diamond-shaped region of Figure 22.3 (see Prandtl 1920). The extremum theory of Hill (1958) applied by Neimark (1968), and direct solution by Rosenfeld (1959) show that deformation does occur in this field if the material has at least infinitesimal strain hardening.

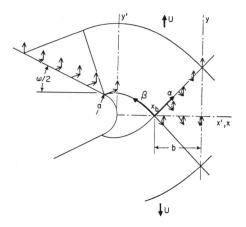

FIGURE 22.4. Flow field for a blunted groove.

A quantitative understanding of such crack growth requires a fracture criterion, which in turn will involve the stress-and-strain histories of material elements along the line in front of the groove.

22.2. Analysis for Core Region

22.2.1. Stress

In the diamond-shaped core region $-b < x < 0$, the normal stress components, $\sigma_x, \sigma_y, \sigma_z$, are given in terms of the yield strength in shear k and the semiangle of the groove $\omega/2$ (see, for example, Wang 1954):

$$\sigma_{x,y,z} = 2k\left[\left(0, 1, \frac{1}{2}\right) + \frac{\pi}{2} - \frac{\omega}{2}\right]. \tag{22.1}$$

22.2.2. Strain and Displacement Increments

Neimark (1968) has shown that displacements vary nearly linearly across the core region, corresponding to uniform strain there. Along the plane $y = 0$, the strain increments and displacement increments are then given in terms of the extensions $\pm dU$ of the rigid parts by

$$d\varepsilon_{x,y} = \mp dU/b, \tag{22.2}$$

$$du_x = \int_0^x d\varepsilon_x \, dx = -x \, dU/b. \tag{22.3}$$

22.2.3. Flank Angle

The change in flank angle is calculated most conveniently by considering coordinate axes fixed in the upper end of the specimen. The tangential displacement along the boundary of the logarithmic spiral region is

$\sqrt{2dU}$, giving a component dU normal to the flank. Since this component varies linearly along the flank, and since distances along the flank are equal to distances along the x axis in the core, the angular change in the flank angle is given by

$$d\omega/2 = -dU/2b. \tag{22.4}$$

For small displacements of specimens with a large diamond-shaped region, the change in flank angle can be neglected.

22.2.4. Coordinates of Particles

A fracture criterion in general depends on the stress-and-strain history followed by a particular particle as it approaches the crack. Within the core region, the coordinates of a point initially at x_0 can be found by integration of Equation 22.3, noting that $dx = du_x$ and taking a steady rate of decrease of the core region $db/dU < 0$:

$$dx = -x\, dU/b; \quad x = x_0\left(1 + \frac{U - U_0}{b_0}\frac{db}{dU}\right)^{-1/(db/dU)}. \tag{22.5}$$

22.3. Analysis for Root Region

For simplicity in determining the coordinates of a particle and the stress-and-strain history in the logarithmic spiral region, choose the x', y' coordinates of Figure 22.4, which are fixed at the tip of the groove.

22.3.1. Stress

By appropriate change of variable and correction of sign, the stress distribution can be obtained, for example, from Equation 19 of Wang (1954):

$$\text{for} \quad 0 < x' < a(\exp \omega/2 - 1),$$

$$\sigma_{x,y,z} = 2k\left[\left(0, 1, \frac{1}{2}\right) + \ln\left(1 + \frac{x'}{a}\right)\right]. \tag{22.6}$$

22.3.2. Displacement and Strain Increments

In the region of logarithmic spirals, the displacement distribution turns out to be identical with that given by Wang for a field without a diamond-shaped core. This identity follows from the Geiringer equations, which show that the straight characteristics in the fan-shaped regions above and below the logarithmic spiral transmit the normal components of displacements without change from the rigid region to the boundary of the logarithmic spiral region. Wang expressed the displacement distribution as a series of Bessel functions and also graphically. Along $y = 0$, in terms of the curvilinear coordinate β of Figure 22.4, which is

$$2\beta = \frac{\pi}{2} - \frac{\omega}{2} - \ln\left(\frac{x'}{a} + 1\right) \quad \text{for} \quad \omega < \pi \quad \text{and}$$

$$\frac{x'}{a} + 1 < \exp(\pi/2 - \omega/2), \tag{22.7a}$$

the displacement rate relative to the center is

$$\frac{du_x}{dU} = (1 + 2\beta)I_0(2\beta) + 2\beta \sum_{p=1}^{\infty} (-1)^p \frac{4p}{4p^2 - 1} I_{2p}(2\beta). \tag{22.7b}$$

This displacement rate and the corresponding strain rates, obtained by numerical differentiation over eight intervals, are plotted in Figure 22.5.

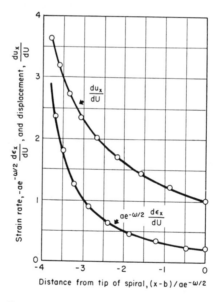

FIGURE 22.5 Strain and displacement rates in the logarithmic spiral region.

22.3.3. *Root radius*

Taking the deformed shape of the root of the groove as approximately a circular arc, one can choose the arc either (see Figure 22.6)

a. to match the normal displacements of the surface at the flank and at the root, or

b. to match the displacements of the particles initially at the point of tangency and at the root.

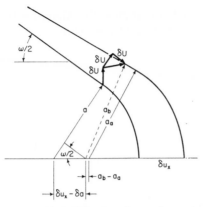

FIGURE 22.6. Effective change in root radius.

Since the angle between the radii a_a and a_b of the two approximations is only of the order of the variables $\delta\omega/2$, $\delta U/a$, and the relative difference in length of the two radii, $(a_a - a_b)/a$, the length difference itself will be a second order quantity. The change from the initial radius is then found from the change in radii to the flank:

$$\delta a = \delta U \cos \omega/2 + \delta U - (\delta u_x - \delta a)\sin \omega/2$$

$$\frac{da}{dU} = \frac{1 + \cos \omega/2 - \sin \omega/2(du_x/dU)_{x'=0}}{1 - \sin \omega/2} \qquad (22.8)$$

The inward motion of the root of the groove, $(dx/dU_{x'=0})$, is found from Figure 22.5, and the resulting value of da/dU is plotted in Figure 22.7.

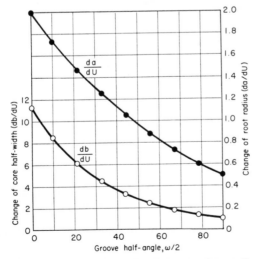

FIGURE 22.7. Rate of change of root radius a and core dimension b with total extension $2U$.

22.3.4. Core Width

The change in the core half-width b arises not only from the inward displacement at the tip of the groove, found from Figure 22.5 or Equations 22.7, but also from the increase in the size of the logarithmic spiral region due to the increase in root radius, and the decrease due to increasing flank angle, $d(\omega/2)/dU = 1/b$:

$$\frac{db}{dU} = \left(\frac{du_x}{dU}\right)_{x'=0} + \frac{d[a \exp(\pi/2 - \omega/2) - a]}{dU},$$

or

$$\frac{db}{dU} = \left(\frac{du_x}{dU}\right)_{x'=0} + \frac{da}{dU}[\exp(\pi/2 - \omega/2) - 1] - \frac{a}{2b}\exp(\pi/2 - \omega/2)$$

$$(22.9)$$

For relatively sharp grooves ($a/b \ll 1$), which will be of most interest in crack propagation, the last term may be neglected. The rate of core shrinkage then depends only on $\omega/2$ and is therefore nearly constant during extension. This function is also plotted in Figure 22.7.

22.3.5. Coordinates

Since strain increments and stress depend on x'/a, successive values of x'/a following a particle should be found. Recasting the displacement rate in terms of the rate of change of x'/a and the radius rate da/dU,

$$\frac{du_{x'}}{dU} = \frac{da}{dU}\frac{du_{x'}}{da} = \frac{da}{dU}\frac{dx'}{da} = \frac{da}{dU}\frac{d(x'/a) + (x'/a)(da/a)}{da/a}.$$

The displacement rate relative to the notch root is then expressed in terms of that in the x coordinate system, which can in turn be found from Equations 22.7:

$$\frac{du_{x'}}{dU} = \frac{du_x}{dU} - \left(\frac{du_x}{dU}\right)_{x'=0}. \qquad (22.10)$$

Combining and rearranging,

$$\int_{\exp(\pi/2-\omega/2)-1}^{x'/a} \frac{(da)/(dU)\, d(x')/(a)}{(du_x)/(dU) - [(du_x)/(dU)]_{x'=0} - x'/a\, da/dU} = \ln\frac{a}{a_0},$$

$$(22.11)$$

where a_0 is the tip radius at which the point in question was first swept over by the logarithmic spiral region. The integral of Equation 22.11, obtained by a trapezoidal rule over the intervals shown, is plotted in Figure 22.8. For convenient plotting the blunting is expressed as a_0/a; a particle follows a path downward to the left, as indicated by the arrow.

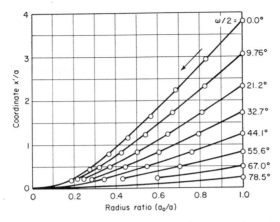

FIGURE 22.8. Motion of a hole during blunting. a_0 is the radius when the spiral reaches the hole.

22.4. Analysis of Rate of Crack Growth

With the stress-and-strain history known, a fracture criterion can be applied. From the observations described in the Introduction, consider the mechanism to be the growth and coalescence of holes. Take equally spaced cylindrical holes, parallel to the leading edge of the crack, and lying directly ahead of the crack in its plane, as shown in Figure 22.9.

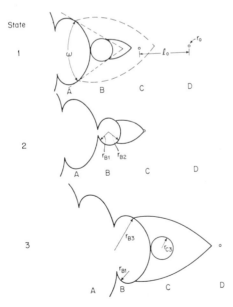

FIGURE 22.9. Growth of hole C to coalescence. $\omega = 65.4$ deg.

These are similar to the assumptions of Johnson (1968), who treated only the first increment of crack growth. Such a configuration might correspond to one stage in the grain boundary fracture of 7075–T6 aluminum-zinc alloy. To simplify the analysis further, neglect interactions of the holes with each other and with the logarithmic spiral region induced by the blunted crack. The rate of growth of these holes depends on the two transverse components of stress, σ_x and σ_y, the equivalent stress and strain, $\bar{\sigma}$ and $\bar{\varepsilon}$, and on the strain-hardening exponent n in the equation $\bar{\sigma} = \sigma_1 \bar{\varepsilon}^n$ approximately according to McClintock, (1968a), Equation 23,

$$\frac{d \ln r/r_o}{d\bar{\varepsilon}} = \frac{\sqrt{3/2}}{(1-n)} \sinh\left[(1-n)\frac{\sigma_x + \sigma_y}{2\bar{\sigma}/\sqrt{3}}\right] + \varepsilon_x + \varepsilon_y. \tag{22.12}$$

(A similar expression has recently been obtained by Tracey 1968). For the plane plastic strain considered here, and for the stress in the logarithmic spiral region from Equation 22.6,

$$\bar{\varepsilon} = -2\varepsilon_x/\sqrt{3}; \ \varepsilon_x + \varepsilon_y = -\varepsilon_z = 0$$

Combining, the hole growth rate is

$$\frac{d \ln r/r_o}{d\varepsilon_x} = -\frac{1}{1-n} \sinh\{(1-n)[1 + 2\ln(1 + x'/a)]\}. \tag{22.13}$$

The steady-state eccentricity is given in terms of semimajor and semiminor axes r_y and r_x approximately by McClintock (1968a) Equation 24,

$$m = \frac{r_y - r_x}{r_y + r_x} = \frac{\sigma_y - \sigma_x}{\sigma_y + \sigma_x} = \frac{1}{1 + 2\ln(1 + x'/a)} \tag{22.14}$$

The maximum eccentricity occurs when the hole reaches the root, when $x'/a = r_{C3}/r_{B3}$ (see Figure 22.9) and is of the order of 1/2 to 1/3. Necking between holes and between the holes and the root of the crack will tend to decrease the eccentricity; for simplicity take it to be zero (circular holes).

The problem now is to calculate the growth of the holes and their change in spacing as functions of the flank angle, in order to find what original ratio of inclusion diameter to spacing led to a given flank angle. Two special cases may be considered, depending on whether or not as one hole coalesces with the root of the crack, its logarithmic spiral region extends to the next nucleus. First consider the simpler case, when it does not, as shown in Figure 22.9. It is easiest to analyze the problem by starting with the final state and working backward to the entry of the hole to the plastic zone, state 2, and the moment of coalescence of the previous hole, state 1. Radii of individual holes are denoted by letters A, B, etc., and the radius of the root of the crack by that of the hole that formed it.

The radius ratio at coalescence is related to the crack half-angle $\omega/2$ by

$$\sin \omega/2 = \frac{r_{B3} - r_{C3}}{r_{B3} + r_{C3}} \; ; \frac{r_{C3}}{r_{B3}} = \frac{1 - \sin \omega/2}{1 + \sin \omega/2}$$

$$= \frac{r_{B1}}{r_{A1}} = \frac{r_{B1}}{r_{B3}}. \tag{22.15}$$

Here is the point at which the analysis differs significantly from Johnson's (1968); unless very large reversed plastic strains occur in the unloading flanks, which seems unlikely in view of the results of Chitaley (1968) for mode III, the ratio of the two radii requires a flank angle that in turn requires fully plastic flow.

At coalescence the coordinate of the hole C relative to the root of the B is

$$\frac{x'_{BC3}}{r_{B3}} = \frac{x'_{AB1}}{r_{A1}} = \frac{r_{C3}}{r_{B3}}. \tag{22.16}$$

The increase in radius required to take a hole from the tip of the logarithmic spiral region to x'/a is found from Figure 22.8 at the appropriate value of x'/a and $\omega/2$:

$$\frac{r_{B2}}{r_{B3}} = \frac{a_o}{a} \left(\frac{x'}{a} = \frac{r_{C3}}{r_{B3}}, \frac{\omega}{2} \right). \tag{22.17}$$

The displacements of the ends of the specimen required for these changes in radii are found from Figure 22.7:

$$\Delta U = \frac{1}{da/dU} \Delta a. \tag{22.18}$$

The advance of the root of the crack is found from Figure 22.5 at $x_r - b = -a \exp(\pi/2 - \omega/2) - a$

$$\Delta u_x = \left(\frac{du_x}{dU} \right)_{x'=0} \Delta U. \tag{22.19}$$

The initial spacing of the holes is determined from the fact that during a cycle in which the logarithmic spiral region picks up one hole, it advances due to blunting by the amount

$$\Delta b = \frac{db}{dU} \Delta U, \tag{22.20}$$

where db/dU is found from Figure 22.7. The region retreats due to coalescence at state 3 by the amount

$$r_{B3} \left[\exp\left(\frac{\pi - \omega}{2} \right) - 1 \right] - r_{C3} \left[\exp\left(\frac{\pi - \omega}{2} \right) + 1 \right]$$

At the same time, the material in the core region (neglecting strain in the core) is advancing by ΔU. The amount of material overtaken in one cycle, which is the initial spacing of the holes, is

$$\frac{l_o}{r_{C3}} = \left(\frac{db}{dU} - 1\right)\frac{\Delta U}{r_{C3}} - \frac{r_{B3}}{r_{C3}}\left[\exp\left(\frac{\pi - \omega}{2}\right) - 1\right]$$

$$+ \exp\left(\frac{\pi - \omega}{2}\right) + 1. \tag{22.21}$$

Next consider the case shown in Figure 22.10, where when one hole coalesces with the crack tip (hole B in state 2), the next hole, C, is already

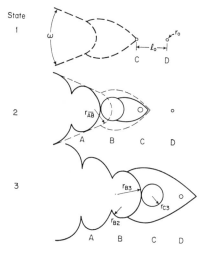

FIGURE 22.10. Growth of hole C to coelescence. $\omega = 42.4$ deg.

within the logarithmic spiral region. The radius ratio at coalescence is given again by Equation 22.15 except that coalescence now occurs at state 2 rather than state 1. At coalescence the coordinate of the next hole relative to the root of the crack is found by first noting the coordinate of C at state 3 from Equation 22.16, turning to Figure 22.8 to determine the radius ratio required to take this point to the tip of the plastic zone,

$$\frac{r_{oBC3}}{r_{B3}} = \frac{a_o}{a}\left(\left(\frac{x'}{a}\right)_{max} = \frac{r_{C3}}{r_{B3}}\right),$$

and then calculating the corresponding radius ratio for state 2:

$$\frac{r_{oBC3} = r_{oBC2}}{r_{B2}} = \frac{r_{oBC3}}{r_{B3}}\frac{r_{B3}}{r_{B2}} \tag{22.22}$$

For very small included angles ω, a number of such blunting and coalescing steps are required for the point C to come in from the tip of the logarithmic spiral region at state 1. Rather than considering these steps individually, it may be a satisfactory approximation to assume the root radius to be constant at the mean value

$$r_{\overline{AB}} = \frac{r_{A2} + r_{B2}}{2} \tag{22.23}$$

The corresponding coordinate of the hole C relative to the average coordinate of the root of the crack turns out to be

$$\frac{x'_{\overline{ABC2}}}{r_{\overline{AB}}} = \frac{x'_{BC2} + r_{B2}}{r_{AB}} \tag{22.24}$$

With the geometry determined, the growth of the hole is found by integrating Equation 22.13 for the hole growth per unit strain, the strain per unit displacement from Figure 22.5, the displacement per unit increase in root radius from Figure 22.7 and the rate of increase in coordinate of a particle with respect to root radius from Figure 22.8:

$$\ln \frac{r_{C2}}{r_{C1}} = \int_{x'_{\overline{ABC2}}/r_{\overline{AB}}}^{\exp(\pi/2-\omega/2)-1} \frac{d\ln(r/r_o)}{d\varepsilon_x} \frac{a\exp(-\omega/2)\,d\varepsilon_x}{dU} \exp(\omega/2)$$

$$\times \frac{dU}{da}\left[\frac{da}{a} \Big/ d\left(\frac{x'}{a}\right)\right] d\left(\frac{x'}{a}\right) \tag{22.25}$$

Likewise the increase in spacing between the two holes B and C can be found from

$$\ln \frac{BC_1}{BC_2} = \int_{x'_{\overline{ABC2}}/r_{\overline{AB}}}^{\exp(\pi/2-\omega)/2-1} \left(a\exp(-\omega/2)\frac{d\varepsilon_x}{dU}\right)\exp(\omega/2)$$

$$\times \frac{dU}{da}\left[\frac{da}{a} \Big/ d\left(\frac{x'}{a}\right)\right] d\left(\frac{x'}{a}\right) \tag{22.26}$$

These two integrals were evaluated by a trapezoidal rule over their range of integration, since the approximations in the analysis did not warrant a more precise calculation.

The flank angle is plotted in Figure 22.11 as a function of the initial ratio of hole spacing to diameter. The spacing ratios for large flank angles are underestimates, since the hole growth due to the logarithmic spiral subtended by the radius of hole A, shown dashed in Figure 22.9, was neglected. This approximation is of no concern, however, because included angles of greater than 90 deg would require such widely spaced inclusions that their volume fraction would be parts per million. The vanishing of the flank angle at a finite hole spacing would not mean a complete loss of

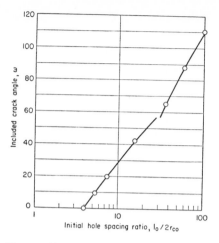

FIGURE 22.11. Dependence of crack angle on initial hole spacing.

ductility; a finite crack-tip displacement is still associated with the process of growing the hole to the size of the tip radius. This would not be a steady-state process, however, unless the spacings and mean diameters of the holes increased in the same very particular and unnatural manner. Actually, for included angles of less than 30 deg, the holes become so closely spaced within the logarithmic spiral region that it would be quite in error to neglect interactions[2]. One should instead treat the fracture process as one requiring the gradual deterioration in average strength of a porous region, which exerts progressively less normal traction σ_{nn} on its surroundings as it is extended by a displacement discontinuity $[u_y]$. That is, a traction-displacement boundary condition of the type proposed by Barenblatt (1962) would have to be applied to the plastic half-spaces on either side of the crack:

$$\sigma_{nn}([u_y]). \tag{22.27}$$

22.5. Steady-State Conditions

In order to observe the crack propagation angle as a nearly steady state, it is necessary that the hole-spacing ratio be relatively unaffected by strain in the core before the logarithmic spiral region reaches a point. For example, suppose that it is desired to have the crack-growth angle affected by less than 10 percent as the crack growth sweeps over N hole nuclei. The required ligament width can then be obtained from

$$\frac{b}{l_o} = N\left(\frac{dU}{da}\frac{\Delta a}{a}\right)\frac{r_{C3}}{l_o}\frac{|d \ln r/d\varepsilon_x|}{(d(\ln r)/d\omega)\,\Delta\omega} \tag{22.28}$$

[2] One may also have to include elastic effects.

where the factor

$\dfrac{dU}{da}$ is found from Figure 22.7

$\dfrac{\Delta a}{a}$ is found from Equations 22.15 or 22.17

$\dfrac{r_{c3}}{l_o}$ is found from Equations 22.21 or 22.26

$d \ln r/d\varepsilon_x$ is found from Equation 22.13

$d(\ln r)/d\omega$ is found from Figure 22.11.

Evaluation of Equation 22.28 shows that for a 10 percent variation in flank angle as the crack travels over ten holes, the required ligament half-thickness increases from 70 to 200 initial hole spacings as the flank angle drops from 110 deg to 20 deg. Expressed another way, such a steady state will be found over only 5 to 15 percent of the ligament thickness.

22.6. Discussion and Conclusions

What do experiments tell? Experiments have been run on plane-strain grooved specimens of OFHC copper, 1100–F aluminum (McClintock, 1968b), and more recently annealed 6061–0 aluminum alloy. In all cases the crack angle changed rapidly, as expected. The copper was so ductile that it started as nearly a 90-deg crack, terminating in a double-shear crack. The aluminum formed a flat-bottomed tip, cracking from the corners until 45-deg shear could occur on a single plane across the remaining section. The 6061–0 alloy started with an included flank angle of 45 deg to 60 deg, and then changed to a zigzagging, sharper crack, as shown in Figure 22.12.

What about higher-strength, less ductile materials? One would expect the cracking angle to be smaller, but if so, fracture should occur by sheets of holes growing gradually in front of the crack. Such gradual growth seems not to be observed by sectioning. Perhaps the explanation is very simple. Perhaps what we are really seeing is not pure tensile fracture (mode I), but a zigzagging, 45 deg fracture under combined shear and tension (modes I and II). Here high-strain regions in the fans to either side of the tip of the crack could cause rapid crack growth even without appreciable blunting, and along planes at ± 45 deg the triaxiality is as high as directly in front of the crack.

Reexamination of the stereo scanning micrographs of Figure 22.2 shows the presence of such zigzagging at all scales of observation. Even the centers of the holes in Figure 22.1 show some zigzagging. Also,

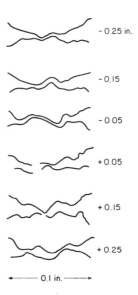

FIGURE 22.12. Section of a 30 deg grooved 6061-0 Aluminum specimen at fracture. Initial ligament 0.125 in. by 1.25 in.

FIGURE 22.13. Cup and cone fracture of 3-in. diameter bar of medium carbon steel (Miklowitz 1950). Note scale of roughness.

Figure 22.13, from Miklowitz (1950), shows a tensile fracture which many viewers at first judge to be from a small specimen. Actually the original diameter was 3 in., indicating that the scale of the roughness in the central region is set more nearly by the size of the specimen than by the micro-structure. Here again zigzagging seems dominant. The theory for hole growth in shear bands by McClintock, Kaplan, and Berg (1966) could be strengthened by incorporating a fully plastic analysis based on extending the solutions of Green (1954) for junctions under combined shear and tension to a row of holes. The fully plastic strain distribution for a zig-zagging notch remains to be determined.

And so I must acknowledge my incomplete mastery of Orowan's pre-cepts. This rather complicated analysis does predict the sharpening of cracks in ductile fracture, and gives promise of applying to sheets of inclusions such as are found in some grain boundaries and in transverse tests on rolled materials. On the other hand, more considered observations at the beginning might have suggested a more appropriate model for the majority of fractures of interest, namely, zigzagging under combined tension and shear.

My first acknowledgement is to Professor Orowan for his help and inspiration over the years, especially in seeking the essence of the problem, based on what is actually happening at the appropriate scale of observa-tion, whether it be atomic or macroscopic. I am also indebted to Robert Leonard, James Joyce, Charles Mahlmann, Miss Cheryl Cretin, and Miss Sandra Grant, whose cheerful and last-minute cooperation made this manuscript possible. Finally, I wish to thank the National Science Founda-tion under Grant No. GK–1875X, without whose financial support this group could not have been brought together.

REFERENCES

Barenblatt, G. I., 1962, *Adv. in Appl. Mech.*, **7**, New York: Academic Press, p. 55.

Birkle, A. J., R. P. Wei, and G. E. Pellissier, 1966, *Trans. Amer. Soc. Metals*, **59**, 981.

Bluhm, J. I., and R. J. Morrissey, 1965, *Proc. First Int'l Conf. on Fracture*, Sendai, **3**, 1739.

Burghard, H. C., Jr., and D. L. Davidson, 1965, *Proc. First Int'l Conf. on Fracture* Sendai, **2**, 571.

Chitaley, A. D., 1968, *Elastic-Plastic Mechanics of Cracks with Growth*, Sc.D. thesis, Dept. Mech. Eng., M.I.T., Cambridge, Mass.

Green, A. P., 1954, *J. Mech. Phys. Sol.*, **2**, 197.

Griffith A. A., 1920, *Phil. Trans.*, *Roy. Soc.*, **A221**, 163.

Griffith, A. A., 1924, *Proc. First Int'l Conf. Appl. Mech.*, Delft, p. 55.

Hill, R., 1958, *J. Mech. Phys. Sol.*, **6**, 236.

Johnson, M. A., 1968, *Large Scale Geometry Changes at a Crack Tip*, M. S. thesis, Div. of Eng., Brown University, Providence, R. I.

Lee, E. H., 1952, *J. Appl. Mech.*, **19**, 331.

Lee, E. H., and A. J. Wang, 1952, *Proc. Second U.S. Nat. Cong. Appl. Mech.*, p. 489.

McClintock, F. A., 1961, *Welding J. Res. Suppl.*, **26**, 202.

McClintock, F. A., 1968a, *J. Appl. Mech.*, **35**, 363.

McClintock, F. A., 1968b, *Int. J. Fracture Mech.*, **4**, 101.

McClintock, F. A., S. M. Kaplan, and C. A. Berg, Jr., 1966, *Int. J. Fracture Mech.*, **2**, 614.

Miklowitz, J., 1950, *J. Appl. Mech.*, **17**, 159.

Neimark, J. E., 1968, *J. Appl. Mech.*, **35**, 111.

Orowan, E., 1955, *Welding J. Res. Suppl.*, **34**, 157s.

Prandtl, L., 1920, *Nachr. Königl. Gesell. Wiss. Göttingen, Math-phys. Kl.*, p. 74.

Rosenfeld, R. L., 1959, "The Initial Deformation of a Rigid-Plastic Work-Hardening Notched Tensile Specimen in Plane Strain," *Res. Memo. 20*, Fatigue and Plasticity Lab. Dept. of Mech. Eng., M.I.T., Cambridge, Mass.

Tracey, D. M., 1968, *On the Large Ductile Expansion of Cavities*, M. S. thesis, Div. of Eng., Brown University, Providence, R. I.

Wang, A. J., 1954, *Quart. Appl. Mech.*, **11**, 427.

ABSTRACT. It is attempted to obtain the fatigue crack-propagation law for solids, such as metals and polymets, on the basis of nucleation theory. The results lead to Paris's law for the case of mild steel, and suggest that for the case of polymers it may be reasonable to take the maximum stress of the cycle in the stress-intensity factor instead of the stress range.

The possibility is shown that the effects of both the loading cycle itself and the loading frequency can be included.

23. A Kinetic Approach to Fatigue Crack Propagation

T. YOKOBORI

23.1. Introduction

As far as the approach is concerned, it may be useful to divide *S–N* relationships into two areas: the range of finite fatigue life, and the endurance limit, since we believe the mechanisms may be different in these two areas. A unified engineering theory for metal fatigue (Yokobori 1963, and Yokobori, Yoshida, and Ishimura 1969) proposed previously concerns mainly the endurance limit, being based on the conditions of not initiating or not propagating some critical crack. With respect to the range of finite fatigue life, we shall give a kinetic approach whether it be microscopic or macroscopic. The nucleation theory of metal fatigue was previously proposed from this viewpoint (Yokobori, 1955).

Since that time, need for a theory has arisen to explain the following features: (1) the fatigue crack propagation law, and (2) frequency dependence of fatigue life. Thus in the present paper an attempt is made to extend the theory to explain these features.

23.2. Extension of Fatigue Crack as a Nucleation Process

Let us consider the nucleation of a microcrack near the tip of an already initiated crack. According to the nucleation theory of fracture

327

of Yokobori (1952, 1955), the expression for the rate of the initiation of an unstable crack is given by[1]

$$I = \frac{VZkT}{h} \exp\left[-\frac{\Delta f^*}{kT} - \frac{A}{q^4 S^4 kT} + \frac{vq^2 S^2}{2EkT} \right], \tag{23.1}$$

where

$$A = \pi^3 \rho^3 E^2 / 6(1 - v^2)^2 \tag{2.32}$$

and

S = applied stress
E = Young's modulus
v = Poisson's ratio
ρ = specific surface energy
Z = concentration of nucleation sites
V = the volume of the specimen
Δf^* = activation energy associated with the separation of a pair of atoms as the edge of the crack moves between them
v = effective volume created by the separation of a pair of atoms
q = stress concentration at the nucleation sites,

and h, k, T have their customary meaning in rate theories.
Putting

$$B = \frac{v}{2E} \tag{23.3}$$

and

$$S_o = \left(\frac{A}{B}\right)^{1/6}, \tag{23.4}$$

then the term $(vq^2 S^2 / 2E) - (A/q^4 S^4)$ in the parentheses in the right-hand side of Equation 23.1 is written as follows:

$$\frac{vq^2 S^2}{2E} - \frac{A}{q^4 S^4} = (B^2 A)^{1/3} \left[\left(\frac{qS}{S_o}\right)^2 - \frac{1}{(qS/S_o)^4} \right]. \tag{23.5}$$

For the range of $qS/S_0 < 1$, the following approximation is valid:

$$\left(\frac{qS}{S_o}\right)^2 - \frac{1}{(qS/S_o)^4} \doteqdot 6.5 \log \frac{qS}{S_o}. \tag{23.6}$$

---

[1] It might be more reasonable to assume αqS instead of $vq^2 S^2 / 2EkT$, where α = the effective volume of a unit process. However, using this assumption, the type of the expression $I \propto (S/S_0)/1/mkT$ is obtained approximately (Yokobori 1951, 1952) except that S_0 and m are different constants from those in Equations 23.4 and 23.8, respectively.

Then the right-hand side of Equation 23.5 becomes (Yokobori 1952, and Yokobori and Kitagawa 1967)

$$\frac{vq^2S^2}{2E} - \frac{A}{q^4S^4} = \frac{1}{m}\log\left(\frac{qS}{S_0}\right), \tag{23.7}$$

where

$$\frac{1}{m} = 6.5(B^2A)^{1/3} = 2.26\pi\left(\frac{v}{1-v^2}\right)^{2/3}\rho. \tag{23.8}$$

Substituting Equations 23.7 and 23.8 into Equation 23.1, we get

$$I = \frac{VZkT}{h}\left(\frac{qS}{S_0}\right)^{1/mkT}\exp\left(-\frac{\Delta f^*}{kT}\right). \tag{23.9}$$

The above expression holds for uniformly increasing stress. In the case of fatigue testing, the applied stress oscillates as

$$S = \sigma_m + \sigma_a \sin 2\pi\omega t, \tag{23.10}$$

where

σ_a = stress amplitude
σ_m = mean (static) stress
ω = stress frequency per unit time.

Thus μ, the crack nucleation rate per cycle, is obtained by integrating Equation 23.9 over the range of time from 0 to $1/\omega$ as follows:

$$\mu = \frac{VZkT}{h}\exp\left(-\frac{\Delta f^*}{kT}\right)\left(\frac{1}{S_0}\right)^{1/mkT}\int_0^{1/\omega}(qS)^{1/mkT}\,dt. \tag{23.11}$$

In the region adjacent to the tip of the main crack, it may be assumed that strain hardening has occurred and the local stress near the tip of the crack may be assumed to be higher than the initial yield point. We assume that the microcrack will initiate within this region, the width of which is ε, as shown in Figure 23.1. The stress σ at the point a distance x from the tip of the main crack is[2]

$$\sigma = \sqrt{\frac{c}{2x}}(\sigma_m + \sigma_a \sin 2\pi\omega t) \tag{23.12}$$

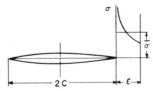

FIGURE 23.1. Stress distribution near the tip of the main fatigue crack.

[2] The effect of elastic-plastic stress distribution near the crack tip is calculated in another paper (Yokobori and Ichikawa 1968).

and then, the average stress $\bar{\sigma}$ over the region of width ε is given as follows:

$$\bar{\sigma} = \frac{1}{\varepsilon} \int_0^\varepsilon \sigma \, dx = 2\sqrt{\frac{c}{2\varepsilon}}(\sigma_m + \sigma_a \sin 2\pi\omega t). \tag{23.13}$$

23.2.1. The Case of Iron and Steel

For the case of iron and steel, it has been known (Frost and Dugdale 1957; Yokobori et al. 1967) that mean (static) stress σ_m has no influence on the fatigue crack propagation. Hence Equation 23.13 reduces to

$$\bar{\sigma} = 2\sqrt{\frac{c}{2\varepsilon}}\sigma_a \sin 2\pi\omega t. \tag{23.14}$$

Substituting $\bar{\sigma}$ of Equation 23.14 into qS in Equation 23.11, and integrating, we get μ. For convenience, the averaged value $(2/\pi)\sigma_a$ of $\bar{\sigma}$ over the range 0 to $1/2\omega$ of time was used in the integration. Furthermore, the half cycle corresponding to compression is assumed to make no contribution to crack nucleation. Then from Equations 23.11 and 23.14, we get

$$\mu = \frac{M}{\omega}(\sqrt{c}\,\sigma_a)^{1/mkT}, \tag{23.15}$$

where

$$M = \frac{VZkT}{2h}\left[\exp\left(-\frac{\Delta f^*}{kT}\right)\right]\left(\frac{4}{\pi}\sqrt{\frac{1}{2\varepsilon}}\frac{1}{S_o}\right)^{1/mkT}. \tag{23.16}$$

Let us assume when a microcrack nucleates within the region of width ε at the tip of the main fatigue crack, the latter extends by joining the microcrack. Then $\bar{N}$, the average number of cycles for extension of the main fatigue crack, is given by

$$\bar{N} = \frac{1}{\mu} = \frac{\omega}{M}(\sqrt{c}\,\sigma_a)^{-1/mkT}. \tag{23.17}$$

23.2.2. The Case of Aluminum Alloys and High Polymers

In the case of aluminum alloys (Frost and Dugdale 1957) and high polymers (Yokobori and Sasahira 1968) it has been observed that the mean (static) stress has an effect upon the fatigue crack propagation rate. In this case we must use Equation 23.13 instead of Equation 23.14. The usual data concern the case of $\sigma_m/\sigma_a \geq 1$ for the axial fatigue test. Thus substituting $\bar{\sigma}$ of Euqation 12.13 into qS in Equation 23.11, and integrating with an approximation of a rectangular stress wave for the sinusoidal stress wave for convenience, we get

$$\mu = \frac{M_o}{\omega}\left\{\left[\sqrt{c}\left(\sigma_m + \frac{2}{\pi}\sigma_a\right)\right]^{1/mkT} + \left[\sqrt{c}\left(\sigma_m - \frac{2}{\pi}\sigma_a\right)\right]^{1/mkT}\right\}. \tag{23.18}$$

The second term is negligible in comparison with the first term in the parentheses of the right-hand side of Equation 23.18. Then Equation 23.18 reduces to

$$\mu = \frac{M_o}{\omega}\left[\sqrt{c}\left(\sigma_m + \frac{2}{\pi}\sigma_a\right)\right]^{1/mkT},$$
(23.19a)

or approximately

$$\mu = \frac{M_o}{\omega}[\sqrt{c}(\sigma_m + \sigma_a)]^{1/mkT},$$
(23.19b)

where

$$M_o = M\left(\frac{\pi}{2}\right)^{1/mkT},$$

and we get

$$\overline{N} = \frac{1}{\mu} = \frac{\omega}{M_o}\left[\sqrt{c}\left(\sigma_m + \frac{2}{\pi}\sigma_a\right)\right]^{-1/mkT}.$$
(23.20)

23.3. Fatigue Crack-Propagation Law Based on the Concept of Nucleation Theory

It is assumed that fatigue life consists of incubation periods, as shown schematically in Figure 23.2. Then the fatigue crack-propagation

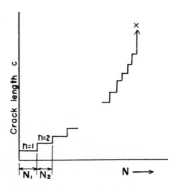

FIGURE 23.2. Schematic illustration of fatigue-crack propagation, half length of main crack versus repeated cycles N. $n =$ the ordering number of incubation period; for example, $n = i$ means the ith incubation period (ith stage).

rate per cycle, dc/dN, is represented as

$$\frac{dc}{dN} = \frac{dc}{dn}\Big/\frac{dN}{dn},$$
(23.21)

where n = the repeated cycles for each incubation period (stage). It is assumed that each incubation period corresponds to a nucleation process, such as considered in Section 23.2. Assuming that the increase of crack length after each incubation period is independent of the number of cycles N, and denoting this increase by ε, then

$$\frac{dc}{dn} = \varepsilon. \tag{23.22}$$

At the ith stage (ith incubation periods),

$$\frac{dN}{dn} = N_i. \tag{23.23}$$

Substituting Equations 23.22 and 23.23 into Equation 23.21,

$$\frac{dc}{dN} = \frac{\varepsilon}{N_i}. \tag{23.24}$$

N in Equations 23.17 and 23.20 corresponds to N_i at the ith stage, and μ in Equations 23.15 and 23.19 corresponds to $\mu_i = 1/N_i$. Thus Equation 23.24 becomes

$$\frac{dc}{dN} = \varepsilon\mu_i. \tag{23.25}$$

23.3.1 The case of Iron and steel

Substituting Equation 23.15 into Equation 23.25, we get

$$\frac{dc}{dN} = \frac{\varepsilon M}{\omega} \left(\sqrt{c}\,\sigma_a\right)^{1/mkT} \tag{23.26}$$

or

$$\frac{dc}{dN} = \frac{\varepsilon M}{\omega} \left(\frac{\Delta K}{2}\right)^{1/mkT}, \tag{23.27}$$

where ΔK = stress-intensity factor corresponding to the stress range. Equation 23.27 gives the fatigue crack-propagation law. If we assume $v = 0.28$, $\rho = 10^2$ erg/cm^2, and $v = 8 \times 10^{-24}$ cm^3, then the value of $1/mkT$ at room temperature is calculated as about 7. The experimental data of Yokobori and Nambu (1968) show that $\log(dc/dN)$ is linear against $\log \Delta K$, as shown in Figure 23.3, which is in accordance with Paris's (1964) law

$$\frac{dc}{dN} = C_1(\Delta K)^\delta, \tag{23.28}$$

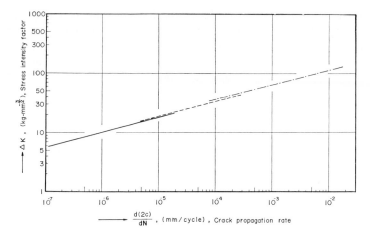

FIGURE 23.3. The experimental data for the fatigue crack-propagation rate, $\log (d(2c)/dN)$ plotted against the stress-intensity factor ΔK, corresponding to the range of stress, that is, $\sqrt{c} (2\sigma_a)$, for low-carbon steels. The solid line (Yokobori, and Nambu 1968) is for fatigue microcrack (the crack length ranging from 13 microns to 300 microns), the dashed line (Yokobori, et al. 1967) and the dash-dot line (Paris 1964) are for fatigue macrocracks.

where he assumes $\delta = 4$. If we assume that the μ_i increase (Yokobori, Kawagishi, and Yoshimura 1969) during each incubation period instead of assuming constant values, then the averaged value $\bar{\mu}_i$ over each incubation period should be used. Then, as shown by Yokobori and Yoshimura (1968), $1/2mkT$ should be used instead of $1/mkT$, and the value of $1/2mkT$ at room temperature becomes 3.5, being in good agreement with experimental data.

Equations 23.26 or 23.27 show no influence of ferrite grain size on the fatigue crack-propagation rate, which is in good agreement with the experimental data of Yokobori et al. (1967).

23.3.2 The case of Polymers
Substituting Equation 23.19 into Equation 23.25, we get

$$\frac{dc}{dN} = \frac{\varepsilon M}{\omega} \left[\sqrt{c} \left(\sigma_m + \frac{2}{\pi} \sigma_a \right) \right]^{1/mkT} \tag{23.29}$$

or

$$\frac{dc}{dN} \bigg/ \frac{\varepsilon M}{\omega} (K_m)^{1/mkT}, \tag{23.30}$$

where $K_m = \sqrt{c}(\sigma_m + \sigma_a) =$ the stress-intensity factor corresponding to the maximum stress, but not to the range of stress amplitude. In the case of polymers it can be seen by comparing Figure 23.4b (Yokobori and

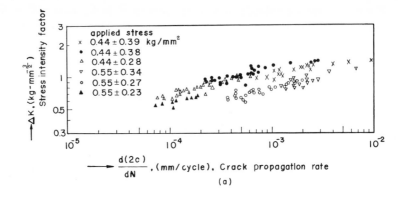

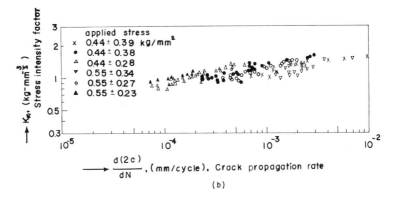

FIGURE 23.4 a and b. The experimental data for the fatigue crack-propagation rate, log $(d(2c)/dN)$ plotted against stress intensity factor ΔK and K_m respectively, for a polyethylene specimen (Yokobori and Sasahira 1969). K_m corresponds to the maximum stress of cycle, that is, $\sqrt{c}(\sigma_m + \sigma_a)$.

Sasahira 1969) with Figure 23.4a that Equation 23.30 is in better agreement with the data than is either Equation 23.27 or Equation 23.28.

23.4. S–N Relationships

The whole process of fatigue fracture may be divided into the initiation and the propagation process of the main fatigue crack. The theory described in Sections 23.2 and 23.3 deals with the propagation process of the main crack. Denoting the mean number of cycles to the initiation of the main crack and the number of cycles from its initiation until its propagation by $\bar{N}_o$ and $\bar{N}_p$, respectively, then the total life $\bar{N}_t$ is written as:

$$\bar{N}_t = \bar{N}_o + \bar{N}_p. \tag{23.31}$$

For higher stress amplitude, it may be assumed $\bar{N}_p \gg \bar{N}_o$, that is $\bar{N}_t \approx \bar{N}_p$. On the other hand, for iron and steel Equation 23.26 is rewritten as

$$\frac{dc}{c^{1/2mkT}} = \frac{\varepsilon M}{\omega} \sigma_a^{1/mkT} \, dN_p. \tag{23.32}$$

It may be considered that when the size of the main fatigue crack reaches a critical value c_f, catastrophic static fracture, say, by the Griffith mechanism will occur. Thus for higher stress amplitude the mean fatigue life $\bar{N}_t$ will be obtained by integrating Equation 23.32 with respect to c from c_o to c_f as follows:

$$\int_{c_o}^{c_f} \frac{dc}{c^{1/2mkT}} = \frac{\varepsilon M}{\omega} \sigma_a^{1/mkT} \bar{N}_t, \tag{23.33}$$

where c_o = size of a critical crack which will eventually lead to final fracture. Integrating Equation 23.33, we get

$$\bar{N}_t = \frac{2\omega mkTR}{\varepsilon M} \sigma_a^{-1/mkT}, \tag{23.34}$$

where

$$R = \frac{1}{1 - 2mkT} (c_o^{-(1/2mkT)+1} - c_f^{-(1/2mkT)+1}). \tag{23.35}$$

Equation 23.34 represents the S–N curve. This equation is in agreement with experiments (Yokobori, et al. 1968).

The theoretical curve is as shown in Figure 23.5 (Yokobori and Sasahira

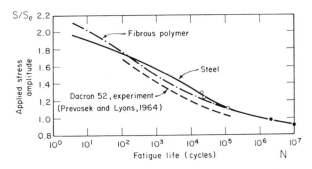

FIGURE 23.5. Comparison of theoretical S–N curves of solids with the data. Theoretical curves were calculated assuming the whole process as one simple process (Yokobori 1955), but the trend will be similar based on the present theory, except that the parameters were different. The solid line was calculated for mild steel (Yokobori and Sasahira 1967), and the dash-dotted line was calculated (Prevosek and Lyons 1964) for fibrous polymers. The Se-fatigue strength is calculated at 10^6 repeated cycles. Data for mild steel are from Yokobori (1953).

1967) if we assume the whole process as only one nucleation process using Equation 23.1 instead of the approximation of Equation 23.9. The result will, however, also be similar when we take the present treatment using Equation 23.1 instead of Equation 23.9 and assuming the parameters used in Yokobori and Sasahira (1967). It is to be noted that the convex part in the low-cycle range dos not appear when $\alpha q S$ is assumed instead of $v q^2 S^2 / 2 E k T$ in Equation 23.10. Thus the feature of the part of the S–N curve in the low-cycle range will be considered characterized by another mechanism than the present one.

In the range of smaller stress amplitude, $\bar{N}_0$ is not negligible when compared with $\bar{N}_p$ in Equation 23.31. On the other hand, the initiation process may also be considered as a nucleation process, and, therefore, $\bar{N}_0$ can be written as

$$\bar{N}_0 = \omega H, \tag{23.36}$$

where d may be a function of $\sqrt{d\sigma_a}$, (Yokobori and Kawasima 1969) that is $H = H(\sqrt{d\sigma_a})$, where $d =$ average grain diameter. This can also be inferred from the experimental trend of the influence of grain size on fatigue life (Yokobori and Kawasima 1969) at least in the smaller range of stress amplitude.

23.5. The Effects of Load Cycling and of Load Frequency on Fatigue Life in Terms of Repeated Cycles

It is known empirically that fatigue life in terms of cycles to failure does not depend on the frequency ω of cycling, Kennedy (1962); Head (1956). We will consider the problem as follows. To be in accordance with this empirical fact, it is necessary, at least phenomenologically, to take into account the requirement of load cycling, that is, that the next stage of propagation is impossible unless the load is cycled. Thus let us assume that

$\bar{N}_i =$ the averaged repeated cycles required for the ith propagation occurring after the conditions or the states capable for the ith propagation have been attained, and

$\bar{N}_i^* =$ the averaged repeated cycles required for attaining the conditions or the states capable for the $(i + 1)$ th propagation after the conditions for the ith propagation have been attained.

Then we get (Yokobori, Kawagishi, and Yoshimura 1969),

$$\bar{N}_*^i = [\bar{N}_i] + 1, \tag{23.37}$$

where the symbol $[\chi]$ stands for the Gaussian symbol, that is, the maximum integer not bigger than χ. If n = the total number of the stages of propagation, then the total mean fatigue life $\bar{N}_t$ is given by

$$\bar{N}_t \simeq \sum_{i=1}^{n} \bar{N}_i^* = \sum_{i=1}^{n} \{[\bar{N}_i] + 1\}$$

$$= \sum_{i=1}^{n} [\bar{N}_i] + n. \tag{23.38}$$

The first term $\sum_{i=1}^{n}[\bar{N}_i]$ in Equation 23.38 may be replaced by $\bar{N}_p$ expressed by Equation 23.34, and therefore, in the case $\bar{N}_t \simeq \bar{N}_p$ we get

$$\bar{N}_t = \frac{2\omega mkTR}{\varepsilon M} \sigma_a^{-/1mkT} + n. \tag{23.39}$$

It can be understood from Equation 23.39 that when the frequency is high, the fatigue life is proportional to the load frequency, and, on the other hand, when the frequency is very low, $\bar{N}_t \simeq n$; that is, the fatigue life will not depend on frequency, and in the intermediate range of frequency, fatigue life in terms of repeated cycles depends on both loading frequency and loading cycles.

23.6. Conclusions

The fatigue crack propagation for solids such as metals and polymers was studied as a nucleation process. The following conclusions were reached.

1. The theory leads to a fatigue crack-propagation law similar to that of Paris for the case of mild steel, and suggests that for the case of polymers it may be reasonable to use the maximum stress of the cycle in the stress-intensity factor instead of the stress range.

2. The theory predicts S–N relationships which are in good accordance with experimental data.

3. The theory offers a possibility of explaining the temperature dependence of the fatigue strength and the size effect.

4. By combining the effect of stress or strain cycling with the present nucleation model, the possibility was shown of getting the theoretical formula for fatigue life which includes both the effects of the cyclic load and the effect of the frequency.

REFERENCES

Frost, N. E., and D. S. Dugdale, 1957, *J. Mech. Phys. Sol.*, **5**, 182.

Head, A. K., 1956, *J. Phys. Soc. Japan*, **11**, 468.

Kennedy, A. J., 1962, *Creep and Fatigue in Metals*, London: Oliver and Boyd, p. 326.

Paris, P., 1964, in J. Burke et al. (eds.), *An Interdisciplinary Approach-Fatigue*, Syracuse University Press, p. 107.

Prevosek, D., and J. W., Lyons, 1964, *J. Appl. Phys.*, **35**, 3152.

Yokobori, T., 1951, *Proc. First Japan Natl. Cong. Appl. Mech.*, JSME, p. 97.

Yokobori, T., 1952, *J. Phys. Soc. Japan*, **7**, 44.

Yokobori, T., 1953, *J. Phys. Soc. Japan*, **8**, 265.

Yokobori, T., 1955, *J. Phys. Soc. Japan*, **10**, 368.

Yokobori, T., 1963, *Tech. Rep. Tohoku Univ.* **27**, 143.

Yokobori, T., and M. Kitagawa, 1967, *Semi-Interntl. Symp. Exp. Mech.*, JSME, 2, p. 173.

Yokobori, T., and J. Sasahira, 1967, *J. Japanese Soc. Str. Fract. Materials*, **2**, 40.

Yokobori, T., M. Tanaka, H. Hayakawa, T. Yoshimura, and S. Sasahira, 1967 *Rep. Res. Inst. Str. Fract. Materials*, (Tohoku Univ.), **3**, 39.

Yokobori, T., and M. Ichikawa, 1968, *Rep. Res. Inst. Str. Fract. Materials*, (*Tohoku Univ.*), **4**, 45.

Yokobori, T., and T. Kawasima, 1969, Revue Roumaine des Sciences Techniques— Serie de Metallurgie: *J. Japan. Soc. Str. Fract. Materials*, **4**, No. 1.

Yokobori, T., and M. Nambu, 1968, in *Proc. Third Conf. Dimensioning and Str. Calculations*, Hungarian Acad. Sci., p. 321.

Yokobori, T., M. Kawagishi, and T. Yoshimura, 1969, in P. L. Pratt (ed.), *Fracture 1969*, London, Chapman and Hall, p. 803.

Yokobori, T., and S. Sasahira, 1969, to be published.

Yokobori, T., M. Yoshida, and N. Ishimura, 1969, to be published.

ABSTRACT. A composite consisting of stiff elastic fibres reinforcing a soft ductile matrix is considered. If the strain-hardening characteristics of the matrix under fatigue conditions is known, the number of cycles required to produce a given amount of plastic strain in the matrix can be deduced. Alternatively the number of cycles required to achieve a given flow stress in the matrix can be found. Expressions are derived for these quantities, and the applicability of them to the prediction of the fatigue lives of composites is discussed.

24. Fatigue of the Matrix in a Fibre-Reinforced Composite

A. KELLY and M. J. BOMFORD

24.1. Introduction

Most theories of the strength and modulus of composites predict the strength or elastic moduli in terms of the behavior of the two components (Kelly and Davies 1965, Broutman and Krock 1967, American Society for Metals 1964). In this paper we attempt to apply this approach to the fatigue of an aligned fibrous composite. It is important to do this because the answer to the question, whether or not reinforcement of a matrix with strong fibres has in fact altered the fatigue resistance of the matrix, is not immediately apparent from the results of a fatigue test on a composite. An attempt at a simple quantitative measure of the change of fatigue resistance has been proposed by Baker (1966). This is valuable but will apply only in a region where the Coffin law is obeyed by the matrix (Coffin and Tavernelli 1959).

A theoretical derivation of the type given here has been attempted by Courtney and Wulff (1966). Experimental results on the fatigue of reinforced metals have been reported by Baker (1966), Baker and Cratchley (1964), Ham and Place (1966), Forsyth, George, and Ryder (1964), Morris and Steigerwald (1967), and results for glass reinforced plastics are summarized by Davis, McCarthy, and Schurb (1964).

339

24.2. Model of the Composite

The composite is assumed to consist of straight, strong, continuous elastic fibres, aligned parallel to the applied stress, in a soft ductile matrix. The matrix is assumed to be isotropic and to have well defined cyclic hardening properties. The strains in the two components are assumed to be always equal.

The composite is assumed to be undergoing a constant maximum stress fatigue test, and the stress is great enough for the matrix to yield plastically in each cycle. The fibre is assumed to be elastic, with Young's modulus E_f, and the matrix with Young's modulus E_m yields at a stress σ_{ym} *in a particular cycle*. For the sake of mathematical simplicity, it is assumed that the matrix will not work-harden in one particular cycle, but rather the hardening of the matrix is accounted for by an increase in the yield point of the matrix in the next cycle. The Bauschinger effect is ignored. The stress on the composite is calculated directly from the rule of mixtures.

A schematic representation of the curve for tension-zero cycling appears in Figure 24.1. At the beginning of a cycle the applied stress is zero, but

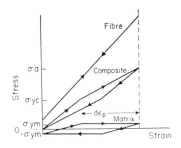

FIGURE 24.1.

the matrix is in compression under a stress $-\sigma_{ym}$, and this is balanced by a tensile stress in the fibres equal to $\sigma_{ym}(1 - V_f)/V_f$, where V_f is the volume fraction of fibres. As the load is increased from zero the composite deforms elastically (with a Young's modulus of $E_c = E_f V_f + E_m(1 - V_f)$) until it yields in tension at a stress σ_{yc}. At this point the tensile stress on the matrix is $+\sigma_{ym}$, and thus from Figure 24.1 and the assumption of equal strain it is obvious that

$$\frac{\sigma_{yc}}{E_c} = \frac{2\sigma_{ym}}{E_m}. \tag{24.1}$$

After yield the composite continues to deform until the maximum applied stress, σ_a, is reached. During this stage the modulus of the composite is given by $E'_c \simeq E_f V_f$, remembering that no work-hardening of the matrix occurs within a cycle.

A similar progression of events occurs in the second half of the cycle when the applied stress decreases from σ_a to 0.

In a half-cycle therefore the plastic strain, $d\varepsilon_p$, of the matrix is deduced from Figure 24.1 as

$$d\varepsilon_p = \frac{\sigma_a - \sigma_{yc}}{E_f V_f} = \frac{\sigma_a - 2\sigma_{ym}(E_c/E_m)}{E_f V_f}. \tag{24.2}$$

Similarly, if a composite is fatiguing in tension-compression between applied stresses $\pm\sigma_a$, it can be shown by similar arguments that the plastic strain per half cycle is

$$d\varepsilon_p = \frac{2\sigma_a - \sigma_{yc}}{E_f V_f} = \frac{2\{\sigma_a - \sigma_{ym}(E_c/E_m)\}}{E_f V_f}. \tag{24.3}$$

If the strain hardening per cycle is small it follows that the plastic strain in di half cycles is given from Equations 24.2 and 24.3, respectively, by

$$d\varepsilon_p = \frac{\sigma_a - 2\sigma_{ym}(E_c/E_m)}{E_f V_f} \cdot di \tag{24.4}$$

and

$$d\varepsilon_p = \frac{2\sigma_a - 2\sigma_{ym}(E_c/E_m)}{E_f V_f} \cdot di. \tag{24.5}$$

24.3. Number of Cycles to Attain a Given Flow Stress of a Given Plastic Strain

It is assumed (1) that the matrix behaves in the composite as it does if the fibres are not present; and (2) that the matrix has a well-defined flow stress-cumulative plastic strain curve under fatigue conditions of the form $\sigma_{ym} = C\varepsilon_p^n + \sigma_m$. We first calculate the number of cycles to attain a given flow stress of the matrix. This is done in Sections 24.3.1.1 and 24.3.1.2 below. We next find the number of cycles required to produce a given total plastic strain in the matrix, as discussed in Sections 24.3.2.1, 24.3.2.2, and 24.3.2.3.

From the basic Equations 24.4 and 24.5 above, and from assumption (2), the number of cycles in each case can be calculated by integrating the relevant differential equation between appropriate limits. These limits are: the lower limit, $\sigma_{ym} = \sigma_m$ (or $\varepsilon_p = 0$) at $i = 1$ (the beginning of the test), and the upper limit $\sigma_{ym} = \sigma_{mf}$ (or $\varepsilon_p = \varepsilon_{pf}$) at $i = 2N(X)$ at the end of the test, where i is the number of half cycles, σ_m is the yield stress of the matrix in the first cycle, and σ_{mf} is the yield stress of the matrix after $N(X)$ cycles. The symbol (X) will be used for identifying the various N's during the different calculations and for reference, later in the paper, to the various assumed rates of hardening. Two types of hardening, linear and parabolic, are considered here as these are often observed approximately, enabling analytic expressions to be obtained.

24.3.1. Stress Criterion for Failure

24.3.1.1. LINEAR FATIGUE HARDENING

We take

$$\sigma_{ym} = K\varepsilon_p + \sigma_m, \tag{24.6}$$

where K is the strength coefficient for linear hardening.

(i) For tension-zero fatigue we obtain from Equations 24.6 and 24.4

$$d\varepsilon_p = \frac{d\sigma_{ym}}{K} = \frac{\sigma_a - 2\sigma_{ym}(E_c/E_m)}{E_f V_f} \cdot di$$

and integrating this gives

$$2N(1) - 1 = \frac{E_f V_f E_m}{2KE_c} \ln\left\{\frac{\sigma_a - 2\sigma_m(E_c/E_m)}{\sigma_a - 2\sigma_{mf}(E_c/E_m)}\right\}$$

(ii) Similarly, from Equation 24.5 for push-pull conditions we obtain

$$2N(2) - 1 = \frac{E_f V_f E_m}{2KE_c} \ln\left\{\frac{\sigma_a - \sigma_m(E_c/E_m)}{\sigma_a - \sigma_{mf}(E_c/E_m)}\right\}$$

24.3.1.2. PARABOLIC FATIGUE HARDENING

We take

$$\sigma_{ym} = K'\varepsilon_p^{1/2} + \sigma_m, \tag{24.7}$$

where K' is the strength coefficient for parabolic hardening. Then from Equation 24.7,

$$\frac{d\sigma_{ym}}{d\varepsilon_p} = \frac{K'^2}{2\sigma_{ym}}$$

(i) Using Equation 24.4, we obtain

$$\frac{2\sigma_{ym}\, d\sigma_{ym}}{K'^2} = \frac{\sigma_a - 2\sigma_{ym}(E_c/E_m)}{E_f V_f}\, di,$$

which, when integrated, gives

$$2N(3) - 1 = \frac{2E_f V_f}{K'^2}\left\{\sigma_a\left(\frac{E_m}{2E_c}\right)^2 \ln\left\{\frac{\sigma_a - 2\sigma_m(E_c/E_m)}{\sigma_a - 2\sigma_{mf}(E_c/E_m)}\right\} - \frac{E_m}{2E_c}(\sigma_{mf} - \sigma_m)\right\}$$

(ii) Similarly, from Equation 24.5 we obtain for push-pull conditions

$$2N(4) - 1 = \frac{2E_f V_f}{K'^2}\left\{2\sigma_a\left(\frac{E_m}{2E_c}\right)^2 \ln\left\{\frac{\sigma_a - \sigma_m(E_c/E_m)}{\sigma_a - \sigma_{mf}(E_c/E_m)}\right\} - \frac{E_m}{2E_c}(\sigma_{mf} - \sigma_m)\right\}$$

24.3.2. *Strain Criterion for Failure*

24.3.2.1. NO FATIGUE HARDENING IN THE MATRIX

We take $\sigma_{ym} = \text{constant} = \sigma_m$.

(i) Equation 24.4 can be integrated directly to give

$$2N(5) - 1 = \frac{E_f V_f \varepsilon_{pf}}{\sigma_a - 2\sigma_m(E_c/E_m)},$$

where ε_{pf} is the cumultative plastic strain in the matrix after $N(X)$ cycles.

(ii) Similarly, from Equation 24.5 for push-pull fatigue

$$2N(6) - 1 = \frac{E_f V_f \varepsilon_{pf}}{2\{\sigma_a - \sigma_m(E_c/E_m)\}}.$$

24.3.2.2. LINEAR FATIGUE HARDENING IN THE MATRIX

(i) Tension-zero cycling fatigue. From Equations 24.4 and 24.6 we obtain

$$di = \frac{E_f V_f}{\sigma_a - 2(K\varepsilon_p + \sigma_m)E_c/E_m} \cdot d\varepsilon_p,$$

which, when integrated, gives

$$2N(7) - 1 = \frac{E_f V_f E_m}{2KE_c} \ln\left\{\frac{\sigma_a - 2\sigma_m(E_c/E_m)}{\sigma_a - 2(K\varepsilon_{pf} + \sigma_m)E_c/E_m}\right\}.$$

(ii) Push-pull fatigue. Similarly, from Equation 24.5,

$$2N(8) - 1 = \frac{E_f V_f E_m}{2KE_c} \ln\left\{\frac{\sigma_a - \sigma_m(E_c/E_m)}{\sigma_a - (K\varepsilon_{pf} + \sigma_m)E_c/E_m}\right\}.$$

24.3.2.3. PARABOLIC FATIGUE GARDENING IN THE MATIX

(i) Tension-zero fatigue. From Equations 24.4 and 24.7 we obtain

$$di = \frac{E_f V_f}{\{\sigma_a - 2(K'\varepsilon_p^{1/2} + \sigma_m)E_c/E_m\}} d\varepsilon_p,$$

which, when integrated, gives

$$2N(9) - 1 = \frac{E_f V_f E_m^2}{2K'^2 E_c^2}\left\{\left(\sigma_a - 2\frac{E_c}{E_m}\sigma_m\right)\right.$$
$$\left. \times \ln\left\{\frac{\sigma_a - 2\sigma_m E_c/E_m}{\sigma_a - 2(K'\varepsilon_{pf}^{1/2} + \sigma_m)E_c/E_m}\right\} - 2K'\varepsilon_{pf}^{1/2}\frac{E_c}{E_m}\right\}.$$

(ii) Push-pull fatigue. Similarly, from Equations 24.5 and 24.7 we obtain

$$2N(10) - 1 = \frac{E_f V_f E_m^2}{2K'^2 E_c^2}\left[2\left(\sigma_a - \sigma_m\frac{E_c}{E_m}\right)\right.$$
$$\left. \times \ln\left\{\frac{\sigma_a - \sigma_m E_c/E_m}{\sigma_a - (\sigma_m + K'\varepsilon_{pf}^{1/2})E_c/E_m}\right\} - 2K'\varepsilon_{pf}^{1/2}\frac{E_c}{E_m}\right].$$

24.4. Application of the Model to Real Composites

In Section 24.3 expressions were derived for the number of cycles required to produce a given flow stress or a given total plastic strain in the matrix. If failure occurs in the matrix when a given well-defined flow stress is reached, or after a given well-defined total plastic strain, then these equations can be used to predict "lives" of the composite under fatigue conditions, provided that the matrix in a composite behaves exactly as it does in an independent fatigue test on the matrix, which subjects the matrix to the same regimen of axial strain as the fatigue test on the composite.

This is a rather poor assumption in fact because during fatigue of metals the hardness or yield stress is rather constant over a large part of the life before failure. Similarly, only during fatigue at high stresses (short lives) is there a well-defined relation between total strain and failure. With these reservations in mind we shall assign a failure stress σ_{mf}, and a total plastic strain to failure ε_{pf}, to the matrix, and calculate "lives" from the equations in Section 24.3.

Of course the separate expressions derived in Section 24.3 for the number of cycles to reach either a certain cumulative plastic strain or a certain flow stress of the matrix can be made identical if the matrix really obeys either Equation 24.6 or Equation 24.7 during fatigue (see the equations for $N(9)$ and for $N(3)$). We have derived separate expressions because in reality the conditions under which the matrix fails when a certain flow stress is reached may not be the same as those under which failure occurs after a certain cumulative plastic strain. It follows that the work-hardening curves will not necessarily be the same for the two conditions of failure.

To evaluate the expressions for $N(1) \ldots N(10)$ and to plot curves of N versus σ_a, we have chosen to take the constants appropriate to continuous tungsten wires reinforcing a copper matrix at room temperature because this has been extensively studied for its tensile strengthening properties (see, for example, Kelly and Davies 1965). Also, it will allow us to compare our theoretical results with those of Courtney and Wulff (1966). The values of the various parameters used are listed in Table 24.1.

The parameters in Table 24.1 are assumed constant. Experimentally it is found that all the values of K, K', σ_{mf}, and ε_{pf} vary with the applied stress σ_a. The parameters K and K' are derived from the curves of Kemsley and Paterson (1960). These are for cycling at constant plastic strain, but we shall use them for the present case and shall ignore the orientation dependence of the curves for different positions of the tensile axis in the unit triangle.

Kemsley and Paterson's curves show that the hardening characteristics

TABLE 24.1.

Parameter	Value	Reference
E_f	60.10^6 psi	—
E_m	17.10^6 psi	—
K	10^3 psi	Kemsley and Paterson (1960)
K'	10^3 psi	Kemsley and Paterson (1960)
σ_m	10^3 psi	Bomford (1968)
σ_{mf}	45.10^3 psi	Courtney and Wulff (1966)
ε_{pf}	10	Coffin and Tavernelli (1959)

are neither purely linear nor purely parabolic. Usually the curve starts in a parabolic manner and quite soon changes over to linear hardening; a more complicated treatment should take these points into account.

The value of the stress for failure σ_{mf} in Table 24.1 is probably rather high, and as such overestimates the "lives" for our calculations, but we use it because it is the value used by Courtney and Wulff and it will be useful for comparison later. This value for σ_{mf} is approximately the value observed at the tips of fatigue cracks in copper calculated from microhardness measurements by Bomford (1968). The value for the cumulative plastic strain for failure, ε_{pf}, is an estimate from Coffin and Tavernelli (1959). Of course, the cumulative plastic strain to failure in a fatigue test will increase as the life increases (Coffin's law) for high-strain fatigue, and a similar law holds for long-life fatigue. However, for the sake of mathematical simplicity, ε_{pf} is held constant.

The "Titan" computer was used to calculate the σ_a versus N curves, and graphs of these curves are shown in Figures 24.2 – 24.11.

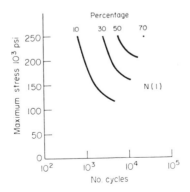

FIGURE 24.2.

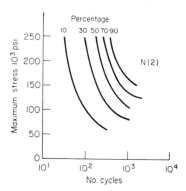

FIGURE 24.3.

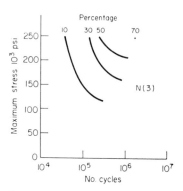

FIGURE 24.4.

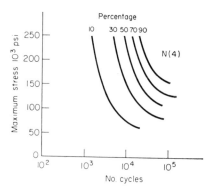

FIGURE 24.5.

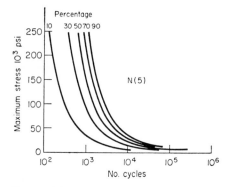

FIGURE 24.6.

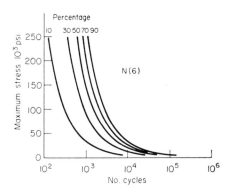

FIGURE 24.7.

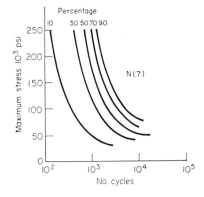

FIGURE 24.8.

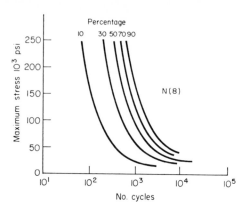

FIGURE 24.9.

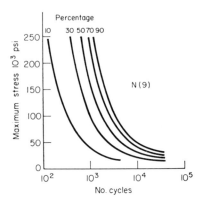

FIGURE 24.10.

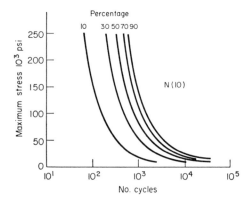

FIGURE 24.11.

24.5. Discussion

The curves shown in Figures 24.2–24.11 are somewhat similar in shape to conventional S–N curves for metals. The curves are drawn for $\sigma_a < 250,000$ psi (83% of the breaking stress of the fibres). The assumed volume fraction of fibres are given as percentages against each curve. The number of cycles required to attain a given flow stress or a given cumulative plastic strain becomes infinite in all cases for applied stresses 5,000 psi less than the last point plotted on the curves. These "fatigue limits" are easily deduced from the expressions in Section 24.3. For example the expression for $N(9)$ becomes infinite if

$$\sigma_a = \frac{2E_c}{E_m}(\sigma_m + K'\varepsilon_{pf}^{1/2}),$$

thus, using the values in Table 24.1, a composite with $V_f = 0.50$ has a fatigue limit of 18.8×10^3 psi.

The "lives" predicted are shorter than those found experimentally (Bomford 1968). One reason for this is that the rate of hardening decreases during the life. Experimentally, as we observed above, the hardness reaches a plateau and thereafter for quite a large percentage of the life keeps at a constant value until just before failure occurs.

In the description of the model composite the fibres were assumed to be infinitely strong. Tungsten wires have a tensile strength of about 300,000 psi. If the stress on the tungsten necessary to produce the relevant fatigue stress were greater than 300,000 psi, the composite will fail by another mode—namely, the tensile fracture of some of the weaker fibres which will lead to internal stress concentrations from which fatigue cracks can initiate. This will lead to a reduction in the fatigue life.

The lives in tension-zero fatigue are always greater than the relevant ones in push-pull. In the latter the plastic strain per cycle will always be greater for a certain applied stress.

Inspection of the σ_a–N curves shows that the number of cycles to failure always increases as the volume fraction of fibres increases. This is contrary to the findings of Courtney and Wulff (1966) at the long-life end of the curves. Their calculations predict a decrease of the fatigue limit with an increase in volume fraction of fibres. The model cannot lead to this and it is the converse of what is observed experimentally by Morris and Steigerwald (1967).

Our equations usually predict much longer lives than those of Courtney and Wulff. This is because the hardening rates that they use ($K = 10^5$ psi and $K' = 4.6.10^4$ psi) are much higher than ours, and are relevant to unidirectional tensile deformation, not to a cyclic deformation test.

The Bauschinger effect has been ignored in this analysis and the difference in Poisson's ratio between the fibre and matrix neglected. This latter difference will lead to a transverse stress set up in the composite and of course fatigue of the matrix under complex stresses will have different characteristics from uniaxial fatigue.

It has been assumed that the matrix in the composite work hardens exactly as does the bulk matrix. This is found experimentally in simple tension provided the interfibre spacing is greater than about 10μ; with spacings less than this the hardening properties of the matrix alter considerably (Lilholt 1968).

Finally, in a volume of this nature, it should be remarked that the composite studied here bears a superficial resemblance to a notional material introduced by Orowan (1939) when considering the fatigue of metals. This he considered to consist of plastic regions in an elastic matrix. The plastic regions became *unloaded* as a result of the plastic flow. In our composite the elastic region (fibre) and plastic region (matrix) are strained in parallel so that no unloading of the matrix occurs.

M. J. Bomford is grateful to the Ministry of Technology for a personal maintenance grant. Both authors are indebted to Mr. N. H. Macmillan for writing the computer programmes.

REFERENCES

American Society for Metals, 1964, *Fibre Composite Materials*, Metals Park, Ohio: ASM.

Baker, A. A., 1966, *Appl. Mat. Res.*, **5**, 210.

Baker, A. A., and D. Cratchley, 1964, *Appl. Mat. Res.*, **3**, 215.

Bomford, M. J., 1968, PhD Thesis, University of Cambridge.

Broutman, L. J., and R. M. Krock, 1967, *Modern Composite Materials*, Reading, Mass: Addison-Wesley.

Coffin, L. F., and J. F. Tavernelli, 1959, *Trans. AIME*, **215**, 794.

Courtney, T. H., and J. Wulff, 1966, *J. Mats. Sci.*, **1**, 383.

Davis, J. W., J. A. McCarthy, and J. N. Schurb, 1964, *Mat. Des. Eng.*, **60** No. 7, 87.

Forsyth, P. J. E., R. W. George, and D. A. Ryder, 1964, *Appl. Mat. Res.*, **3**, 223.

Ham, R. K., and T. Place, 1966, *J. Mech. Phys. Sol.*, **14**, 276.

Kelly, A., and G. J. Davies, 1965, *Met. Rev.*, **10**, 1.

Kemsley, J. S., and M. S. Paterson, 1960, *Acta Met.*, **8**, 453.

Lilholt, H., 1968, private communication.

Morris, A. W. H., and E. A. Steigerwald, 1967, *Trans. AIME*, **239**, 730.

Orowan, E., 1939, *Proc. Roy. Soc.*, **A171**, 79.

4. Geology

ABSTRACT. In an Andradean fluid, having a stress strain-rate law $(\sigma/\sigma_0)^n = (\varepsilon/\varepsilon_0)$ and a density ρ, the characteristic distance of diffusion $\delta^{(v)}$ of the effect of a fixed boundary velocity $|u_0|$ is

$$\delta^{(v)} = \left[\left(\frac{n+1}{\rho n^2} \right) \left(\frac{\sigma_0}{\varepsilon_0^{1/n}} \right) |u_0|^{(1-n)/n} \right]^{n/(n+1)} t^{n/(n+1)},$$

and the characteristic distance of diffusion $\delta^{(\tau)}$ of the effect of a fixed boundary shear stress $|\tau_0|$ is

$$\delta^{(\tau)} = \left[\frac{2\sigma_0^n}{n\rho\varepsilon_0} |\tau_0|^{(1-n)} \right]^{1/2} \sqrt{t}.$$

The temporal character of the diffusion of a fixed boundary stress behaves as $\sqrt{t}$ just as in the case of a Newtonian fluid. Possible applications of these observations to problems of geophysics are suggested.

25. Diffusion of Boundary Disturbances into Andradean Fluids

C. A. BERG

25.1. Introduction

One important mechanical parameter governing fluid systems is the charactersitic distance of diffusion of a boundary disturbance. In a Newtonian fluid having kinematic viscosity v, the characteristic distance δ of diffusion of a disturbance in time t is

$$\delta = \sqrt{vt}, \tag{25.1}$$

as may be seen either by dimensional analysis of the Navier-Stokes equations or by examination of explicit solutions.

The characteristic distance of diffusion of disturbances can often reveal the essential physical behavior of fluid systems, without recourse to specialized mathematical techniques (e.g., Langharr 1942, Schlicting 1960). In non-Newtonian fluids, where the exact analysis may be prohibitively difficult, a study of the diffusion of boundary disturbances can give clues to essential physical behavior. One such case is the kinetics of deformation and thermal convection in the mantle of the earth which poses very complex problems of non-Newtonian fluid dynamics. The earth's mantle is a polycrystalline aggregate at high temperature and may be assumed to undergo viscous flow as an Andradean fluid. In order to gain

353

insight into the essential physics of the kinetics of viscous flow in the mantle, the diffusion of boundary disturbances in an incompressible Andradean fluid will be examined.

25.2. Boundary Disturbances in a Viscous Fluid

The first, and simplest, nontrivial study of the spread of boundary disturbances through a viscous fluid was the solution of Stokes in 1851 for the flow in a Newtonian fluid near a flat plate which is suddenly accelerated from rest at time $= 0$ and moves in its own plane with a constant velocity u_0. Figure 25.1 shows the accelerated plate and adjacent

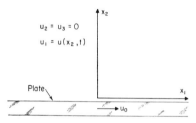

FIGURE 25.1. A flat plate lying adjacent to a semi-infinite body of Andradean fluid; the plate is suddenly accelerated in the direction of the x_1 axis.

fluid. The velocity u in the fluid is parallel to the velocity of the plate and depends upon the coordinate x_2 normal to the plate and time t according to

$$u/u_0 = \operatorname{erfc}(x_2/2\sqrt{vt}), \tag{25.2}$$

where erfc is the complementary error function. The role of $\sqrt{vt}$ as the characteristic distance of diffusion in a Newtonian fluid is clear from Equation 25.2. If the plate in Stokes' original problem had been subjected to a constant shear stress τ_0 beginning at $t = 0$ and persisting thereafter, the linearity of all governing equations would lead to a shear stress τ distribution in the fluid of the same type as Equation 25.2; i.e., $\tau/\tau_0 = \operatorname{erfc}(x_2/2\sqrt{vt})$. In a Newtonian (i.e. linear) fluid a disturbance consisting of a fixed boundary stress diffuses in just the same way as does the effect of fixed velocity. The response of Andradean fluids to boundary disturbances consisting of fixed velocities and fixed stress will now be examined using the situation first studied by Stokes (1851), the flow in the vicinity of a suddenly accelerated flat plate, as a specific model.

25.3. Andradean Fluids

When a polycrystalline material at high temperature (e.g., the mantle of the earth) undergoes viscous deformation in pure shear, the shear stress is related to the strain rate by

$$\sigma/\sigma_0 = (\varepsilon/\varepsilon_0)^{1/n}, \tag{25.3}$$

where the exponent $n > 1$ is a function of temperature, and σ_0 and ε_0 are characteristic parameters of the material. The representation 25.3 for steady nonNewtonian flow of crystalline materials is due to Andrade (1911). Common values of the exponent n reported for polycrystalline structural metals at typical service temperatures vary from 3 to 5; however, for a given material the exponent n is a strong function of temperature, approaching unity as the temperature approaches the melting point and approaching infinity as the temperature approaches absolute zero. The shape of the stress strain-rate curve 25.3 for pure shear deformation may be compared with that of a Newtonian fluid. In pure shear of a Newtonian fluid the shear stress is related to the shear strain rate by

$$\sigma/\sigma_0 = \varepsilon/\varepsilon_0, \tag{25.4}$$

where σ_0/ε_0 is the Newtonian viscosity μ of the fluid. Figure 25.2 shows the stress strain-rate curves for a Newtonian and an Andradean fluid

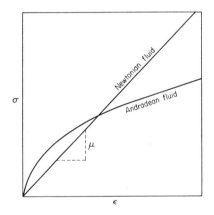

FIGURE 25.2. Comparison of Andradean and Newtonian fluid stress (σ) strain-rate (ε) behavior. The slope $(d\sigma/d\varepsilon)$ of the curve for the Andradean fluid is infinite at the origin.

having $n = 3$. The slope of the stress strain-rate curve of the Andradean fluid is infinite at the origin and gradually vanishes as $\varepsilon \to \infty$. The slope of the Newtonian curve is equal to the constant viscosity μ. Since the characteristic distance of diffusion of a boundary disturbance in a Newtonian fluid behaves as $\sqrt{vt}$, with $v = \mu/\rho$, one expects that in the case of Andradean fluids a disturbance which causes small strain rates will propagate rapidly and that a disturbance which causes large strain rates will propagate slowly. Since the slope of the Andradean stress strain-rate curve is singular at the origin, it is not possible to linearize the constitutive relation of the Andradean fluid at vanishingly small strain rates or to assign an "effective viscosity" to the fluid in the small strain-rate range.

25.4. Calculations

The mode of flow to be studied here is illustrated in Figure 25.1. The flat plate and the adjacent fluid are initially at rest. At the time $t = 0$ the plate is accelerated in the x_1 direction. The conditions of the acceleration will be either fixed velocity at, or fixed shear stress acting on, the plate for $t > 0$.

The velocity in the incompressible fluid is parallel to the velocity of the plate; thus the velocity components fluid are

$$u_1 = u(x_2, t); \qquad u_2 = u_3 = 0. \tag{25.5}$$

The components of strain rate ε_{ij} associated with a given velocity field (u_i) are $\varepsilon_{ij} = u_{i,j} + u_{j,i}$; according to Equations 25.3 the only nonvanishing strain rates are

$$\varepsilon_{12} = \varepsilon_{21} = u_{1,2}. \tag{25.6}$$

To obtain the stress components in the fluid, note that in plane flow (Equation 25.5) of an incompressible material the local deformation is always pure shear. In the present case, the directions of maximum shear strain rate are parallel to the x_1 and x_2 axes. One may use Equation 25.3 to describe the deviatoric stress components acting at any point in the incompressible fluid; for computational convenience the following form of this relation is used:

$$\sigma_{12} = (\sigma_0 / \varepsilon_0^{1/n}) \varepsilon_{12}^{1/n}. \tag{25.7}$$

In Equation 25.7 it will be understood that the branch of $(\varepsilon_{12}^{1/n})$ will be taken so as to make the sign of σ_{12} and ε_{12} agree. The remaining off-diagonal components of deviatoric stress vanish because of fluid isotropy. Since the x_1 and x_2 axes lie in the direction of maximum shear strain rate, the normal stress components σ_{11} and σ_{22} will be equal. It is assumed that whatever body force field is present can be supported by the fluid in static equilibrium — i.e. by a distribution of pure hydrostatic stress p_0. This hydrostatic stress, which contains an arbitrary uniform pressure p_0, may be superposed upon the stress field in the fluid so that the body forces are identically supported by p_0, and so that σ_{11} and σ_{22} vanish at a given point. The equilibrium conditions are thus reduced to

$$\sigma_{22,2} \equiv \sigma_{11,2} \equiv \sigma_{33,3} = 0. \tag{25.8}$$

and

$$\sigma_{11,1} + \sigma_{12,2} = \rho[\partial u / \partial t]. \tag{25.9}$$

At large distances from the plate (i.e., as $x_2 \to \infty$) the fluid is assumed to be either at rest or in a uniform state of shear. Thus $\sigma_{11,1}$ vanishes as $x_2 \to \infty$. Since σ_{11} depends only upon x_1, and σ_{11} and σ_{22} now vanish at one point, then σ_{11} and σ_{22} vanish everywhere. According to Equation 25.8, σ_{33} depends upon x_1 and x_2, but it does not necessarily vanish. The "normal stress effects" prevent simultaneous elimination of σ_{11}, σ_{22}, and σ_{33} at any point in the fluid by the simple superposition of a hydrostatic stress.

The stress σ_{11} is now eliminated from the equilibrium equation to yield the equation governing the distribution of velocity $u_2 = u(x_2, t)$ in the fluid

$$(u_{,22})^{(1-n)/n} = n(\varepsilon_0^{1/n}/\sigma_0)\rho \, \partial u/\partial t. \tag{25.10}$$

The two cases considered here are (1) fixed velocity of the plate for $t > 0$, and (2) fixed shear stress acting on the plate for $t > 0$; the plate is assumed to be at rest for $t \leq 0$. For case 1 the auxiliary conditions governing $u(x_2, t)$ at the plate $(x_2 = 0)$ are

$$u(0, t) = 0, \, t \leq 0$$
$$u(0, t) = u_0, \, t > 0. \tag{25.11}$$

Fixing the stress $\sigma_{12}(0, t)$ acting on the plate is equivalent to fixing the velocity gradient at the plate surface; the condition for fixed stresses acting at the plate surface (case 2) is

$$u_{,2}(0, t) = p_0, \, t > 0. \tag{25.12}$$

If the fluid is at rest at large distances from the plate, the appropriate conditions are

$$u(x_2, t) \to 0, \qquad x_2 \to \infty$$
$$u_{,2}(x_2, t) \to 0, \qquad x_2 \to \infty. \tag{25.13}$$

If the fluid is in a uniform state of translation at large distances from the plate, one has

$$u(x_2, t) \to u_{(\infty)}, \qquad x_2 \to \infty, \tag{25.14}$$

and if the fluid is in a uniform state of shearing flow at infinity, one has

$$u_{,2}(x_2, t) \to p_\infty, \qquad x_2 \to \infty. \tag{25.15}$$

25.5. Diffusion of Boundary Disturbances Consisting of Fixed Boundary Velocities

The velocity distribution in an Andradean fluid adjacent to a flat plate which is suddenly accelerated to a fixed velocity u_0 with the fluid at infinity being at rest (the boundary conditions of Equations 25.11 and 25.13) has been given earlier (Berg 1967). In order to avoid complications in choosing appropriate branches of $(\varepsilon_{12})^{1/n}$ in Equation 25.7, it is assumed that ε_{12} is positive. With $\varepsilon_{12} = u_{,2} > 0$, and with the fluid at infinity being at rest, the plate velocity (u_0) must be negative (i.e., the plate moves to the left). However, the restriction $u_0 < 0$ is used only for convenience and carries no physical importance. Dividing Equation 25.10 by the absolute value of the prescribed plate velocity $|u_0|$, and using the dimensionless velocity $\tilde{u} = u/|u_0|$, lead to a solution of Equation 25.10 in the similarity parameter

$$\beta = x_2 \left(\frac{(n+1)}{n^2} \cdot \frac{\sigma_0 \, |u_0|^{(1-n)/n}}{\rho \varepsilon_0^{1/n}} \cdot t \right)^{-n/(n+1)}, \tag{25.16}$$

the transformed version of Equation 25.10 being

$$(\tilde{u}')^{(1-n)/n} \tilde{u}'' + \tilde{u}' = 0, \tag{27.17}$$

where ( )′ represents $d(\ )/d\beta$. The solution of Equation 25.17, subject to the conditions $\tilde{u}(0) = -1$ and $\tilde{u}(\infty) = 0$, is

$$\tilde{u} + 1 = \int_0^\beta \left[\frac{n-1}{n} \right] [\alpha^2 + \beta_0^2(n)]^{n/(1-n)} \, d\alpha. \tag{25.18}$$

with $\beta_0(n)$ determined so that $\tilde{u}(\infty)$ vanishes.

The values of $\beta_0(n)$ required to satisfy the vanishing of $\tilde{u}$ at infinity for several values of the exponent n are shown in Figure 25.3. The correspond-

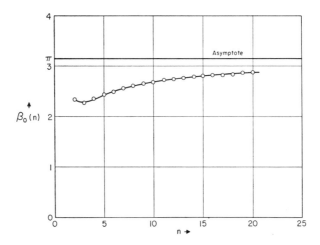

FIGURE 25.3. The behavior of $\beta_0(n)$ as a function of n; $\beta_0(n)$ was determined (numerically) so that u would vanish at infinity.

ing velocity profiles are shown in Figure 25.4, and the values of $\tilde{u}(\beta_0(n))$ are shown in Figure 25.5. At $\beta = \beta_0$ the ratio of local fluid velocity to plate velocity $\tilde{u} = u(\beta_0)/|u_0|$ is of the order unity (this ratio approaches

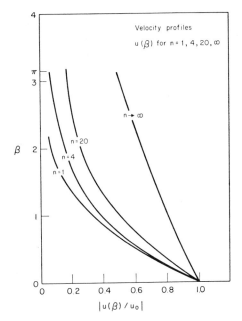

FIGURE 25.4. Profiles of velocity in the Andradean fluid, having exponent n, lying adjacent to a flat plate suddenly accelerated to the velocity u_0.

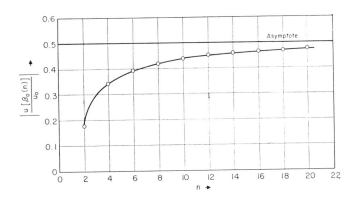

FIGURE 25.5. The velocity ratio $|u(\beta_0(n))/u_0|$ at the point $\beta = \beta_0(n)$, as a function of the Andradean exponent (n) of the fluid.

$1/2$ as $n \to \infty$), as is the value of $\beta_0(n)$; $\beta_0(n) \to \pi$, as $n \to \infty$. The parameter β of Equation 25.16 provides a characteristic distance of diffusion of the "fixed velocity" boundary disturbance into the previously undisturbed fluid; that is, if one sets β equal to unity the value of x_2 for a given value of t provides a measure of the distance over which the effect of the boundary disturbance has penetrated the fluid. The dependence of this characteristic

distance of diffusion upon the Andradean exponent n of the fluid is shown
by Equation 25.16. If n is 1— the Newtonian case — the similarity parameter
β is $x_2/\sqrt{2vt}$. As $n \to \infty$ the Andradean fluid approaches nonhardening
rigid-plastic behavior, and $\beta \sim x_2/t$, which gives a diffusional behavior
approaching wave propagation. However, in the case $n \to \infty$ the apparent
wave speed of the disturbance approaches zero, and the solution of
Equation 25.18 as $n \to \infty$ gives plastic slip flow at the plate (Berg 1967).
In summary, Equation 25.16 shows that the characteristic distance $\delta^{(v)}$ of
diffusion of a disturbance caused by a fixed velocity applied suddenly to
the boundary of an Andradean fluid initially at rest is

$$\delta^{(v)} = \left[\left(\frac{n+1}{pn^2} \right) \left(\frac{\sigma_0}{\varepsilon_0^{1/n}} \right) |u_0|^{(1-n)/n} \right]^{n/(n+1)} \cdot (t)^{n/(n+1)}. \tag{25.19}$$

25.6. Diffusion of Disturbances Caused by Fixed Boundary Stresses

Prescribing the shear stress acting at the plate in Figure 25.1 is
equivalent to prescribing the gradient of velocity u_2 at the plate. Differentiating Equation 25.10 with respect to x_2 and using p to denote $u_{,2}$ yields
the equation governing the distribution of velocity gradient in the fluid:

$$\left(\frac{1-n}{n} \right) p^{(1-2n)/n}(p_{,2})^2 + p^{(1-\;)/n}p_{,22} = \frac{n\rho\varepsilon_0^{1/n}}{\sigma_0} p_{,t}. \tag{25.20}$$

In order to determine how the velocity gradient (i.e., stress) diffuses into
an Andradean fluid adjacent to a flat plate to which a fixed boundary stress
is suddenly applied, the entire system being initially at rest, one solves
Equation 25.20 subject to the conditions of Equations 25.12 and 25.13.
Dividing Equation 25.20 by the absolute value of the prescribed velocity
gradient $p_0 > 0$ at the plate, and using the dimensionless velocity gradient
$\tilde{p} = p/p_0$, one seeks a similarity solution of Equation 25.20 in a parameter
$\hat{\beta}$ given by

$$\hat{\beta} = x_2/(\{2\sigma_0/(n\rho\varepsilon_0^{1/n}p_0^{(n-1)/n})\}^{1/2} > t), \tag{25.21}$$

the reduced form of Equation 25.20 being

$$(1-n)/n(\tilde{p})^{(1-2n)/n}(\tilde{p}')^2 + (\tilde{p})^{(1-n)/n}(\tilde{p})'' + \hat{\beta}\tilde{p}' = 0. \tag{25.22}$$

In the similarity parameter $\hat{\beta}$ time always appears as $\sqrt{t}$, just as in the
Stokes (1851) solution for suddenly accelerated flow in Newtonian fluids.
Leaving aside for the moment the question of satisfying Equations 25.12
and 25.13, one sees that the transient response of an Andradean fluid to a
fixed boundary stress is radically different from the transient response of
the same fluid to a fixed boundary velocity. The characteristic distance of
diffusion of the effect of a fixed boundary velocity Equation 25.19 is proportional to $t^{n/(n+1)}$, whereas the characteristic distance of diffusion of the

effect of a fixed boundary stress is, according to Equation 25.21, proportional to $\sqrt{t}$, just as in a Newtonian fluid.

To solve Equation 25.22 it is convenient to use $f(\hat{\beta})$ defined by

$$\tilde{p}(\hat{\beta}) = e^{f(\hat{\beta})}. \tag{25.23}$$

Through Equation 25.23 the equation of equilibrium Equation 25.22 becomes

$$f'' + \hat{\beta} f' e^{(n-1)f/n} + (f')^2/n = 0, \tag{25.24}$$

subject to the conditions

$$f(0) = 0, \quad f(\infty) \to -\infty. \tag{25.25}$$

Instead of attempting directly to solve Equation 25.23 subject to 25.25, consider the case in which f and f' are both prescribed at the origin, say

$$\begin{aligned} f(0) &= 0, \quad \text{and} \\ f'(0) &= -a < 0. \end{aligned} \tag{25.26}$$

One could construct a power series solution of Equations 25.24 and 25.26 in some neighborhood of the origin. We consider two extreme cases. First, if $|f'(0)| \ll 1$, the term $(f')^2/n$ in 25.26 may be neglected. Since $f(0) = 0$ and $f'(0)$ is small, there will be a neighborhood of the origin in which f is small compared with unity. In this neighborhood Equation 25.24 may be replaced by

$$f'' + \hat{\beta} f' = 0, \tag{25.27}$$

the solution of which is

$$f = -a \int_0^\beta e^{-\alpha 2/n} \, d\alpha. \tag{25.28}$$

The integral in Equation 25.28, the error function of $\hat{\beta}$, is bounded. Thus if a is sufficiently small, $|f|$ will be small for all $\beta \geqslant 0$. The integral 25.30 does not provide that $f(\infty) \to -\infty$, as is necessary if the fluid is initially at rest. According to Equation 25.28, f approaches the value $-a\sqrt{\pi/2}$ and Equation 25.28 thus represents the response of the fluid to an increment of velocity gradient p_0 at the plate when a uniform velocity gradient $p = p_0 e^{-a\sqrt{\pi/2}}$ initially existed in the fluid; the response of the Andradean fluid in this case is just the same as a Newtonian fluid having a viscosity corresponding to the slope of the Andradean stress strain-rate curve at the initial value of (uniform) strain rate in the fluid. Although Equation 25.28 does not provide the response of the initially undisturbed fluid to a suddenly applied boundary stress, it does show that similarity solutions of the type assumed do exist.

Consider a second case in which $f(0) = 0$ and $f'(0)$ is once more negative and very large in magnitude

$$f'(0) < 0; |f'(0)| \gg 1. \tag{25.29}$$

In this case the behavior of f near the origin is dominated by the term involving $(f')^2$ in Equation 25.24, and in this case an approximate solution is given by[1]

$$f(\hat{\beta}) = n \log (na - \hat{\beta})/na. \tag{25.30}$$

Because of the occurrence of the logarithmic singularity the condition that the fluid be at rest at infinity is not met. Figure 25.6 show curves of

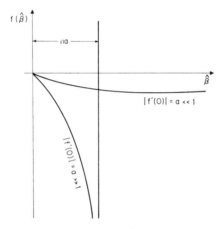

FIGURE 25.6. The behavior of $f(\hat{\beta})$ as a function of $\hat{\beta}$ when $|f'(0)| \ll 1$ and when $|f'(0)| \gg 1$; in both cases $f'(0) < 0$.

$f(\hat{\beta})$ for the cases $|f'(0)| \ll 1$ and $|f'(0)| \gg 1$. At any time, for the case $|f'(0)| \gg 1$, the velocity gradient $\tilde{p} = e^{f(\hat{\beta})}$ vanishes at a finite distance from the plate at $\beta = na$. The (possible) physical significance of such a solution will not be examined here; if physically realizable, flows of this type entail a "diffusional shock" at $\beta = na$, with the fluid in the region $\beta > na$ being at rest. This matter is deferred for later consideration. At present we note that for sufficiently small values of $|f'(0)|$, f is finite at infinity, and for sufficiently large values of $|f'(0)|$ the integral of Equation 25.24 has a singularity which prevents the solution curve from reaching infinity on the $\hat{\beta}$ axis.

The diffusion of a "fixed stress" disturbance into an Andradean fluid initially at rest requires an integral of Equation 25.24 for which $f(\beta) \to -\infty$

[1] The continuation (Equation 25.30) of $f(\hat{\beta})$ away from the origin is based upon retaining only the terms f'' and $(f')^2$ in 25.24. This continuation, if initially justified, is justifiable on the strip of the $\hat{\beta}$ axis from the origin to the singularity $0 \le \hat{\beta} \le na$.

as $\hat{\beta} \to \infty$; for some intermediate value of the initial slope $f'(0)$ this condition will be met. Since the limiting approximate integrals above arise either because the initial curvature $f''(0)$ of $f(\beta)$ is very small and becomes smaller, or because the initial curvature is very large and grows larger — with the consequent singularity at $\hat{\beta} = na$, one supposes that on the branch of the integral, on which $f(\hat{\beta}) \to -\infty$ as $\hat{\beta} \to \infty$, the curvature is small compared with the slope of the curve. One then expects that on the desired branch of $f(\hat{\beta})$, at large values of $\hat{\beta}$ the term f'' in Equation 25.26 will be negligible, and $f(\hat{\beta})$ should behave as

$$f(\hat{\beta}) \sim \left(\frac{n}{1-n}\right) \log \hat{\beta}. \tag{25.31}$$

According to 25.31 $f''/f' \sim 1/\hat{\beta}$, so that as $\hat{\beta} \to \infty$ the assumption that f'' in 25.24 is small compared with the remaining terms is justifiable, and 25.31 does provide the correct behavior, at large distances from the plate; the velocity gradient $\tilde{p}$ at large distances from the accelerated plate in the initially undisturbed Andradean fluid should decay as $\hat{\beta}^{-n/n-1}$

Curves of $f(\hat{\beta})$, each meeting the condition $f(0) = 0$ and having various values of initial slope $f'(0) < 0$, were constructed numerically. For an Andradean exponent $n = 3$, it was found that the value of $f'(0)$ required to meet the condition $f(\hat{\beta}) \to -\infty$, which provides that the fluid be undisturbed at large distances from the plate, lay in the interval $-1.02250 < f'(0) < -1.02000$.

In view of the foregoing, $\hat{\beta}$ of Equation 25.20 may be used to estimate the characteristic distance of diffusion of a disturbance caused by a fixed shear stress suddenly applied to the flat boundary of an Andradean fluid that is initially at rest; the characteristic distance of diffusion $\delta^{(\tau)}$ in this case is given by

$$\delta^{(\tau)} = \{2\sigma_0/(n\rho\varepsilon_0^{1/n}p_0^{(n-1)/n}\}^{1/2}\sqrt{t},$$

where $p_0 > 0$ is the value of the velocity gradient (strain rate) corresponding to the applied stress. $\delta^{(\tau)}$ may be expressed in terms of the shear stress τ_0 applied to the boundary as follows:

$$\delta^{(\tau)} = \left\{\frac{2\sigma_0^n}{n\rho\varepsilon_0}\tau_0^{(1-n)}\right\}^{1/2}\sqrt{t}. \tag{25.32}$$

25.7. Discussion

In an Andradean fluid the temporal character of the distance of diffusion of a boundary disturbance depends upon the type of disturbance. For fixed boundary velocities the distance of diffusion is proportional to $t^{n/(n+1)}$; for fixed boundary stresses the characteristic distance of diffusion is always proportional to $\sqrt{t}$, as in a Newtonian fluid. Recalling the

constitutive relation of the Andradean fluid (Equation 25.7) the characteristic diffusion length of a fixed boundary stress (Equation 25.32) may
be interpreted as

$$\delta^{(\tau_0)} = \left[\frac{1}{\rho} \left(\frac{\partial \sigma_{12}}{\partial \varepsilon_{12}} \right)_{\sigma_{12} = \tau_0} t \right]^{1/2},$$

so that an effective viscosity, being the slope of the Andradean stress
strain-rate curve at the value of the applied boundary stress, determines
the characteristic length of diffusion. The outcome, that $\delta^{(\tau_0)} \sim \sqrt{''v'' \, t}$
with $''v''$ being the effective kinematic viscosity, is a strikingly simple result
which might have been guessed in advance on the basis of dimensional
arguments, but which, nevertheless, seems not to have been stated precisely heretofore. Because the basic temporal character of diffusion of
disturbances from boundaries to which a fixed loading is applied is the same
in both Newtonian and Andradean fluids, it would be reasonable to expect
that fluid dynamic phenomena, such as the onset of convective instability,
the kinetics of which are controlled by viscous diffusion of disturbances
away from boundaries on which fixed, or slowly varying stresses act, will
be highly similar in Andradean and Newtonian fluids. It has been suggested (Orowan 1965) that since the mantle of the earth most probably
flows as an Andradean fluid — which in its stress-strain-rate behavior is
quite different from a Newtonian fluid — the convective stability criterion
developed for Newtonian fluids are probably not at all applicable to convection in the mantle. However, because of the very small thermal diffusivity
of the mantle a locally superheated region will not cool down very rapidly
and its buoyancy will exert an approximately constant loading on the
surrounding cooler mantle material. As the superheated mass moves
upward under its buoyancy the effect of its motion will diffuse into the
surrounding cooler mantle material, entraining that material in the motion
of the superheated mass; the total mass of material which the buoyancy
force must accelerate will thus increase. If the rate of entrainment of surrounding material by viscous diffusion is extremely rapid the motion of
the superheated mass is quickly retarded, whereas a slow rate of viscous
diffusion will allow the upward motion of the superheated mass to accelerate. The buoyancy force acting on the superheated mass applies a slowly
varying stress to the surrounding cooler material (since the accelerations
in mantle motion are small). Thus, whether the mantle behaves as an
Andradean or Newtonian fluid will have little influence on the kinetics
of the spread of the viscous disturbances into the surrounding mantle.
As a first estimate, the prediction of thermal convective stability or instability based upon Newtonian flow should be applicable to the case at
hand, even though the mantle may flow as an Andradean fluid, provided
that one uses as the viscosity in the stability parameters the slope ($\partial\sigma/\partial\varepsilon$)

of the Andradean flow curve evaluated at the stress level that the buoyancy of the superheated mass applies to its cooler surroundings.

Many helpful discussions with Professor E. Orowan of the mechanical problems of mantle motion are gratefully acknowledged.

REFERENCES

Andrade, E. N. da C., 1911, *Proc. Roy. Soc.*, **A84**, 1.

Berg, C. A., 1967, in *Proc. NATO Advanced Study Institute*, Newcastle, New York: Wiley, in press.

Berg, C. A., 1968, in *Proc. Fifth International Congress on Rheology*, Kyoto, in press.

Langharr, H., 1942, *J. Appl. Mech.*, **9**, A55.

Norton, F. H., 1929, *Creep of Steel at High Temperatures*, New York: McGraw-Hill.

Orowan, E., 1965, *Phil. Trans. Roy. Soc.*, **A258**, 284.

Schlicting, H., 1960, *Boundary Layer Theory*, New York: McGraw-Hill.

Stokes, G. G., 1851, *Cambr. Phil. Trans.*, **9**, 8.

ABSTRACT. A satisfactory model of thermally induced creep motion in the earth's upper mantle is most readily obtained if the two following properties are taken into account: (a) concentration of radioactivity in continental plates entails horizontal temperature gradients whose dynamic effects are not negligible; (b) stresses in a crystalline solid can be divided into statically relaxable and nonrelaxable ones. In the upper mantle, below about 100 km depth, where the melting point is approached though not reached, the static relaxation times can become smaller than the times required to produce appreciable strains; creep then begins to approximate viscous flow. A qualitative and crude but rather coherent and contradiction-free model of uppermantle convection becomes possible. Its main feature is a preponderance of horizontal displacements as compared to models which would consist of a number of "cells."

26. Pattern of Convective Creep in the Earth's Mantle

W. M. ELSASSER

The large and regular displacements of ocean bottom which have been discovered in recent years (e.g., Vine 1966) leave little doubt that the primary agency of large-scale tectonic phenomena near the earth's surface is some sort of convection, no doubt primarily thermal in origin, although chemical effects will play a secondary role. Up to now the approximate equality of the mean continental flow and the mean oceanic heat flow has always been the chief argument against upper-mantle convection and in favor of an essentially vertical differentiation of crustal material from the underlying mantle. With any such simple picture gone, the question arises for the theoretician all over again, to design patterns that make dynamic sense of creep motion in the upper mantle. The guiding principle must again be found in the fact that by all observational evidence a large fraction of the heat leaving the surface of the continents is produced by radioactive heat sources located in the continental plates themselves; the usual estimates of the fraction generated in the continents range from two-thirds to three-quarters of the total heat flow. The radioactive heat produced in the upper mantle is undoubtedly smaller than this by a rather substantial numerical factor. We are here not as yet concerned with quantitative pictures of

367

observed large-scale geological phenomena; these, if they ever can be provided, are still very far away. For simplicity, let us think of an upper
mantle with negligible internal radioactive heat development in which the
continents swim as radioactive hot-plates. It has rather early been suspected
(Bullard et al. 1956) that heat must be carried upward under the oceans by
convection if the equality of heat flows between continents and oceans is
to prevail. Recently, I have put this argument into a more mathematical
form (Elsasser 1967) by showing that the material of the ocean bottom
cannot be in static equilibrium unless the heat flows are equal — which is
tantamount to postulating convection in the upper mantle under the oceans.
One would not have advanced such views except for the remarkable empirical evidence which shows that ocean bottoms are not very old but have
been renewed by mantle material flowing out laterally from the suboceanic
ridges (e.g., Hess 1962).

The condition for vertical instability that may lead to convective overturn
is well known: if the temperature gradient dT/dp (assuming a chemically
homogeneous substance) exceeds the adiabatic value, the stratification is
unstable and energy is gained by its overturn. The adiabatic gradient in
the upper mantle is usually estimated as about 0.3–04. deg/km, on sound
thermodynamical grounds; this is some 50 to 80 times smaller than the
temperature gradient near the top of the mantle as inferred directly from
the heat flow and known thermal conductivities. Hence, the upper mantle
is mechanically very unstable indeed (and so, by the way, are the continental
plates). It does not follow, however, that there will be actual overturn. The
reason is that the mantle, being a crystalline solid, moves by creep, and
creep is a notoriously nonlinear phenomenon. At moderately large stresses
the creep rate increases exponentially with increasing stress, but since an
exponential does not go to zero for vanishing argument, this is often replaced by the Ree-Eyring approximation, which expresses the strain-rate
as $\dot{\varepsilon} = \sinh(\sigma/\sigma_0)$. Even this approximation turns out rather unsatisfactory
at low stresses, where it would predict a linear dependence of strain rate
upon stress; what has actually been found in the observations is a power
law. Thus a law, $\dot{\varepsilon} \sim \sigma^4$, was found in a number of laboratory experiments.[1]

The strong dependence of strain rate upon stress means that certain
strain patterns are not likely to occur, namely, those that involve large
strain rates and correspondingly very large stresses. Conspicuous among
these are cell-type closed circulation patterns of limited extent. Since it
seems practically certain that the layer of very steep temperature gradients
and correspondingly extreme instability cannot reach deeper into the

[1] It is well known that fine-grained polycrystalline metals at very low stress show a
linear creep rate-stress law resulting from grain boundary sliding or Nabarro-Herring
diffusional creep. It is not at all certain, however, what role these mechanisms may play
in the deformation of the earth's mantle which must possess a particularly large grain
size.

mantle than a few hundred kilometers, the much-depicted type of convection "cell" would necessarily have to have sharp corners, that is, large shear strains at corners, and correspondingly very large stresses there. Since the geological evidence does not indicate the existence of horizontally limited "cells" but rather a very widespread, almost world-wide system of correlations, it is necessary to look for those features of the upper mantle that lead to a pronounced alignment of motions along the horizontal, taking place in a comparatively thin layer of the upper mantle.

In recent years the author has tried to study a principal feature of the following kind: the steep vertical temperature gradient itself makes the material rapidly softer, the deeper one goes down into the upper mantle. There are laboratory experiments by Towle and Riecker (1969) which show that a suitably defined laboratory "shear strength" in creep depends roughly exponentially on a "reduced" temperature, T/T_m, where T_m is the melting point, this being true both on variations of T itself and on variations of T_m with hydrostatic pressure. A widely adopted model of the earth assumes that most radioactive material is concentrated in the upper mantle. Such a model is highly plausible on physical arguments that we cannot discuss here; it is rather obvious that in it the temperature gradient in the top mantle will be steep but farther down the temperature curve will flatten out. This temperature distribution has its closest approach to the melting point at moderate depth, say below 100 km; it gives rise to the Gutenberg layer of lowered seismic wave velocity. The corresponding increase of softness with temperature should be very pronounced. As a consequence of this, extensive pieces of the topmost mantle can move horizontally like solid plates. This has been demonstrated by Morgan (1968) by his analysis of macrotectonic data.

For some time now I have considered (Elsasser 1966, 1967), if in a somewhat speculative fashion, the mechanical effects of thermal stratification of the upper mantle. I felt consistently, however, that an essential element was still lacking. I believe now that this missing element can be found in the *horizontal* temperature gradients which result from the concentration of radioactivity in the continental plates. This leads to a constant cooling of the suboceanic mantle, whereas the cooling of the continental plates and their underlying mantle will be much less, or may even be considered as negligible. In order to show that horizontal temperature gradients do have significant effects, consider schematically two fluids of equal volume and fixed densities — ρ and $\rho + \Delta\rho$, respectively — contained in a rectangular box of horizontal area A and height $2h$. Let the heavier fluid first be on top of the lighter one, the two being separated by a horizontal plane boundary. On overturning the configuration so that the light fluid comes to lie on top, the potential energy gained is clearly $2ghA\,\Delta\rho$. Again, let the fluids be horizontally next to each other, separated by a vertical wall, and let the terminal position be as before, with the light fluid on top of the heavier one;

the potential energy gained is now $ghA \Delta\rho$, one-half of the previous value. But then, horizontal creep displacements in a shallow trough are fairly readily effected because they do not involve the turning of sharp angles, requiring excessively high stresses.

In his geophysical writings, E. Orowan has repeatedly pointed out that the material of the mantle is likely to be much softer than that of the continental plates. He attributes this to the presence of small amounts of "fluxing" agents in the upper mantle, outstanding among them water. Even minute quantities of water are known from laboratory experiments to enhance creep tremendously. On the other hand, the continental plates are desiccated and correspondingly hard. If, therefore, the suboceanic mantle cools substantially without much change in the temperature of the continents, the corresponding creep pattern ought to be primarily a down-flow of material from the ocean bottom toward the layer underlying the continents. Next, continuity requires that the material come up again somewhere, and experience shows that it rises under the submarine ridges. One might thus expect that the oceans are the primary places of convective upwelling. But there are no rises in the largest of oceans, the Pacific, except for the East Pacific Rise confined to one corner of it. The Darwin Rise apparently was once an incipient submarine ridge, but it does not seem to have developed a mechanism of lateral spreading.

One of the great problems in the convection of a solid in creep is its *initiation*. Since the strain rate-to-stress relationship is so highly nonlinear, all motions are strongly concentrated where the stresses are large. It goes almost without saying that, with a horizontally stratified oceanbottom, the maximum stresses are at or near the continental margins. As Orowan (1958) has discussed, the continents, if in isostatic — that is, vertical — equilibrium, still have an uncompensated horizontal "spreading" stress. Furthermore, if the ocean bottoms do cool, the isothermal surfaces in the upper mantle will dip downward toward the oceans, giving rise to stresses of thermal origin Downflow of mantle material takes place all around the margin of the Pacific Ocean, directed toward the continents; the over-whelming majority of earthquakes are found concentrated in these regions (see Gutenberg and Richter, 1965, Figures 3 and 4). Once such a downflow has started it can maintain itself by virtue of the pull of the sinking material, which is colder than the average mantle material at the same level. We have good reason to believe that the oceanic crust, being quite thin, will first be dragged down with the sinking mantle material by friction. Only at suffi-cient depth do the stresses of bouyancy become strong enough to permit the separation of crustal from mantle material and the rising of the crustal material, which is then added to the roots of the mountain chains. Orowan (1967) has pointed out that the familiar phenomenon of island arcs also has its origin in a downflow of mantle material of the ocean bottom, with the convex tide of the arc facing upstream. One need not, however, believe, as Orowan does, that these phenomena correspond necessarily to compressive horizontal stresses; for reasons into which we cannot enter here it seems plausible that the negative buoyancy of the descending material is sufficient to drive the process once it has been initiated. In this case the stresses above and behind the descending material would be essentially tensile.

Many geologists have remarked that since the material at the edges of the Pacific sinks, the ocean's area must decrease steadily. This decrease is balanced by the creation of new ocean bottom off the ridges. The three principal oceans other than the Pacific, namely, the Atlantic, Indian, and Arctic Oceans, all have major ridges running through them; thus if the ocean-bottom material is created by sideways flow from ridges, the bottom of these oceans is likely to be younger and the material at some depth warmer than under the Pacific. The main question is now, what are the principal stress and flow patterns that underlie (literally) the observed surface phenomena?

The initiation process of the ridges may well have been substantially different from the later, steady-state process. One event that is hard to overlook, if one admits the occurrence of drift at all, is the breakup of the original Gondwanaland into its present fragments, South America, Africa, India, Australia, and Antarctica. Even if these fragments did not form an altogether coherent continental mass to begin with, they must certainly have been much closer together than they are now in order to account for the remnants of the Permian glaciation. How could a process of upwelling have started so that a huge mass of continental type did break apart spontaneously? The simplest assumption we can make is that the contraction of the Pacific Ocean created a tensile stress in the continents. Once the continental plates have cracked open and a certain vertical displacement has taken place in the mantle, the material in the ridge is hotter than the surrounding mantle at the same level. If, then, the material can flow laterally off the top of the ridges by gravity sliding, the ridge-forming process can become self-sustaining.

The most spectacular feature of certain ridges is no doubt their median position between continental margins; this applies in particular to the Midatlantic Ridge. But according to Menard (1966), only a moderate fraction of the overall length of ridges is found in this median position; a more widespread and, as Menard thinks, even more characteristic property of ridges is exhbiited by their tendency to keep an approximately constant distance from shorelines. Africa is the prime example of this; it is surrounded by ridges on three sides. In the case of Antarctica, fully surrounded by ridges, their distance to the shores is slightly more uneven. This ridge-shore relationship requires an explanation in physical terms. A suggestive idea to be found in the literature is that there is a down-current of mantle material under, for example, Africa. This author feels unable to accept such a model.

The eastern part of Africa is covered by a series of rift valleys which clearly indicate upwelling (not downwelling), and in the remainder of the continent no surficial evidence of a downdraft is at all apparent. This leaves us apparently with only one explanation: *The ridges recede progressively from the shorelines* as the material sliding sideways off the ridges generates new ocean bottom. Such an explanation requires that the updrafts corresponding to the ridges can and do *migrate* within the material of the mantle. In the language of hydrodynamics, the ridges are dynamic rather than purely substantial features. Further thought indicates that no other model is likely to satisfy the topological requirements for the appearance of new ocean bottom at some places of the globe and a compensating disappearance at others.

One view that emerges from such studies is that pronounced vertical motions occur only in limited regions: upward motion mainly under well-marked submarine ridges, and downward motion near trenches; the rest of the near-surface material moves horizontally as solid plates. Now if there is generalized motion on a spherical surface, a principle of relativity applies: a state of rest exists only by definition of some specific body as being in that state. As an example of this, consider the Southern Hemisphere and let South Africa be " at rest." An observer located there sees the Midatlantic Ridge move toward the west with a certain velocity, say v; he sees the Carlsberg Ridge in the Indian Ocean moving toward the east with about the same velocity. He sees South America moving westward with about $2v$, and he would see Australia and the archipelago located to the north of it moving eastward, also with a velocity $2v$ except for one complication: the Java Trench is one of the three trenches which are not part of the Pacific margin; under it, ocean-bottom material must slope down and move toward the northeast. This will express itself in a southwestward movement of the continental mass. To the observer situated in South Africa this appears as a reduction of the eastward velocity of this land mass.

A conspicuous feature of macrotectonics is the difference in symmetry between the sinking and rising motions. The sinking motion is slanting and associated with a shear zone marked by strong seismicity. The rising under ridges, on the other hand, appears to have complete bilateral symmetry in all features so far observed. In the case of a straight ridge, this implies a vertical plane of symmetry along the axis. The simplest explanation of the difference between descending and ascending motions is in terms of a *vertical variation* of material properties arising from the rapid increase of temperature with depth. In this view, the shape of a vertical current is mainly determined by the depth at which it originates.

Observationally, the downdrafts resemble the fracture and sliding of a crystalline solid. While the updrafts approximate features of plastic flow (Orowan 1965) rising to some extent like blocks rather than in the form of a smooth velocity distribution, they do lack the sharply defined fracture zones, punctuated by earthquakes, of the downdrafts and show such remarkable bilateral symmetry that in fact they resemble viscous flow at least as much as plastic flow. We should here remember that there is no laboratory equivalent to motions of such extreme slowness; therefore, we must next emphasize a significant feature of crystalline solids; this is *stress relaxation*. Consider first the fact that in a solid in creep, the stresses are related not only to the static strains but also to the strain rates. An elementary example of stress relaxation is the following: Consider a vertical cylinder of material compressed by a load on top of it but confined in a rigid enclosure so that it cannot expand laterally. Simple elastic calculations show that there are horizontal stresses which are, however, smaller than the vertical stress. If we wait a long enough time there will be rearrangements of the crystal inhomogeneities until finally the stress would be uniform in all directions, i.e., would be hydrostatic. Here is the prototype of a *statically relaxable stress*. More precisely, the differences from uniform, hydrostatic stress are relaxable; the mean stress — the linear average of the principal stresses, if uniform throughout the body — is not. Now if a body is deformed *infinitely slowly* we may assume that the statically relaxable stresses can never get large and so may be neglected. A crystalline solid would approach the behavior of a viscous fluid, in which there are three effective types of stress-induced effects: (a) hydrostatic compression, (b) its gradient, which gives rise to the circulation-generating force term $\nabla p/\rho$ in the equations of motion, and (c) the viscous shear forces in the Stokes-Navier equations.

Numerical values of static relaxation times are difficult to come by. There is no clearcut evidence that the time of the Fennoscandian uplift, of order 10^4 years, or similar empirical geological times, can be interpreted simply in terms of static relaxation times. So as to avoid a lengthy and probably sterile discussion, let us just say that static relaxation times in the mantle may conceivably vary between 10^3 years and 10^6 years, depending on the material and especially on T/T_m, which latter is a function of depth; we have discussed this last dependence already. In our model, radioactivity is assumed concentrated in the upper mantle. This results in a very steep temperature gradient in the upper mantle, going over into a very shallow gradient farther down. The melting point as a function of depth can be estimated by means of the Clapeyron equation and is found to rise rather steadily at the rate of a few deg/km. Thus there must be a region of closest approach of the actual temperature to the melting point. One may estimate this region to lie somewhat below 100 km, although it is quite possible that the temperature remains tolerably close to the melting point for several hundred km farther down; thereafter, as one goes still deeper, the melting point should again rise substantially above the local temperature.

Let us now assume that there is a layer in the upper mantle where the statically relaxable stress components can relax fast enough so that they have little influence upon the dynamics of the flow. The symmetry of flow under ridges would then be an expression of a statically relaxed, nearly viscous behavior. Next, the fact that the fault planes of the descending material are often marked by earthquakes down to several hundred km depth, whereas all earthquakes associated with ridges are extremely shallow, should be taken as the expression of a corresponding temperature difference: the rising material is appreciably hotter, the sinking material appreciably cooler than typical mantle material at the same level.

The layer of closest approach to the melting point should also be the layer in which the "return flow" takes place, that is, the flow which carries the material from the regions of down draft to the regions of updraft On constructing a return flow pattern one must take into account the existence of two major phase transformations in the upper mantle, as rather conclusively established by Anderson (1967). The first transformation occurs at about 400 km depth; the olivine here changing to a spinel-type structure. The second transformation occurs at around 750 km depth; here the silicates break up into an assembly of oxides whose crystal structures are almost certainly cubic. Verhoogen (1965) has demonstrated on thermo-dynamic arguments that a vertical convective current can pass across a phase boundary if in homogeneous material. Now in a composite material, such as the mantle, there is perhaps no sharply defined phase boundary but a transition layer of finite thickness. One further possibility which should be kept in mind is that within such a transition layer the "viscosity" is lowered. This would have an effect upon the horizontal components of the motion, since it might make a motion resembling horizontal sliding possible. So far this is of course only a speculative concept, but its signifi-cance cannot be ignored even at the present stage.

One can adduce rather good arguments for the view that convective motions in the mantle do not (or nor appreciably) extend below the 750 km phase-transformation layer. Certainly, the melting point must jump there and the corresponding reduction in T/T_m must substantially increase the "hardness" for creep. If radioactivity is concentrated in the upper mantle, a process of extrusion of light material must have happened at an earlier stage of the earth's history, when the lower mantle was warmer than it is now. The absence of fluxing agents in the lower mantle which this implies, taken together with a small vertical temperature gradient, makes the lower mantle a quite unfavorable place for creep convection. There are obser-vations which show that deep-focus earthquakes are not found below about 700 km depth; these lend some modest degree of empirical support to this presumption. This would leave for convection only a shallow "tray" whose depth is less than 2 percent of the earth's circumference. We have so far not seen observational evidence to indicate that convection goes much deeper, although conditions may have been different in the earth's remote past.

REFERENCES

Anderson, D. L., 1967, *Science*, **157**, 1165.

Bullard, E. C., A. E. Maxwell, and R. Revelle, 1956, *Adv. Geophys.*, **3**, 153.

Elsasser, W. M., 1966, in P. M. Hurley (ed.), *Advances in Earth Science*, Cambridge Mass.: M.I.T. Press, p. 461.

Elsasser, W. M., 1967, *J. Geophys. Res.*, **72**, 4768.

Elsasser, W. M., 1969, in S. K. Runcorn (ed.), *The Application of Modern Physics to the Earth and Planetary Interiors*, Newcastle, NATO Institute, New York: Wiley.

Gutenberg, B., and C. F. Richter, 1965, *Seismicity of the Earth*, New York: Hafner.

Hess, H. H., 1962, in *Petrological Studies*, A Volume to Honor A. F. Buddington, Geol. Soc. Amer., p. 599.

Menard, H. W., 1966, in *Physics and Chemistry of the Earth*, New York: Pergamon, vol. 6, p. 317.

Morgan, W. J., 1968, *J. Geophys. Res.* **73**, 1959.

Orowan, E., 1958, "Mechanical Problems of Geology," Lecture Notes, Calif. Inst. of Tech., unpublished.

Orowan, E., 1965, *Philos. Trans. Roy. Soc.*, **A258**, 284.

Orowan, E., 1967, *Geophys. J. Roy. Astr. Soc.*, **14**, 385.

Towle, L. C., and R. E. Riecker, 1969, *Science*, **163**, 41.

Verhoogen, J., 1965, *Phil. Trans. Roy. Soc.*, **258**, 276.

Vine, F. J., 1966, *Science*, **154**, 1405.

ABSTRACT. The von Mises requirement of five independent slip systems as a necessary condition for ductility poses an especially acute difficulty in achieving ductility in rocks. The number of independent slip systems available in several common rock-forming minerals (including quartz, calcite, and dolomite) is calculated from slip systems revealed in experiments. In many cases, the von Mises criterion cannot be met and the observation of macroscopic ductility must be sought in other factors such as (1) some relaxation of the requirement of five independent slip systems when heterogeneous strain is allowed, especially in case of fine-scale deformation bands or kinks; (2) deformation by mechanisms other than intracrystalline plasticity, especially by cataclastic deformation or by diffusion creep.

27. The Ductility of Rocks

M. S. PATERSON

27.1. Introduction

Ductility is one of the most important factors in the widespread use of metals as materials of construction, especially where tensile loads are to be borne. In some cases, as in plastic forming processes, it is exploited in a major way. In other cases, it may be a secondary but vital factor in permitting the inherent strength of metals to be exploited through redistribution of stress by local plastic flow in places of stress concentration, without fracture being initiated. Thus there has grown to be a considerable literature on the brittleness and ductility of metals, to which Professor Orowan has made notable contribution (Orowan 1945, 1948, 1955, 1959, and others). In recent years, in the search for materials to meet the widening demands of technology, studies have been made of the possibility of achieving some ductility in other classes of materials, such as ceramics (Parker et al. 1958; Parker 1963). Here we shall draw attention to a less well-known field, the deformation of rocks, in which studies of ductility are also of importance. However, as a preliminary, we need to reexamine the concept of ductility in some detail in order to clarify the use of the term in this more general context.

Etymologically, ductility carries the notion of capacity for being "led

377

or drawn" (Shorter Oxford English Dictionary) and its early technical use is for capacity to be drawn into wire. In modern metallurgical writing, ductility usually means capacity for undergoing permanent change of shape without fracturing; however, it often also carries the implication of a capacity for plastic deformation by the particular mechanisms of crystal plasticity, slip, and twinning. The latter is not an essential part of the concept when it is applied in a macroscopic or phenomenological way; in using the term in this paper we retain only the central notion of capacity for permanent change of shape without gross fracturing. The mechanism of deformation is then a matter for elucidation by detailed studies in particular cases.

That ductility is not an invariable property of a material but one depending on the particular circumstances of test or observation is especially well demonstrated in the case of rocks. In common experience they are very brittle materials. Yet from geological observation they appear often to have undergone extensive permanent deformation in nature, possibly very slowly, and laboratory experiment has now established conditions of temperature, pressure, and chemical environment under which a number of rocks can be plastically deformed relatively rapidly (Griggs and Handin 1960; Paterson 1968). It should be emphasized that almost always in these situations all three principal stresses are compressive.

In discussing the ductility of rocks, scale assumes a greater importance than in the case of metals because of the wider range of scale of the various sorts of heterogeneity that have to be taken into account in rocks. In a simple case such as a uniform body of marble or quartzite, the situation may be similar to that which is usual in metals, the deformation appearing to be pervasively distributed throughout the body when viewed on any scale coarser than the scale of the spacing of slip bands within grains. However, it is more usual for there to be several constituent minerals, which are often segregated, for example into bands; in such a case it is still possible that a deformation can occur which is pervasive on all scales larger than the scale of the mineral grains but is it more likely that deformation will be concentrated, perhaps almost exlusively, within certain bands, or even take place mainly by sliding at the boundaries between bands. Then, it is only valid to speak of the rock as a whole as being ductile if it is being viewed on a scale sufficiently large that individual bands are no longer resolved. Alternatively one may be able to refer to the ductile behaviour of certain bands or domains within the body of rock. Such considerations are fundamental in structural geology.

In terms of scale, the mechanism of a deformation is revealed when the body is examined on a fine enough scale to show up a characteristic heterogeneity in the deformation. For example, a body of rock may on one scale of observation be undergoing permanent deformation in an apparently uniform manner, so that it can be said to be ductile. Examination on a finer scale may reveal that the mechanism of the deformation is slip within the grains, or may show that there is fine-scale fracturing and relative movement on the fractures ("cataclasis"). Alternatively, it may show that deformation is concentrated in certain polycrystalline domains in which further details are resolved on going to a still finer scale.

The main concern of this paper is to review the nature of the mechanisms of deformation that are likely to be of importance in the various circumstances in which rocks display ductility. In particular, we shall examine the extent to which slip in the constituent minerals can contribute to the ductility, in the light of present knowledge of slip systems in minerals, and discuss what other mechanisms may be important in supplementing a predicted inadequacy of slip systems in cases where a substantial degree of ductility is nevertheless observed.

27.2. The von Mises Problem

We focus attention on an individual crystal grain in a polycrystalline aggregate undergoing plastic deformation and demand that no discontinuities in displacement occur at the grain boundary, that is, that no opening up or sliding occurs between the grains. This requirement is usually embodied in the more restrictive assumption that the grain undergoes the same homogeneous deformation as the assemblage of other grains in its neighbourhood, which is equivalent to assuming that the individual grain can undergo any arbitrarily specified deformation, since ductility implies no restriction on the nature of the deformation. It is also assumed that there is no volume change. Then, in order to achieve this deformation by slip, the crystal must have at least five independent slip systems available (von Mises 1928, p. 179; Taylor 1938; Bishop 1953). In this context, each slip system is viewed as giving a simple shear and it is "independent" if the shear cannot be achieved by any combination of other available slip systems. The criteria for establishing the independence of slip systems have been set out by von Mises (1928), Groves and Kelly (1963), and Kocks (1964), and are restated in the Appendix.

The von Mises requirement has been taken into account in many discussions of the mechanism of deformation of polycrystalline materials (for example, Taylor 1938, 1956; Bishop 1953; Livingston, and Chalmers 1957; Kocks 1958, 1964; Groves and Kelly 1963; Copley and Pask 1965; and Pratt 1967) and it has been generally acknowledged as a necessary (although not sufficient) condition for ductility. Thus at atmospheric pressure the NaCl structure materials are observed to be brittle in the polycrystalline form at the temperatures at which either only $\{110\}$ $\langle \bar{1}10 \rangle$ or only $\{100\}$ $\langle 011 \rangle$ slip systems are known to be active, but markedly ductile behaviour

is found at higher temperatures at which it is known that both systems are
operative or that pencil glide is occurring in the $\langle 110 \rangle$ direction; only in
the latter cases are five independent slip systems available (Groves and
Kelly 1963). The well-known ductility of polycrystalline face-centered and
body-centered cubic metals is also consistent with the availability of five
independent slip systems, while the more limited ductility of polycrystalline
hexagonal close-packed metals at temperatures at which single crystals are
very ductile can be related to the inadequacy of the easy basal slip and the
need for activity on other slip systems (Groves and Kelly 1963; Tegart
1964). The situation in ice is similar to the latter (Tegart 1964).

It is therefore of interest here to examine the situation in common rocks.
As a background to this, Table 27.1 lists slip systems that have been
reported in minerals of trigonal symmetry together with the numbers that
are independent among equivalent systems or various combinations of
them, since it happens that several of the more important rockforming
minerals are trigonal (the method used in deducing these numbers is
illustrated in the Appendix). We now discuss the various rock types in turn.

27.2.1. *Marble and Limestone*

The $\{01\bar{1}2\}$ $\langle 0\bar{1}11 \rangle$ twinning that occurs so easily in calcite, even at room
temperature and pressure, is seen from Table 27.1 to be insufficient alone
to satisfy von Mises' requirement for ductility, giving at most three independent systems when all are oriented so that the sign of the resolved shear
stress is favourable for twinning and when the strains are within the limits
set by complete twinning. However, experiments on single crystals under
high confining pressure have demonstrated the occurrence of slip on the
systems $\{10\bar{1}1\}$ $\langle \bar{1}012 \rangle$ and $\{02\bar{2}1\}$ $\langle \bar{1}012 \rangle$ over a wide range of temperatures (Turner, Griggs, and Heard 1954; Griggs, Turner, and Heard 1960;
Turner and Heard 1965; Borg and Handin 1967), and there have also been
indications of $\{10\bar{1}0\}$ $\langle \bar{1}012 \rangle$ slip, especially under constrained conditions
(Turner and Heard 1965; Paterson, and Turner 1968). Table 27.1 shows
that the $\{02\bar{2}1\}$ $\langle \bar{1}012 \rangle$ slip alone, or a combination of any two of these
three slip modes, suffices to give five independent slip systems and so
explain the observed ductility under high-pressure conditions. It is also
seen that when crystals are favourably oriented for twinning, the combination of twinning and any one of the above slip modes gives the
equivalent of five independent slip systems. The latter situation is relevant
to those experiments on Yule marble in which the applied stress was
oriented with respect to the preferred orientation of the crystals so as to
favour twinning in a large proportion of them (Griggs and Millar 1951;
Griggs et al. 1951, 1953; Turner et al. 1956); a major role of the twinning
is apparent from microexamination and from the relative low-stress strain
curves but, as noted above and earlier pointed out by Handin and Griggs
(1951), twinning alone could not give a homogeneous deformation.

TABLE 27.1. Slip systems in trigonal crystals

Slip systems	Symbol	Multiplicity	Number independent	Possible Examples
(0001)⟨1̄21̄0⟩	basal	3	2	Quartz, dolomite
{101̄0}⟨1̄21̄0⟩ or {101̄0}[0001]	m(a) or m(c)	3	2	Quartz
{101̄0}⟨1̄123⟩	m(c + a)	3	2	,,
{101̄1}⟨1̄123⟩ or {011̄1}⟨1̄21̄0⟩	r or z, (c + a)	3	3	,,
(0001)⟨1̄21̄0⟩ + {101̄0}⟨1̄21̄0⟩	basal + m(a)	6	4	,,
{101̄0}⟨1̄21̄0⟩ + {101̄0}[0001]	m(a) + m(c)	6	4	,,
(0001)⟨1̄21̄0⟩ + {101̄0}⟨1̄21̄0⟩ + {101̄0}[0001]	basal + m(a) + m(c)	9	4	,,
(0001)⟨1̄21̄0⟩ + {101̄0}⟨any⟩ + {101̄0}⟨any other⟩	basal + m (any two)	9	4	,,
{101̄1}⟨1̄123⟩ + (0001)[1̄21̄0] or {101̄0}[0001]	r + basal or m(c)	6	5	,,
{101̄1}⟨1̄123⟩ + {101̄0}⟨1̄21̄0⟩	r + m(a)	6	$5(\gamma^2 \neq \tfrac{3}{2})$*	,,
{101̄1}⟨1̄123⟩ + {101̄0}⟨1̄123⟩	r + m(c + a)	6	$5(\gamma^2 \neq \tfrac{1}{2}$ or $\tfrac{3}{2})$	,,
{011̄1}⟨1̄123⟩	r + z	6	$5(\gamma^2 \neq \tfrac{1}{2}$ or $\tfrac{3}{4})$	,,
(0001)⟨1̄21̄0⟩ + {101̄0}⟨1̄21̄0⟩ + one of {101̄1}⟨1̄123⟩	basal + m(a) + one r	7	5	,,
{022̄1}⟨01̄14⟩	f	3	3	Dolomite twinning
(0001)⟨1̄21̄0⟩ + {022̄1}⟨01̄14⟩	basal + f	6	5	Dolomite
{011̄2}⟨01̄11⟩	e	3	3	Calcite twinning
{101̄1}⟨1̄012⟩	r	3	3	Calcite
{1̄21̄0}⟨1̄012⟩	a	3	2	,,
{022̄1}⟨1̄012⟩	f	6	5	,,
{011̄2}⟨01̄11⟩ + {101̄1}⟨1̄012⟩	e + r	6	$5(\gamma^2 \neq \tfrac{3}{2})$	,,
{011̄2}⟨01̄11⟩ + two of {022̄1}⟨1̄012⟩	e + two f	5	5	,,
{011̄2}⟨01̄11⟩ + {1̄21̄0}⟨1̄012⟩	e + a	6	5	,,
{101̄1}⟨1̄012⟩ + two of {022̄1}⟨1̄012⟩	r + two f	5	$5(\gamma^2 \neq \tfrac{3}{8})$	,,
{101̄1}⟨1̄012⟩ + {1̄21̄0}⟨1̄012⟩	r + a	6	$5(\gamma^2 \neq \tfrac{3}{8})$	,,
{1̄21̄0}⟨1̄012⟩ + three of {022̄1}⟨1̄012⟩	a + three f	6	$5(\gamma^2 \neq \tfrac{3}{8})$	,,

*$\gamma = c/a$

27.2.2 Dolomite

The only slip mode that has been demonstrated in dolomite single crystals is (0001) $\langle \bar{1}2\bar{1}0 \rangle$, which can contribute two independent slip systems (Table 27.1). In orientations favourable to $\{02\bar{2}1\}$ $\langle 0\bar{1}14 \rangle$ twinning, the equivalent of five independent systems is available, within the limit of strain set by complete twinning, from a combination of these modes of deformation, both of which have been observed in experimentally deformed dolomite rock where the temperature and the orientation of a large proportion of the grains are favourable to twinning (Turner et al. 1954; Handin and Fairbairn 1955; Griggs, Turner, and Heard 1960). Ductility is also observed under conditions unfavourable to twinning, which requires that other mechanisms of deformation have been involved to supplement the basal slip. However, the range of ductile behaviour in dolomite rock is much more limited than that in calcite rock, consistent with the greater difficulty of operating sufficient slip systems to satisfy von Mises' requirement in the former.

27.2.3. Quartzite

Quartz is normally very resistant to plastic deformation but under suitable conditions indications of slip have been observed on many combinations of planes and directions (Carter, Christie, and Griggs 1964; Christie, Griggs, and Carter 1964, 1966; Christie and Green 1964; Heard and Carter 1968). Basal slip in the direction of the a axis and $\{10\bar{1}0\}$ slip in the direction of the c axis are the only systems that have so far been studied in detail, others being inferred from lamellae observed in thin section. A number of these systems are listed in Table 27.1; several others involving planes inclined to all the axes have been omitted but they can be taken to be similar to those involving $\{10\bar{1}1\}$ planes in respect of the von Mises problem. It is seen that none of the individual slip modes can satisfy the von Mises requirement; moreover, no combination of basal and prismatic slip gives more than four independent slip systems. At least one rhombohedral or pyramidal slip plane must be active in addition in order to give five independent systems, but five independent systems can also be contributed by combination of rhombohedral or pyramidal planes without basal or prismatic planes being involved. Thus, and with the further possibility of pencil glide, there are many ways in which the von Mises requirements can be met in quartz-rich rocks under conditions where at least some of the above slip systems are available.

27.2.4. Micaceous and Similar Rocks

There are many rocks in which the predominant mineral is one of the micas or a similar platy mineral such as talc or pyrophyllite. Single-crystal studies have so far revealed only slip parallel to the cleavage plane in these minerals (for example, Borg and Handin 1966), so that, even if slipping can occur in more than one direction in the slip plane, only two independent slip systems at most are known to be available. Therefore, unless closer study reveals other slip systems, the explanation of observed ductility in these rocks must involve, to an important degree, processes other than uniformly distributed slip. The high degree of preferred orientation in many of the mica-rich rocks may substantially reduce the problem of accommodation between grains in them, but this factor does not apply in material of low preferred orientation such as talc and pyrophyllite, widely used as pressure media in very high pressure apparatus because of their flow properties. A similar problem can be expected to arise in rocks consisting of aggregates of fibrous minerals such as chrysotile serpentine, since a pencil-type of glide parallel to the fibres would give only the equivalent of two independent slip systems.

27.2.5. *Other Rocks*

On present knowledge there are very few, if any, other rocks in which von Mises' criterion for ductility can be fully met. Thus, olivine, an orthorhombic mineral, has been observed to slip on the systems (100) [001], {110} [001], (010) [100], and (100) [010] in high pressure experiments at various temperatures (Raleigh 1965, 1967), but the maximum number of independent systems available from these is three (a combination of {110} [001] and (010) [100] or (100) [010]). Nevertheless, a substantial degree of ductility is observed experimentally in olivine-rich rock, and Raleigh's data on slip systems were in fact derived from polycrystalline specimens, in many of which only one slip plane was dominant. Barite, another orthorhombic mineral, is listed by Handin (1966), quoting early German sources, as having the slip systems (001) [100], {011} [0$\bar{1}$1], (010) [100], and {102} [010], amongst which four independent systems can be found, but barite-rich rock does not appear to have been studied experimentally. On the other hand, the orthorhombic anhydrite rock has been shown to be moderately ductile at high pressure, especially at 300°C (Handin 1953; Handin and Hager 1957, 1958), although fewer slip systems are listed for it (Handin 1966). Also, gypsum rock is readily deformed under relatively low confining pressures (Goguel 1948; Heard and Rubey 1966) in spite of only two independent slip systems (010) [001] and (010) [301], having been reported in this monoclinic mineral (Mügge 1899, quoted by Handin 1966). Further, the low symmetry minerals that are major constituents in other silicate rocks showing limited ductility under experimental conditions have been observed to have only one or two slip systems, for example, enstatite, diopside, kyanite, and some feldspars (Griggs, Turner, and Heard 1960; Raleigh 1965, 1967; Starkey and Brown 1964; Borg and Handin 1966; Riecker and Rooney 1967).

In summary, there are relatively few rocks in which it is known that the minerals can undergo slip on five independent slip systems, even if twinning is taken into account. It is probably significant that these tend to include the more ductile rocks, although a high degree of ductility can also be observed where fewer slip systems are known to be available, as in the case of talc or gypsum. However, there are many other rocks in which a limited degree of ductility has been observed, in all of which the number of known slip systems falls far short of von Mises' requirement. The explanation of the ductility may, in some cases, lie in the activity of slip systems not yet discovered, the total amount of research in this field being still relatively small. However, the low symmetry of many rock-forming minerals makes it unlikely that von Mises' requirement can be fully met in a large number of geologically important rocks. Therefore, we shall now discuss briefly various factors that may contribute to resolving this problem, either by relaxing the von Mises requirement or by providing mechanisms of deformation other than intracrystalline plasticity by gliding processes.

27.3. Factors Reducing the von Mises Problem

Elastic strains have been ignored in the discussion so far. Their contribution may make it possible to provide completely the strain required on the homogeneous strain model in given grains where the strain requirement can already be very nearly met by shearing on an available set of less than five independent slip systems. As pointed out by Pratt (1967), this may be important where the total strains are of the order of the elastic range. However, it is unlikely to be of much importance at larger strains unless there are three or four independent slip systems in such a relationship of orientations that the required strains are nearly met by them in a majority of grains, and even then it is difficult to imagine strains being achieved in this way which exceed the elastic limit at by most perhaps an order of magnitude.

The assumption that the strain is homogeneous on the scale of the grain size is another, probably more important factor, underlying the criterion that the number of independent slip systems in any grain be five. This assumption is not necessary in order to ensure continuity of displacements within the body, as can be demonstrated by visualizing the deformation of a composite body containing hard inclusions which remain coherent with the matrix. Moreover, although it has often proved to be a useful approximation in calculating stress-strain properties, it is abundantly clear that the assumption is not even a strictly tenable one, both because observation reveals marked heterogeneity of deformation within grains and from grain to grain (for example, Barrett and Levenson 1940; Boas and Hargreaves 1948; Urie and Wain 1952) and because the equations of equilibrium cannot be satisfied throughout the body under the condition of uniform strain (Bishop and Hill 1951; Bishop 1954; Taylor 1956; Kocks 1958; Lin 1964; Hill 1965). However, it is not clear to what extent the admitting of heterogeneity of strain can lead to a relaxation of the requirement of five independent slip systems. If three or four independent slip systems were available, one would generally expect that a good approximation to the macroscopic strain could be achieved in a majority of grains while, at the same time, local departures from the macroscopic strain would compensate both for the small amounts by which the macroscopic strain was inattainable in these grains and for the more markedly different strains in the minority of unfavourably oriented grains, the final distribution of strain still being compatible with the requirements of continuity, equilibrium, and the stress-strain properties of the grains. Theoretical investigation of this question is clearly of importance and some indication of the nature of the heterogeneity of strain that might arise is illustrated in the simplified elastic-plastic problem calculated by Lin (1964). However, it seems unlikely that less than three or four independent slip systems could suffice for ductility, even allowing heterogeneous distribution of strain, unless a strong preferred orientation in the grains allowed a favourable macroscopic approximation to single crystal behaviour.

Many features such as deformation bands, kinks, bend-gliding, and local concentrations of slip represent variations of strain within grains which are an aspect of the general heterogeneity of strain just discussed. However, repeated planar concentrations of shearing strain which occur on a sufficiently fine scale can also be viewed as equivalent to additional "slip systems." Finely distributed twinning in calcite and dolomite has already been treated in this way above, but kinking needs special consideration here.

Kinking is a mode of deformation, long known in minerals, which has been considerably studied since Orowan (1942) drew attention to it in metals. It enables a crystal with a unique slip plane to shorten in a direction parallel to that plane, which it cannot do by uniform slip. The kinking can be treated as a simple shear parallel to the kink band boundary or "kink plane," with a direction normal to the "kink axis," that is, the intersection of the kink plane and the slip plane in the undeformed crystal (the geometry of kinking is discussed in more detail by Paterson and Weiss 1966). Then it can be shown, (see Appendix), that an assemblage of kink bands is equivalent to a slip system that is independent of the primary slip system itself, and that if there are two slip directions in the primary slip plane, with each of which kinking is associated, the equivalent of four independent slip systems is available. The addition of more than one orientation of kink associated with each primary slip direction, for example, conjugate kinks, does not increase this number, but if there is kinking associated with a third slip direction in the slip plane the equivalent of five independent slip systems is available. However, the following conditions must be met in order that the kinks can be considered as equivalent to additional slip systems:

1. The kinking must be on a fine scale and suitable mechanisms must be available for accommodation of the strain at the ends (cf. twinning).

2. The kinks and slip must freely interpenetrate, at least when viewed on the scale of the grain (this can also give rise to serious accommodation problems at the intersections).

3. Kinking can contribute only components of shortening parallel to the primary slip plane and lengthening normal to it, not the reverse (cf. the restriction on the sense of shear in twinning).

In spite of these restrictions, kinking can be expected to contribute importantly to the ductility of aggregates in minerals having a unique slip plane, such as mica, talc, pyrophyllite, and kyanite, or of other minerals with substantially less than five independent slip systems, such as olivine; in fact, kinking within grains is often a striking feature in deformed specimens of these materials.

Thus, when account is taken of elastic strain and of heterogeneity of deformation, including kinking and similar deformation features, conditions for ductility based on intracrystalline glide can be substantially met in a much wider range of materials than the criterion of five independent slip systems would suggest. It remains to mention other mechanisms of deformation which may contribute to minor accommodations still needed or which may provide a major alternative to intracrystalline glide.

27.4. Role of Other Mechanisms of Deformation

In metals, change of shape by diffusion of material is thought to be an important mechanism of deformation only in creep experiments at temperatures near the melting point (McLean 1966). However, in the deformation of rocks at high temperatures it could be important in helping to accommodate misfits arising from an insufficient number of slip systems. There is little direct experimental evidence for this but very little work has been done under conditions where it might be expected. On the other hand, diffusion may well be of considerable importance in the natural deformation of rocks, and perhaps even be a predominant mechanism in some cases, since the long geological time scale will tend to favour relatively high temperature phenomena. Since diffusion will occur to some extent in all directions, no limitation on ductility need arise analogous to the effect of an insufficiency of slip systems.[1]

Cataclasis, that is, distributed fracturing and relative movement on the fracture surfaces, is a mechanism that often has to be considered in studies

[1] Climb of dislocations, which involves diffusion, could also be of importance in this context (Webb and Hayes 1967).

on the deformation of rocks, Thus it may be a supplementary mechanism where the strain requirements cannot be fully met by slip processes, either helping to avoid misfit between grains or acting as an accommodation mechanism on the finer scale, as at the ends of slip bands or kinks. The microcracks involved in this do not propagate indefinitely in a way that would lead to gross fracture because the stresses are predominantly compressive in situations where cataclasis is important, as in high pressure experiments or the geological environment of deep burial.

Cataclasis tends to assume major importance, however, at pressures that are not greatly in excess of the brittle-ductile transition pressure, especially when the temperature is relatively low. Thus, there is a whole spectrum of behaviour from fully intracrystalline plasticity, as in marble at sufficiently high pressure to almost complete cataclasis, as in a poorly bonded sandstone at relatively low temperature and moderate pressure. Cataclasis introduces new characteristic macroscopic properties of deformation; notably, frictional effects lead to a marked dependence of the stress-strain curve on the hydrostatic component of the stress, and volume changes can be expected from opening of cracks or changes in the packing of particles in porous rocks (for example, Borg, Friedman, Handin, and Higgs 1960). The latter introduces considerations analogous to those of "critical void ratio" in soil mechanics, and other stability questions involving the nature of the constraints at the boundaries arise. However, these matters have not been extensively explored and elaboration here would be inappropriate.

27.5. Conclusions

Although the von Mises criterion of five independent slip systems is met by a few rock-forming minerals, present knowledge suggests that in general it cannot be satisfied especially in view of the low symmetry of most of the minerals. In many cases, the observed ductility appears to be based on the possibility of partly relaxing von Mises' requirement by taking account of inhomogeneity of deformation, especially of kinking and similar effects, while mechanisms such as elastic or cataclastic deformation or diffusion creep may play a supplementary role. In other cases, cataclasis can be the principal mechanism.

REFERENCES

Aitken, A. C., 1949. *Determinants and Matrices*, Edinburgh: Oliver and Boyd, 6th ed.

Barrett, C. S., and L. H. Levenson, 1940, *Trans. AIME*, **137**, 112.

Bishop, J. F. W., 1953, *Phil. Mag.*, **44**, 51.

Bishop, J. F. W., 1954, *J. Mech. Phys. Sol.*, **3**, 130.

Bishop, J. F. W., and R. Hill, 1951, *Phil. Mag.*, **42**, 414 and 1298.

Boas, W., and M. E. Hargreaves, 1948, *Proc. Roy. Soc.*, **A193**, 89.

Borg, I., M. Friedman, J. Handin, and D. V. Higgs, 1960, in D. Griggs, and J. Handin (eds.), *Rock Deformation*, Geol. Soc. of Amer., Memoir 79, 133.

Borg, I., and J. Handin, 1966, *Tectonophysics*, **3**, 249.

Borg, I., and Handin, J., 1967, *J. Geophys. Res.*, **72**, 641.

Carter, N. L., J. M. Christie, and D. T. Griggs, 1964, *J. Geol.*, **72**, 687.

Christie, J. M., and H. W. Green, 1964, *Trans. Amer. Geophys. Union*, **45**, 103 (abstract).

Christie, J. M., D. T. Griggs, and N. L. Carter, 1964, *J. Geol.*, **72**, 734.

Christie, J. M., D. T. Griggs, and N. L. Carter, 1966, *J. Geol.*, **74**, 368.

Copley, S. M., and J. A. Pask, 1965, *J. Amer. Ceram. Soc.*, **48**, 636.

Goguel, J., 1948, *Introduction a l'Étude Mécanique des Déformations de l'Écorce Terrestre*, Paris, Imprimerie Nationale, 2nd ed.

Griggs, D. T., and J. Handin (eds.), 1960, *Rock Deformation*, Geol. Soc. Amer., Memoir 79.

Griggs, D. T., and W. B. Millar, 1951, *Bull. Geol. Soc. Amer.*, **62**, 853.

Griggs, D. T., F. J. Turner, I. Borg, and J. Sosoka, 1951, *Bull. Geol. Soc. Amer.*, **62**, 1385.

Griggs, D. T., F. J. Turner, I. Borg, and J. Sosoka, 1953, *Bull. Geol. Soc. Amer.*, **64**, 1327.

Griggs, D. T., F. J. Turner, and H. C. Heard, 1960, in D. Griggs and J. Handin (eds.), *Rock Deformation*, Geol. Soc. Amer., Memoir 79, 39.

Groves, G. W., and A. Kelly, 1963, *Phil. Mag.*, **8**, 877.

Handin, J., 1953, *Trans. ASME*, **75**, 315.

Handin, J., 1966, in S. P. Clark (ed.), *Handbook of Physical Constants*, Geol. Soc. Amer. Memoir, 97, 223.

Handin, J., and H. W. Fairbairn, 1955, *Bull. Geol. Soc. Amer.*, **66**, 1257.

Handin, J., and D. T. Griggs, 1951, *Bull. Geol. Soc. Amer.*, **62**, 863.

Handin, J., and R. V. Hager, 1957, *Bull. Amer. Soc. Petroleum Geol.*, **41**, 1.

Handin, J., and R. V. Hager, 1958, *Bull. Amer. Soc. Petroleum Geol.*, **42**, 2892.

Heard, H. C., and N. L. Carter, 1968, *Amer. J. Sci.*, **266**, 1.

Heard, H. C., and W. W. Rubey, 1966, *Bull. Geol. Soc. Amer.*, **77**, 741.

Hill, R., 1965, *J. Mech. Phys. Sol.*, **13**, 89.

Kocks, U. F., 1958, *Acta Met.*, **6**, 85.

Kocks, U. F., 1964, *Phil. Mag.*, **10**, 187.

Lin, T. H., 1964, *J. Mech. Phys. Sol.*, **12**, 391.

Livingston, J. D., and B. Chalmers, 1957, *Acta Met.*, **5**, 322.

McLean, D., 1966, *Rep. Progr. Phys.*, **29**, 1.

von Mises, R., 1928, *Z. angew. Math. Mech.*, **8**, 161.

Nicholas, J. F., 1966, *Acta Cryst.*, **21**, 880.

Orowan, E., 1942, *Nature*, **149**, 643.

Orowan, E., 1945, *Trans. Inst. Eng. Shipbuilders Scotland*, p. 165.

Orowan, E., 1948, *Rep. Prog. Phys.*, **12**, 185.

Orowan, E., 1955, *Welding J.*, **34**, 157s.

Orowan, E., 1959, in B. L. Averbach et al. (eds.), *Fracture*, Cambridge, Mass: M.I.T. Press, p. 147.

Parker, E. R., 1963, in *Mechanical Behaviour of Crystalline Solids*, Nat. Bureau Standards Monograph 59, 1.

Parker, E. R., J. A. Pask, J. Washburn, A. E. Gorum, and W. Luhman, 1958, *J. Metals*, **10**, 351

Paterson, M. S., 1969, in H. Ll. D. Pugh (ed.), *Mechanical Properties of Materials under Pressure*, Elsevier, in press.

Paterson, M. S., and F. J. Turner, 1968, to be published.

Paterson, M. S., and L. E. Weiss, 1966, *Bull. Geol. Soc. Amer.*, **77**, 343.

Pratt, P. L., 1967, *Geophys. J.*, **14**, 5.

Raleigh, C. B., 1965, *Science*, **150**, 739.

Raleigh, C. B., 1967a, *Geophys. J.*, **14**, 45.

Raleigh, C. B., 1967b, in P. J. Wyllie (ed.), *Ultramafic and Related Rocks*, New York: Wiley, p. 191.

Riecker, R. E., and T. P. Rooney, 1967, *Bull. Geol. Soc. Amer.*, **78**, 1045.

Starkey, J., and W. L. Brown, 1964, *Z. Krist.*, **120**, 388.

Taylor, G. I., 1938, *J. Inst. Metals*, **62**, 307.

Taylor, G. I., 1956, in *Proc. Colloq. Deformation and Flow of Solids*, Berlin: Springer, p. 3.

Tegart, W. J. McG., 1964, *Phil. Mag.*, **9**, 339.

Turner, F. J., D. T. Griggs, R. H. Clark, and R. H. Dixon, 1956, *Bull. Geol. Soc. Amer.*, **67**, 1259.

Turner, F. J., D. T. Griggs, and H. Heard, 1954, *Bull. Geol. Soc. Amer.*, **65**, 883.

Turner, F. J., D. T. Griggs, H. Heard, and L. E. Weiss, 1954, *Amer. J. Sci.*, **252**, 477.

Turner, F. J., and H. C. Heard, 1965, *Univ. Calif. Publ. Geol. Sci.*, **46**, 103.

Urie, V. M., and H. L. Wain, 1952, *J. Inst. Metals*, **81**, 153.

Webb, W. W., and C. E. Hayes, 1967, *Phil. Mag.*, **16**, 909.

APPENDIX—INDEPENDENCE OF SLIP SYSTEMS

A general strain can be written in terms of six components, of which only five need be specified independently if volume changes can be neglected. However, in the case of shearing parallel to a given plane, in a given direction, only one parameter, the amount of shear, can be freely varied since the ratio of the strain components remains always the same. Because of this, only one of five arbitrarily specified strain components can be exactly reproduced by a single shearing process. If a second shear parallel to another plane and direction is superposed, then in general two of the strain components can be made to take on any arbitrarily specified values by suitably choosing the amounts of the two shears. Proceeding in this way, it can be seen that any strain having five independently specified components can be only in general produced by simple shearing processes if there are five such independent shears available.

If the strain components of each of n shears, referred to a common set of axes, are written as the rows of a matrix, thus

$$
\begin{bmatrix}
e_{11}^{(1)} & e_{22}^{(1)} & e_{33}^{(1)} & e_{23}^{(1)} & e_{31}^{(1)} & e_{12}^{(1)} \\
e_{11}^{(2)} & e_{22}^{(2)} & e_{33}^{(2)} & e_{23}^{(2)} & e_{31}^{(2)} & e_{12}^{(2)} \\
\vdots & \vdots & \vdots & \vdots & \vdots & \vdots \\
e_{11}^{(n)} & e_{22}^{(n)} & e_{33}^{(n)} & e_{23}^{(n)} & e_{31}^{(n)} & e_{12}^{(n)}
\end{bmatrix},
\tag{27A.1}
$$

then the number of independent shears is given by the rank of this matrix, that is, by the number of columns in its largest square submatrix having a nonzero determinant. The third column is always linearly dependent on the first two because of the constant volume condition, so this number cannot exceed five.

Because of the cumbersome expressions involved in resolving a finite strain into components, it is usual to consider only infinitesimal strain increments, implicitly assuming that any required finite strain can be achieved by a suitable integration. For convenience, we can use the "engineering" expressions for the components of an infinitesimal strain in terms of the displacements. Then, as Bishop (1953) has shown, if a shear of magnitude s occurs parallel to a plane having n_1, n_2, n_3 as the direction cosines of its normal and to a direction having the direction cosines β_1, β_2, β_3, it has strain components

$$
sn_1\beta_1, \; sn_2\beta_2, \; sn_3\beta_3, \; s(n_2\beta_3 + n_3\beta_2), \; s(n_3\beta_1 + n_1\beta_3), \; s(n_2\beta_3 + n_3\beta_2).
$$

The factor s and any other common multiplying factor can be omitted in forming the matrix 27A.1 in testing for independence since they are common to a whole row. We now illustrate the application of this in two cases.

27.A1. Slip in Trigonal Crystals

Using Miller-Bravais indices $(hkil)$ for a plane and $[uvtw]$ for a direction (Nicholas 1966), and expressing the direction cosines in terms of these, it can be shown that the strain components of a simple shear (slip) having this plane and direction are proportional to

$$
3hu, \; (k-i)(v-t), \; 3lw, \; \frac{1}{\sqrt{3}\,\gamma}\{2\gamma^2(k-i)w + 3l(v-t)\},
$$

$$
\frac{1}{\gamma}(2\gamma^2 hw + 3lu), \; \sqrt{3}\{h(v-t) + (k-i)u\}
$$

where γ is equal to the c/a axial ratio.

As an example, we consider the slip modes $\{10\bar{1}1\}$ $\langle\bar{1}012\rangle$ and $\{1\bar{2}10\}$ $\langle\bar{1}012\rangle$, comprising the systems $(10\bar{1}1)$ $[\bar{1}012]$, $(0\bar{1}11)$ $[01\bar{1}2]$, $(\bar{1}101)$ $[1\bar{1}02]$,

$(1\bar{2}10)$ $[\bar{1}012]$, $(\bar{2}110)$ $[01\bar{1}2]$, and $(11\bar{2}0)$ $[1\bar{1}02]$. The matrix 27A.1 is then proportional to

$$
\begin{pmatrix}
-3 & -1 & 4 & \dfrac{1}{\sqrt{3}\,\gamma}(4\gamma^2 - 3) & \dfrac{1}{\gamma}(4\gamma^2 - 3) & -2\sqrt{3} \\[2ex]
0 & -4 & 4 & \dfrac{1}{\sqrt{3}\,\gamma}(-8\gamma^2 + 6) & 0 & 0 \\[2ex]
-3 & -1 & 4 & \dfrac{1}{\sqrt{3}\,\gamma}(4\gamma^2 - 3) & \dfrac{1}{\gamma}(-4\gamma^2 + 3) & 2\sqrt{3} \\[2ex]
-3 & 3 & 0 & \dfrac{1}{\sqrt{3}\,\gamma}(-12\gamma^2) & \dfrac{1}{\gamma}4\gamma^2 & 2\sqrt{3} \\[2ex]
0 & 0 & 0 & 0 & \dfrac{1}{\gamma}(-8\gamma^2) & -4\sqrt{3} \\[2ex]
3 & -3 & 0 & \dfrac{1}{\sqrt{3}\,\gamma}12\gamma^2 & \dfrac{1}{\gamma}4\gamma^2 & 2\sqrt{3}
\end{pmatrix}.
$$

Omitting the third column and the last row gives a 5×5 submatrix which has a determinant equal to the following, apart from constant multiplying factors:

$$
\begin{vmatrix}
1 & -1 & 4\gamma^2 - 3 & 4\gamma^2 - 3 & -1 \\
0 & -4 & -8\gamma^2 + 6 & 0 & 0 \\
1 & -1 & 4\gamma^2 - 3 & -4\gamma^2 + 3 & 1 \\
1 & 3 & -12\gamma^2 & 4\gamma^2 & 1 \\
0 & 0 & 0 & 4\gamma^2 & 1
\end{vmatrix}
= \text{constant } (8\gamma^2 - 3)^2.
$$

(This evaluation is quickly done by the method of "pivotal condensation," Aitken 1949, p. 45). Thus, since we have shown that we can find a 5×5 submatrix with nonzero determinant for all $\gamma^2 \neq 3/8$, then we have established that the above combination of slip systems contains five independent ones, provided that $\gamma^2 \neq 3/8$. If $\gamma^2 = 3/8$, no such submatrix with a nonzero determinant can be found, since the last two columns are then linearly dependent, and so in this case there is no combination of five slip systems that are independent.

27A.2. Kinking

We consider a body having a unique slip plane in which there are one or more slip directions. We choose orthogonal axes, **1, 2, 3**, which are respectively parallel to a slip direction, to the line normal to this slip direction in the slip plane (kink axis), and to the normal to the slip plane.

Then the direction cosines of the slip plane normal are 0, 0, 1 and of this slip direction 1, 0, 0. Another slip direction in the slip plane will have direction cosines which can be written as proportional to u, v, 0 where u and v are suitable numbers. Then kinks associated with these two slip directions will give simple shears parallel to planes with normals having direction cosines respectively proportional to h, 0, k and $1/v$, $1/u$, L where the numbers h, k, and L are fixed by the angles which the two kink planes make with the slip plane. The shear ("slip") directions associated with the kinks will have direction cosines proportional to $1/h$, 0, $-1/k$, and u, v, $-(u^2 + v^2)/uvL$, respectively. Then the matrix 27A.1 formed of the strain components of the four shears, one parallel to each direction in the slip plane and one associated with each kink, will be proportional to

$$
\begin{pmatrix}
0 & 0 & 0 & 0 & 1 & 0 \\
1 & 0 & -1 & 0 & \dfrac{k}{h} - \dfrac{h}{k} & 0 \\
0 & 0 & 0 & v & u & 0 \\
\dfrac{u}{v} & \dfrac{v}{u} & -A & vL - \dfrac{A}{u} & uL - \dfrac{A}{v} & 2
\end{pmatrix}
$$

where $A = u/v + v/u$. Omitting the second and fourth columns gives a submatrix having nonzero determinant, so this combination of shears is equivalent to four independent slip systems. It can also be shown that further kinks associated with the same slip plane and same two slip directions will not increase the number of independent shears. However, a kink associated with a third slip having direction cosines proportional to u', v', 0 will contribute a fifth independent shear, as can be shown by adding a further row to the above matrix of the same form as the fourth row but with primed symbols and evaluating the determinant of the submatrix formed by omitting one of the first three columns.

Author Index

Subject Index

399